LONDON MATHEMATICAL SOCIETY LECTURE NOTE SERIES

Managing Editor: Professor N.J. Hitchin, Mathematical Institute,
University of Oxford, 24–29 St Giles, Oxford OX1 3LB, United Kingdom

The titles below are available from booksellers, or from Cambridge Univ

London Mathematical Society Lecture Note Series. 302

Discrete and Continuous Nonlinear Schrödinger Systems

M. J. Ablowitz
University of Colorado, Boulder

B. Prinari
Università di Lecce

A. D. Trubatch
United States Military Academy, West Point

CAMBRIDGE
UNIVERSITY PRESS

PUBLISHED BY THE PRESS SYNDICATE OF THE UNIVERSITY OF CAMBRIDGE
The Pitt Building, Trumpington Street, Cambridge, United Kingdom

CAMBRIDGE UNIVERSITY PRESS
The Edinburgh Building, Cambridge CB2 2RU, UK
40 West 20th Street, New York, NY 10011-4211, USA
477 Williamstown Road, Port Melbourne, VIC 3207, Australia
Ruiz de Alarcón 13, 28014 Madrid, Spain
Dock House, The Waterfront, Cape Town 8001, South Africa

http://www.cambridge.org

First published 2004

Printed in the United Kingdom at the University Press, Cambridge

Typefaces Times 10/13 pt. and Helvetica   *System* LaTeX 2_ε [TB]

A catalog record for this book is available from the British Library.

Library of Congress Cataloging in Publication Data

Ablowitz, Mark J.
Discrete and continuous nonlinear Schrödinger systems / M.J. Ablowitz, B. Prinari,
A.D. Trubatch.
    p.   cm. – (London Mathematical Society lecture note series ; 302)
Includes bibliographical references and index.
ISBN 0-521-53437-2 (pbk.)
1. Schrödinger equation.   2. Nonlinear theories.   3. Inverse scattering transform.
I. Prinari, B., 1972–   II. Trubatch, A. D., 1968–   III. Title.   IV. Series.
QC174.26.W28A26   2003
530.12′4–dc21   2003048555

ISBN 0 521 53437 2 paperback

Contents

Contents

Preface

Nonlinear systems are generic in the mathematical representation of physical phenomena. It is unusual for one to be able to find solutions to most nonlinear equations. However, a certain physically significant subclass of problems admits deep mathematical structure that further allows one to find classes of exact solutions. Solitons are a particularly important subclass of such solutions. Solitons are localized waves that, in an appropriate sense, interact elastically with each other. They have proved to be extremely interesting to physicists and engineers due, in part, to their localized and stable nature.

This broad field of study is sometimes called "soliton theory" or "integrable systems." This field has witnessed numerous important developments, which have been studied intensively worldwide over the past 30 years. Some of the directions that researchers have pursued include the following: direct methods to find solutions; studies of the underlying analytic structure of the equations; associated Painlevè-type solutions and relevant generalizations; tests to locate integrable systems; studies of the underlying geometric structures inherent in integrable systems; Bäcklund and Darboux transformations, which can be used to produce new classes of solutions; and so on.

In principle, one would like to be able to solve the general initial-value problem associated with these special nonlinear soliton systems. Depending on the boundary conditions under consideration, sometimes this is feasible. Moreover, in some cases, the mathematical representation is constructive, useful, and indeed elegant. The case for which the Cauchy problem admits a complete solution and the qualitative properties of the solution are well understood corresponds to initial data decaying sufficiently rapidly at infinity. Because of the constructive and relatively explicit nature of the solution, this case has received considerable attention. The method of solution is usually referred to as the inverse scattering transform (IST). The IST applies to many interesting nonlinear soliton systems, including nonlinear partial differential equations in $1 + 1$

(one space–one time) and $2 + 1$ (two space–one time) dimensions, nonlinear discrete (difference) evolution equations, and singular integro-differential equations. A discussion of the IST as it applies to many of these cases can be found in various monographs (cf. [6]).

Nevertheless, there are particular problems that, because of their wide applicability, deserve special attention. It is the purpose of this book to investigate one such important set of equations: nonlinear Schrödinger (NLS) systems. NLS systems in continuous media have been studied heavily since the mid-1960s. The continuous scalar NLS equation arises in the prototypical situation governing slowly varying waves of small amplitude (cf. [30]). In the early 1970s, it was shown [91] that the NLS equation governs the long-distance pulse propagation in optical fibers. In the late 1980s and 1990s, it was discovered [131], [122] that vector NLS systems govern the propagation of polarized waves in optical fibers. Today, these NLS equations, with suitable modifications and additional terms, are used routinely to make predictions about the transmission of information in fibers. Without doubt, wave transmission in optical fibers, used for communication purposes, is an application of critical importance.

In recent years there has developed significant interest in the study of nonlinear waves in media that are governed by nonlinear semi-discrete evolution systems (discrete in space–continuous time). Once again there is a class of physically interesting nonlinear systems that includes the so-called discrete NLS equations. These discrete NLS equations reduce to the continuous NLS equation when the discretization parameter vanishes.

For certain scalar- and vector-continuous and discrete NLS systems, the IST method can be applied in an effective and complete way. The initial-value problem, for given data decaying sufficiently rapidly at infinity, can be linearized in terms of integral equations. Multisoliton solutions can be obtained in all cases. This is described in detail in this book. In an appendix we also describe the IST associated with another class of important discrete equations: the Toda lattice [165] and the so-called nonlinear ladder network [121], [98]. While the research literature has many (but not all) of the results obtained here, it requires researchers to access numerous papers. Early fundamental research on NLS systems appears in the papers of Zakharov and Shabat [192], who first analyzed the scalar-continuous NLS equation; Manakov [120], regarding the continuous-vector NLS equation; and by Ablowitz and Ladik [11, 12], who introduced the scalar integrable discrete NLS equation. The vector extension of the discrete NLS system is more recent (cf. [18]). The unified description of these physically interesting integrable discrete and continuous NLS systems within the context of the IST methodology does not appear anywhere else. One of our motivations in writing this book was to develop the direct and inverse scattering formalism based on the Riemann–Hilbert approach. From a pedagogical point of view, readers will find that the structure follows the one laid out for the Korteweg–de Vries equation in the monograph of Ablowitz and Clarkson [6]. Here we have also attempted to include many of the mathematical details, to make the book suitable for students as well as researchers who wish to study this topic.

We gratefully acknowledge support for these studies by the National Science Foundation, the Air Force Office of Scientific Research, and the Colorado Commission on Higher Education.

Chapter 1

Introduction

1.1 Solitons and soliton equations

Ever since the observation of the "great wave of translation" in water waves, by J. Scott Russell in 1834 [146, 147] while he rode on horseback near a narrow canal in Edinburgh, localized (nonoscillatory) solitary waves have been known to researchers studying wave dynamics. Despite Russell's detailed observations, it was many years before mathematicians formulated the relevant equation, now known as the Korteweg–de Vries (KdV) equation that governs those waves (cf. [38, 39, 112]). From the period 1895–1960, the study of water waves was essentially the only application in which solitary waves were found. However, in the 1960s it was discovered that the KdV equation is a relevant model in many other physical contexts, such as plasma physics, internal waves, lattice dynamics, and others. Critically, in their study of the Fermi–Pasta–Ulam lattice equation [75] Zabusky and Kruskal (1965) found that the KdV equation was the governing equation (cf. [189]). Moreover, in a wholly new discovery, Zabusky and Kruskal observed that the solitary waves of KdV are "elastic" in their interaction. That is, the solitary waves pass through one another and subsequently retain their characteristic form and velocity. Zabusky and Kruskal called these elastically interacting solitary waves *solitons*. The work of Gardner, Green, Kruskal, and Miura [82] showed how direct and inverse scattering techniques could be used to linearize the initial-value problem of KdV. The solitons also were shown to correspond to eigenvalues of the time-independent Schrödinger equation. The remarkable discovery of the soliton was the first in a chain of events that culminated in a mathematical theory of solitons in KdV. This work opened a rich vein of research that continues today, more than 35 years later. Further discussion of both the historical background and subsequent development of soliton theory can be found in [21].

1

Subsequent to the development of a mathematical theory of the solitons of KdV, further research revealed that solitons, with their distinctive elastic interactions, arise in numerous important physical systems. These systems are governed, respectively, by a diverse collection of evolution equations that are characterized mainly by the fact that they admit soliton solutions. For example, soliton solutions have been found in a number of nonlinear partial differential equations in $1 + 1$ dimensions (i.e., one space and one time dimension) and $2 + 1$ (i.e., two spatial dimensions and one time dimension). In addition, soliton solutions have been found in semi-discrete (discrete in space, continuous in time) and doubly discrete (discrete in space and time) nonlinear evolution equations and in nonlinear singular integro-differential equations, among others. A survey of some of these can be found in [6]. It should also be noted that there is a four-dimensional system, referred to as the self-dual Yang–Mills (SDYM) equations, that plays an important role in the study of soliton theory or integrable systems. Indeed, the SDYM equations can be viewed as a "master" integrable system from which virtually all other systems can be obtained as special reductions (cf. Atiyah and Ward [25]; Ward [181, 182, 183]; Belavin and Zakharov [29]; Mason et al. [124, 125]; Chakravarty et al. [50, 51]; Maszczyk et al. [126]; Ablowitz et al. [5]).

Researchers in physics and engineering have understood that stable localized solitary waves, even those that do not have the special property of elastic interaction, have many important applications, and their study has led to substantial research in specialized fields (e.g., nonlinear optics) all by itself. Nevertheless, the systems in which the solitary waves interact elastically, the "true" soliton systems, are important special cases. Moreover, there is an intrinsic richness in the mathematical theory of these soliton systems. Accordingly, the field of research associated with integrable systems has grown, developed, and expanded in many directions. One centrally important issue is the method of solution, sometimes referred to as the inverse scattering transform (IST), for these soliton equations. For a number of physically significant equations, the IST can be carried out in an explicit, effective, and illuminative manner. In particular, the IST is a fruitful approach, and it is the basis for the study of the nonlinear Schrödinger systems described in this book.

There are numerous books, review articles, and edited collections (see, for instance, [6, 21, 47, 65, 74, 140]) that delve widely and deeply into the theory of integrable systems. In this book, we give a detailed description of both the IST and the soliton solutions of integrable nonlinear Schrödinger systems, which are mathematically and physically important soliton equations. The collection of systems examined in this book comprises both continuous and semi-discrete systems of equations. As will be described in Section 1.4, these particular systems arise in the modeling of a wide array of physical wave phenomena. In this book, we present most of the known results for these nonlinear Schrödinger systems, as well as some new ones, in a comprehensive, unified framework built with the mathematical machinery of the inverse scattering transform.

1.2 The inverse scattering transform – Overview

The IST is a method that allows one to linearize a class of nonlinear evolution equations. In doing so one can obtain global information about the structure of the solution. In many respects, one can view the IST as a nonlinear version of the Fourier transform.

The solution of the initial-value problem of a nonlinear evolution equation by IST proceeds in three steps, as follows:

1. the forward problem – the transformation of the initial data from the original "physical" variables to the transformed "scattering" variables;
2. time-dependence – the evolution of the transformed data according to simple, explicitly solvable evolution equations;
3. the inverse problem – the recovery of the evolved solution in the original variables from the evolved solution in the transformed variables.

In fact, with the IST machinery one can do more than solve the initial-value problem; one also can construct special solutions of the evolution equation by positing an elementary solution in the transformed variables and then applying the inverse transformation to obtain the corresponding solution in the original variables. In general, the soliton and multisoliton solutions of soliton equations can be constructed in this way. In particular, in the subsequent chapters of this book, we explicitly construct the soliton solutions of four different nonlinear Schrödinger (NLS) systems. Moreover, one can in principle obtain the long-time asymptotics of solutions. In this book, we obtain formulas for the collision-induced phase shifts of solitons in NLS systems, including the polarization shift of solitons in the vector systems, from the asymptotics of the associated scattering data.

An essential prerequisite of the IST method is the association of the nonlinear evolution equation with a pair of linear problems, a linear eigenvalue problem and a second associated linear problem, such that the given evolution equation results as the compatibility condition between them. The pair of linear operators used to construct the associated linear problems is sometimes referred to as a "Lax pair," due to a formulation by Lax [115]. The solution of the nonlinear evolution equation appears as a coefficient in the associated linear eigenvalue problem. For example, in the work of Gardner et al., the solution of the KdV equation is associated with the potential in the linear Schrödinger equation. The eigenvalues and continuous spectrum of this linear eigenvalue problem constitute the transformed variables. The second associated linear problem determines the evolution of the transformed variables.

The associated eigenvalue problem introduces an intermediate stage in both the forward and inverse problems of the IST. In the forward problem, the first step is to construct eigenfunction solutions of the associated linear problem. These eigenfunctions depend on both the original spatial variables and the spectral parameter (eigenvalue). Second, with these eigenfunctions, one determines *scattering data* that are independent of the original spatial variables. In the inverse problem, the first step is the recovery of the eigenfunctions from the (evolved) scattering data. Finally, one recovers the solution in the original variables from these (evolved) eigenfunctions. As noted previously, the evolution of the scattering data is determined by the second associated linear operator and can be computed explicitly.

The properties of the eigenfunctions are key to the formulation of the inverse problem. In general, the solutions of the associated eigenvalue problem also satisfy linear integral equations. For the NLS systems discussed here (as well as other soliton equations in $1 + 1$ dimensions), by using such integral equations one can show that the eigenfunctions are sectionally analytic functions of the spectral parameter. By taking into account the analyticity properties of the eigenfunctions, one can formulate the inverse problem (in particular, the recovery of the eigenfunctions from the scattering data) as a generalized Riemann–Hilbert problem. The Riemann–Hilbert problem is then transformed into a system of linear algebraic–integral equations. Typically in the formulation of the IST, and in particular for the nonlinear Schrödinger systems considered in this book, the scattering data satisfy symmetry relations that are independent of the evolution. These symmetries in the scattering data are essential and must be taken into account in the solution of the inverse problem.

As explained previously, to apply the IST to a nonlinear evolution equation, one must first find a pair of linear operators that can be associated with the nonlinear equation. However, the general method of construction of the Lax pair for a given evolution equation remains on open problem. Nevertheless, for several equations of physical and mathematical interest, such a pair has been found and the IST developed.

In a major step forward, following the works of Gardner et al. [82] and Lax [115], in 1972 Zakharov and Shabat [192] showed that the method also could be applied to another physically significant nonlinear evolution equation, namely, the nonlinear Schrödinger equation (NLS). Subsequently, Manakov extended this approach to the solution of a pair of coupled NLS equations [120]. In fact, Manakov's work applies equally well to a system of N coupled NLS equations, a system that we refer to as vector NLS (VNLS). Using these ideas, Ablowitz, Kaup, Newell, and Segur developed a method to find a rather wide class of nonlinear evolution equations solvable by this technique [10]. In their work they named the technique the inverse scattering transform (IST). Later, Beals and Coifman analyzed the direct and inverse scattering associated with higher order systems of linear operators [28].

The IST has been extended to semi-discrete nonlinear evolution equations (discrete in space and continuous in time) as well as doubly discrete (discrete in both space and time) systems. Flaschka adapted the IST to solve the Toda lattice equation [79], and Manakov used a similar formulation to solve a nonlinear ladder network [121]. Subsequently, Ablowitz and Ladik developed a method to construct families of semi-discrete and doubly discrete nonlinear systems along with their respective linear operator pairs, as required for the solution of the nonlinear systems via the IST [11, 12] (see also [21]). Included in the formulation of Ablowitz and Ladik are an integrable semi-discretization of NLS (which we refer to as integrable discrete NLS, or IDNLS) as well as a doubly discrete integrable NLS. In Chapter 5 we will further extend the IST method to an integrable semi-discretization of the VNLS that was introduced in [18].

While the work mentioned in the preceding paragraphs consists of applications of the IST method to $1 + 1$–dimensional evolution equations, since the early 1980s significant progress has also been made in the extension of the IST approach to $2 + 1$–dimensional systems. For example, the Kadomtsev–Petviashvili equation [100], which is a $2 + 1$–dimensional generalization of the KdV equation, and the Davey–Stewartson equation [59], which is a natural $2 + 1$–dimensional integrable extension of the NLS equation, can be solved via the IST. However, the extension of the IST to such systems is beyond the scope of this book. A review of some developments in the application of the IST to $2 + 1$–dimensional systems can be found in [6].

It should also be noted that IST for periodic and other boundary conditions has been considered, but we will not discuss this here. Additional references can be found in the Bibliography.

1.3 Nonlinear Schrödinger systems

The scalar nonlinear Schrödinger (NLS) equation

$$iq_t = q_{xx} \pm 2 |q|^2 q \qquad (1.3.1)$$

is a physically and mathematically significant nonlinear evolution equation. It results from the coupled pair of nonlinear evolution equations

$$iq_t = q_{xx} - 2rq^2 \tag{1.3.2a}$$

$$-ir_t = r_{xx} - 2qr^2 \tag{1.3.2b}$$

if we let $r = \mp q^*$.

The NLS equation (1.3.1) arises in a generic situation. It describes the evolution of small amplitude, slowly varying wave packets in nonlinear media [30]. Indeed, it has been derived in such diverse fields as deep water waves [190, 31]; plasma physics [191]; nonlinear optical fibers [91, 92]; magneto-static spin waves [194]; and so on. Mathematically, it attains broad significance because it is integrable by the IST [192], it admits soliton solutions, it has an infinite number of conserved quantities, and so on.

We also note that the form of the NLS equation (1.3.1) with a minus sign in front of the nonlinear term is sometimes referred to as the "defocusing" case. The defocusing NLS equation does not admit soliton solutions that vanish at infinity. However, it does admit soliton solutions that have a nontrivial background intensity (called dark solitons) [92, 193]. We will only discuss the IST for functions decaying sufficiently rapidly at infinity.

The vector nonlinear Schrödinger equation,

$$iq_t^{(1)} = q_{xx}^{(1)} + 2\left(|q^{(1)}|^2 + |q^{(2)}|^2\right)q^{(1)} \tag{1.3.3a}$$

$$iq_t^{(2)} = q_{xx}^{(2)} + 2\left(|q^{(1)}|^2 + |q^{(2)}|^2\right)q^{(2)}, \tag{1.3.3b}$$

arises, physically, under conditions similar to those described by the NLS when there are two wavetrains moving with nearly the same group velocities [144, 185]. Moreover, VNLS models physical systems in which the field has more than one component; for example, in optical fibers and waveguides, the propagating electric field has two components that are transverse to the direction of propagation. Manakov [120] first examined equation (1.3.3) as an asymptotic model for the propagation of the electric field in a wageguide. Subsequently, this system was derived as a key model for lightwave propagation in optical fibers (cf. [72], [122], [131], [179]).

In the literature, the system (1.3.3) is sometimes referred to as the coupled NLS equation. This system admits vector–soliton solutions, and the soliton collision is elastic. Moreover, the dynamics of soliton interactions can be explicitly computed [120]. (In [142] a different point of view is discussed.) Vector–soliton collisions are analyzed in Sections 4.3 (continuous) and 5.3 (discrete). In these sections, the elasticity of vector–soliton interactions and the order-dependence of these interactions are described in detail.

Both the VNLS equation (1.3.3) and its generalization,

$$i\mathbf{q}_t = \mathbf{q}_{xx} \pm 2\,\|\mathbf{q}\|^2\,\mathbf{q}, \tag{1.3.4}$$

where $\mathbf{q}$ is an N-component vector and $\|\cdot\|$ is the Euclidean norm, are integrable by the IST. In [120] only the case $N = 2$ is studied, but the extension to more components is straightforward. The N-component equation can be derived, with some additional conditions, as an asymptotic model of the interaction of N wavetrains in a weakly nonlinear, conservative medium (cf. [144]).

In optical fibers and waveguides, depending on the physics of the particular system, the propagation of the electromagnetic waves may be described by variations of equation (1.3.3). Note that the VNLS equation is the ideal (exactly integrable) case. For example, a model with physical significance is [131, 132, 150, 184, 187]

$$iq_t^{(1)} = q_{xx}^{(1)} + 2\left(|q^{(1)}|^2 + B|q^{(2)}|^2\right)q^{(1)} \tag{1.3.5a}$$

$$iq_t^{(2)} = q_{xx}^{(2)} + 2\left(B|q^{(1)}|^2 + |q^{(2)}|^2\right)q^{(2)}, \tag{1.3.5b}$$

which is equivalent to equation (1.3.3) when $B = 1$. However, based on the properties of equations (1.3.5), apparently it is not integrable when $B \neq 1$ (see the discussion in [18]).

The VNLS (1.3.4) has a natural matrix generalization in the system

$$i\mathbf{Q}_t = \mathbf{Q}_{xx} - 2\mathbf{Q}\mathbf{R}\mathbf{Q} \tag{1.3.6a}$$

$$-i\mathbf{R}_t = \mathbf{R}_{xx} - 2\mathbf{R}\mathbf{Q}\mathbf{R}, \tag{1.3.6b}$$

where $\mathbf{Q}$ and $\mathbf{R}$ are $N \times M$ and $M \times N$ matrices, respectively. When $\mathbf{R} = \mp\mathbf{Q}^H$ (here and in the following, the superscript H denotes the Hermitian, i.e., conjugate transpose), the system (1.3.6a)–(1.3.6b) reduces to the single matrix equation

$$i\mathbf{Q}_t = \mathbf{Q}_{xx} \pm 2\mathbf{Q}\mathbf{Q}^H\mathbf{Q}, \tag{1.3.7}$$

which we refer to as matrix NLS or MNLS. The VNLS corresponds to the special case when $\mathbf{Q}$ is an N-component row vector and $\mathbf{R}$ is an N-component column vector, or vice versa. In particular, we obtain the system (1.3.3) when $M = 1$ and $N = 2$.

Both the NLS and the VNLS equations admit integrable discretizations that, besides being used as the basis for constructing numerical schemes for the continuous counterparts, also have physical applications as discrete systems (see, e.g., Aceves et al. [22, 23]; Braun and Kivshar [40]; Christodoulides and Joseph [54]; Claude et al. [55]; Darmanyan et al. [57, 58]; Davydov [60, 61, 62]; Eilbeck et al. [69]; Eisenberg et al. [70, 71]; Flach et al. [76, 77];

Its et al. [97]; Kenkre et al. [102, 103]; Kivshar, Kivshar, and Luther-Davies [106, 107]; Lederer et al. [116, 117]; Malomed and Weinstein [118]; Morandotti et al. [136, 137, 138]; Scott and Macneil [148]; Vakhnenko et al. [173, 174]).

A natural discretization of NLS (1.3.1) is the following:

$$i\frac{d}{dt}q_n = \frac{1}{h^2}(q_{n+1} - 2q_n + q_{n-1}) \pm |q_n|^2 (q_{n+1} + q_{n-1}),   \quad (1.3.8)$$

which is referred to here as the integrable discrete NLS (IDNLS). It is a $O(h^2)$ finite-difference approximation of (1.3.1) that is integrable via the IST and has soliton solutions on the infinite lattice [11], [12]. We note that, if we change the nonlinear term in (1.3.8) to $2|q_n|^2 q_n$, the equation, which is often called the discrete NLS (DNLS) equation, is apparently no longer integrable, and it has been found that in certain circumstances chaotic dynamics results [19]. It should be remarked that the (apparently nonintegrable) DNLS equation arises in many important physical contexts (cf. [22], [23], [40], [54], [70], [71], [76], [104], [107], [116], and [136]–[138]). See also [17], [68], [105] for additional useful references.

Correspondingly, we will consider the discretization of the VNLS given by the following system:

$$i\frac{d}{dt}\mathbf{q}_n = \frac{1}{h^2}(\mathbf{q}_{n+1} - 2\mathbf{q}_n + \mathbf{q}_{n-1}) - \mathbf{r}_n \cdot \mathbf{q}_n (\mathbf{q}_{n+1} + \mathbf{q}_{n-1}) \quad (1.3.9a)$$

$$-i\frac{d}{dt}\mathbf{r}_n = \frac{1}{h^2}(\mathbf{r}_{n+1} - 2\mathbf{r}_n + \mathbf{r}_{n-1}) - \mathbf{r}_n \cdot \mathbf{q}_n (\mathbf{r}_{n+1} + \mathbf{r}_{n-1}), \quad (1.3.9b)$$

where $\mathbf{q}_n$ and $\mathbf{r}_n$ are N-component vectors and $\cdot$ is the inner product. Under the symmetry reduction $\mathbf{r}_n = \mp\mathbf{q}_n^*$ (here and in the following * indicates the complex conjugate), the system (1.3.9a)–(1.3.9b) reduces to the single equation

$$i\frac{d}{dt}\mathbf{q}_n = \frac{1}{h^2}(\mathbf{q}_{n+1} - 2\mathbf{q}_n + \mathbf{q}_{n-1}) \pm \|\mathbf{q}_n\|^2 (\mathbf{q}_{n+1} + \mathbf{q}_{n-1}),   \quad (1.3.10)$$

which, for $\mathbf{q}_n = \mathbf{q}(nh)$ in the limit $h \to 0$, $nh = x$, gives the VNLS (1.3.4). In [18] it was shown that its solitary wave solutions interact elastically and that (1.3.10) admits multisoliton solutions. Thus the expectation was that the discrete vector NLS system (1.3.10) is indeed integrable. We refer to (1.3.10) as the integrable discrete vector NLS (IDVNLS).

An associated pair of linear operators (Lax pair) for the system (1.3.9a)–(1.3.9b) was constructed in [170]. In fact, the Lax pair for the vector system (1.3.10) is a reduction of a *matrix* generalization of the Lax associated with IDNLS. The matrix analog of the vector system (1.3.9) is given by

$$i\frac{d}{d\tau}\mathbf{Q}_n = \mathbf{Q}_{n+1} - 2\mathbf{Q}_n + \mathbf{A}\mathbf{Q}_n + \mathbf{Q}_n\mathbf{B} + \mathbf{Q}_{n-1} - \mathbf{Q}_{n+1}\mathbf{R}_n\mathbf{Q}_n - \mathbf{Q}_n\mathbf{R}_n\mathbf{Q}_{n-1}$$

$$(1.3.11a)$$

$$-i\frac{d}{d\tau}\mathbf{R}_n = \mathbf{R}_{n+1} - 2\mathbf{R}_n + \mathbf{B}\mathbf{R}_n + \mathbf{R}_n\mathbf{A} + \mathbf{R}_{n-1} - \mathbf{R}_{n+1}\mathbf{Q}_n\mathbf{R}_n - \mathbf{R}_n\mathbf{Q}_n\mathbf{R}_{n-1},$$

$$(1.3.11\text{b})$$

where $\mathbf{Q}_n, \mathbf{R}_n$ are $N \times M$ and $M \times N$ matrices, respectively, $\mathbf{A}$ is an $N \times N$ diagonal matrix, and $\mathbf{B}$ is an $M \times M$ diagonal matrix. $\mathbf{A}$ and $\mathbf{B}$ represent a gauge freedom in the definition of the integrable discrete MNLS (IDMNLS) that will be used in the following. In [83], [84] the IST for an eigenvalue problem that is equivalent to the scattering problem considered in [18], [19], [167], [170] had been formulated.

Note that the system (1.3.11a)–(1.3.11b) does not, in general, admit the reduction

$$\mathbf{R}_n = \mp\mathbf{Q}_n^H. \qquad (1.3.12)$$

However, for $N = M$ one can restrict $\mathbf{R}_n$ and $\mathbf{Q}_n$ to be such that

$$\mathbf{R}_n\mathbf{Q}_n = \mathbf{Q}_n\mathbf{R}_n = \alpha_n\mathbf{I}_N, \qquad (1.3.13)$$

where $\mathbf{I}_N$ is the identity $N \times N$ matrix and α_n is a scalar, and with this restriction $\mathbf{R}_n = \mp\mathbf{Q}_n^H$ is a consistent reduction of the system (1.3.11a)–(1.3.11b) that results in the single matrix equation

$$i\frac{d}{d\tau}\mathbf{Q}_n = \mathbf{Q}_{n+1} - 2\mathbf{Q}_n + \mathbf{A}\mathbf{Q}_n + \mathbf{Q}_n\mathbf{B} + \mathbf{Q}_{n-1} \mp \mathbf{Q}_n\mathbf{Q}_n^H(\mathbf{Q}_{n+1} + \mathbf{Q}_{n-1}).$$

$$(1.3.14)$$

Similarly, the IST for (1.3.14) follows the same lines as that for (1.3.11a)–(1.3.11b) with additional symmetry conditions imposed. The additional symmetry (1.3.13) (which has no analog in the continuous case) has essential consequences for the IST, which are discussed in detail in Section 5.2.2.

1.4 Physical applications

As indicated in the previous section, NLS systems have broad application in physical problems. In this section we briefly describe how certain NLS systems arise in nonlinear optics. We choose to discuss nonlinear optics because of its many scientific and technological applications. Here we will only sketch the key ideas behind the derivation of the NLS equations for some of the nonlinear optics applications. Interested readers will be able to find additional details and applications in the cited references. It should also be noted that this section can be read independently from the text describing the IST analysis.

We begin with a discussion of pulse propagation in optical fibers. Among the physical properties of optical fibers, nonlinearity and dispersion are serious sources of signal distortion. Signal loss in fibers was also a major limitation;

however, in the 1980s this was largely overcome due to the development of all-optical amplifiers (cf. [24]).

Dispersion originates from the frequency dependence of the refractive index of the fiber and leads to frequency-dependence of the group velocity; this is usually called group velocity dispersion or simply GVD. Due to GVD, different spectral components of an optical pulse propagate at different group velocities and thus arrive at different times. This leads to pulse broadening, resulting in signal distortion.

Fiber nonlinearity is due to the so-called Kerr effect, where the refractive index depends on the intensity of the optical pulse. In the presence of GVD and Kerr nonlinearity, the refractive index is expressed as

$$n(\omega, E) = n_0(\omega) + n_2|E|^2, \tag{1.4.15}$$

where ω and E represent the frequency and electric field of the lightwave, respectively; $n_0(\omega)$ is the frequency-dependent linear refractive index; and the constant n_2, referred to as the Kerr coefficient, has a value of approximately $10^{-22}\,\mathrm{m}^2/\mathrm{W}$. Even though fiber nonlinearity is small, the nonlinear effects accumulate over long distances and can have a significant impact due to the high intensity of the lightwave over the small fiber cross section. By itself, the Kerr nonlinearity produces an intensity-dependent phase shift that results in spectral broadening during propagation.

In the usual transmission process with lightwaves, the electric field is modulated into a slowly varying amplitude of a carrier wave. Concretely, a modulated electromagnetic lightwave is written as

$$E(z, t) = \mathcal{E}(z, t)e^{i(k_0 z - \omega_0 t)} + c.c., \tag{1.4.16}$$

where c.c. denotes complex conjugation, z the distance along the fiber, t the time, $k_0 = k_0(\omega_0)$ the wavenumber, ω_0 the frequency, and $\mathcal{E}(z, t)$ the envelope of the electromagnetic field.

Hasegawa and Tappert [91] first derived the NLS equation in the context of fiber optics. Detailed derivations can be found in texts (cf. Hasegawa and Kodama [90] and references therein). A simplified derivation is conveniently obtained from the nonlinear dispersion relation:

$$k(\omega, E) = \frac{\omega}{c}\left(n_0(\omega) + n_2|E|^2\right), \tag{1.4.17}$$

where c denotes the speed of light.

A Taylor series expansion of $k(\omega, E)$ around the carrier frequency $\omega = \omega_0$ yields

$$k - k_0 = k'(\omega_0)(\omega - \omega_0) + \frac{k''(\omega_0)}{2}(\omega - \omega_0)^2 + \frac{\omega_0 n_2}{c}|E|^2, \tag{1.4.18}$$

where $'$ represents the derivative with respect to ω and $k_0 = k(\omega_0)$. Replacing $k - k_0$ and $\omega - \omega_0$ by their Fourier operator equivalents $i\partial/\partial z$ and $i\partial/\partial t$, respectively, using $k - k_0 = \frac{\omega}{c} n_0(\omega)$, and letting equation (1.4.18) operate on $\mathcal{E}$ yields

$$i\left(\frac{\partial\mathcal{E}}{\partial z} + k_0'(\omega_0)\frac{\partial\mathcal{E}}{\partial t}\right) - \frac{k_0''(\omega_0)}{2}\frac{\partial^2\mathcal{E}}{\partial t^2} + \nu|\mathcal{E}|^2\mathcal{E} = 0, \qquad (1.4.19)$$

where $\nu = \frac{\omega_0 n_2}{cA_{\text{eff}}}$ and A_{eff} is the effective cross-section area of the fiber (the factor $1/A_{\text{eff}}$ comes from a more detailed derivation that takes into account the finite size of the fiber; the factor $1/A_{\text{eff}}$ is needed to account for the variation of field intensity in the cross section of the fiber). We note that $k_0'(\omega_0) = 1/v_g$, where v_g represents the group velocity of the wavetrain.

In order to obtain a dimensionless equation, it is standard to introduce a retarded time coordinate $t_{ret} = t - k_0'(\omega_0)z = t - z/v_g$ and dimensionless variables $t' = t_{ret}/t_*$, $z' = z/z_*$, and $q = \mathcal{E}/\sqrt{P_*}$, where t_*, z_*, P_* are the characteristic time, distance, and power, respectively. Substituting this coordinate transformation, choosing the dimensionless variables as $z_* = 1/\nu P_*$, $t_*^2 = z_*| - k''(\omega_0)|$, and dropping the prime yields the NLS equation

$$i\frac{\partial q}{\partial z} + \frac{\text{sgn}(-k_0''(\omega_0))}{2}\frac{\partial^2 q}{\partial t^2} + |q|^2 q = 0. \qquad (1.4.20)$$

There are two cases of physical interest depending on the sign of $-k_0''(\omega_0)$. The so-called focusing case occurs when $-k_0''(\omega_0) < 0$; this is called "anomalous" dispersion. The defocusing case is obtained when the dispersion is "normal": $-k_0''(\omega_0) > 0$. In the anomalous case we will see later, in Chapter 2, that equation (1.4.20) has a special soliton solution given by

$$q(z, t) = 2\eta e^{-2i\xi t + 2i(\xi^2 - \eta^2)(z - z_0)}\text{sech}2\eta(t - 2\xi z - t_0) \qquad (1.4.21)$$

(see equation (2.3.88) and note the redefinition of variables and the factor of $\frac{1}{2}$ difference between equations (1.4.20) and (2.3.88)).

Since solitons are stable localized pulses, Hasegawa and Tappert (1973) suggested their use as the "bit" format for the transmission of information in optical fiber systems. Remarkably, in 1980 scientists at Bell Laboratories observed the solitons described by the NLS equations (1.4.19)–(1.4.20) in optical fibers [135]. One of the serious difficulties however was fiber loss. Fortuitously, in the mid-1980s scientists developed all-optical amplifiers (erbium-doped amplifiers: EDFAs, [127, 64]). This development allowed for the transmission of information optically over long distances (e.g., 10,000 km, which is roughly the distance from the United States to Japan).

With damping and amplification included, the NLS equation (1.4.20) takes the form

$$i\frac{\partial q}{\partial z} + \frac{\text{sgn}(-k_0''(\omega_0))}{2}\frac{\partial^2 q}{\partial t^2} + g(z)|q|^2 q = 0, \qquad (1.4.22)$$

where $g(z) = a_0^2 \exp(-2\Gamma z/z_a)$, $0 < z < z_a$ and periodically extended thereafter, with Γ the normalized damping coefficient and a_0^2 determined by $< g > = \frac{1}{z_a}\int_0^{z_a} g(z/z_a)dz = 1$ with $z_a = l_a/z_*$, l_a being the amplifier length. Typically z_a is small, approximately 0.1. Asymptotic analysis shows that, to leading order, $q(z,t)$ still satisfies the NLS equation (1.4.20). Note that since $g(z)$ is rapidly varying, equation (1.4.22) is satisfied with g replaced by its averaged value $< g > = 1$ in the period z_a (cf. [90]).

The introduction of amplifiers, however, introduces small amounts of noise to the system. This in turn causes the temporal position of the soliton to fluctuate. In 1986, Gordon and Haus [88] showed that this fluctuation seriously limits the distance signals could be reliably transmitted. Soliton control mechanisms were introduced in the early 1990s to deal with these difficulties (cf. [108], [129], [128], [134]). Due to these and other issues, all-optical transmission systems in the 1990s employed a nonsoliton format referred to as "non-return-to-zero" (NRZ). In the NRZ format a continuous wave is transmitted over the total time slot of successive "1" bits, whereas in the RZ format, which includes solitons, an individual pulse is transmitted for each "1" bit regardless of the sequence. NRZ pulses have a lower peak intensity than RZ pulses having the same average power within a given bit slot. It should be noted, however, that the peak intensity must be enhanced and the pulse width decreased accordingly as the bit rate increases, since the average power within a bit slot must be maintained at a certain level so that the signal is not buried by amplifier noise. Thus NRZ pulses also suffer from nonlinearity, especially at high bit rates.

Importantly, the NLS equations (1.4.20) and (1.4.22) and related equations with perturbations due to dispersion variations, higher order chromatic dispersion, and nonlinear effects are the central models used in fiber communications, regardless of the transmission format. It is not feasible to numerically simulate Maxwell's equations over the distances required (thousands of kilometers) due to the disparate scales inherent in the equations.

The development of all-optical transmission systems also took a major leap forward in the 1990s with the advent of wavelength-division-multiplexing (WDM) (cf. [89]). WDM is the simultaneous transmission of multiple signals in different frequency (or, equivalently, wavelength) channels. In terms of soliton communications, this means that different solitons will travel at different velocities, thus causing interactions, some of which, such as frequency and timing shifts due to WDM collisions, can be debilitating (cf. [133], [130], [3]). One of the central problems that arises from interactions is unwanted resonant amplifier induced instabilities in adjacent frequency channels ([119], [2]). This phenomenon, called four-wave mixing (FWM), was so serious that researchers began to investigate dispersion-managed (DM) transmission systems. DM systems also reduce other unwanted effects, such as Gordon–Haus and collision-induced timing jitters.

In a dispersion-managed transmission system the fiber is composed of alternating sections of positive (normal) and negative (anomalous) dispersion fibers. The (dimensionless) NLS equation that governs this phenomena is

$$i\frac{\partial q}{\partial z} + \frac{d(z)}{2}\frac{\partial^2 q}{\partial t^2} + g(z)|q|^2 q = 0, \tag{1.4.23}$$

where $d(z)$ is usually taken to be a periodic, large, rapidly varying function of the form $d(z) = \delta_a + \Delta(z)$ with $|\Delta(z)| >> 1$ and having zero average in the period z_a (the period is usually taken to be the same as that of the amplifier). There has been considerable research in the field of dispersion-managed transmission systems (cf. [89] and references therein).

Researchers have shown that equation (1.4.23) admits various types of optical pulses, such as DM solitons [80, 4], quasi-linear modes [9], and so on. Importantly, both types of pulses can be described via a unified framework [9, 8]. We shall not delve into this matter any further here, because it would take us well outside the scope of this book.

In many applications vector NLS systems are the key governing equations. In optical fibers with constant birefringence (i.e., constant phase and group velocities as a function of distance) Menyuk [131, 132] has shown that the two polarization components of the electromagnetic field $\mathcal{E} = (u, v)^t$ that are orthogonal to the direction of propagation, z, along the fiber asymptotically satisfy the following nondimensional equations (assuming anomalous dispersion):

$$i(u_z + \delta u_t) + \tfrac{1}{2}u_{tt} + (|u|^2 + \alpha|v|^2)u = 0 \tag{1.4.24a}$$

$$i(v_z - \delta v_t) + \tfrac{1}{2}v_{tt} + (\alpha|u|^2 + |v|^2)u = 0, \tag{1.4.24b}$$

where δ represents the group velocity "mismatch" between the u, v components of the electromagnetic field and α is a constant that depends on the polarization properties of the fiber. In deriving equation (1.4.24) it is assumed that the electromagnetic field is slowly varying (as in the scalar problem); certain nonlinear (four-wave mixing) terms are neglected in the derivation of equation (1.4.24) because the lightwave is rapidly varying due to large, but constant, linear birefringence. In this context, birefringence means that the phase and

group velocities of the electromagnetic wave in each polarization component
are different.

In a communications environment, due to the distances involved (hundreds
of thousands of kilometers), the polarization properties evolve rapidly and randomly as the lightwave evolves along the propagation distance z. Not only does
the birefringence evolve, but it does so randomly, and on a scale much faster
than the distances required for communication transmission (birefringence polarization changes on a scale of 10–100 m). In this case, researchers (cf. [180],
[72], [122]) have shown that the relevant nonlinear equation is equation (1.4.24)
but with $\delta = 0$ and $\alpha = 1$. Indeed, this is the integrable vector NLS equation
first derived by Manakov (1974) and is studied in Chapter 4 of this book via
the IST method. Hence we see that both the scalar- and vector-continuous NLS
systems arise naturally in the field of nonlinear optical fiber communications.

Finally, we briefly mention some applications in discrete NLS systems. Our
prototype application is coupled nonlinear optical waveguides. In this case we
consider an optical material in which waveguides are "etched" into suitable
optical material and each waveguide is well separated from each other in, say,
the x-direction (or "n"-direction) with propagation occurring in the longitudinal,
say z, direction. (See, e.g., Figure 1.1.)

From Maxwell's equations, we model the governing wave equation for the
electromagnetic field in the x-direction in Kerr nonlinear material to be [188]

$$\Psi_{zz} + \Psi_{xx} + (f(x) + \delta|\Psi|^2)\Psi = 0, \qquad (1.4.25)$$

where $|\delta| << 1$ and $f(x)$ models the linear index of refraction. One now expands
the solution Ψ in terms of a suitable series of functions (cf. [16]); that is, Ψ is

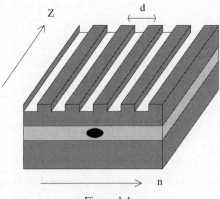

Figure 1.1

expressed in the form

$$\Psi = \sum_{m=-\infty}^{\infty} E_m(Z)\psi_m(x)e^{-i\lambda_0 z}, \qquad (1.4.26)$$

where $Z = \delta z$, $\psi_m = \psi(x - md)$, d is the spacing of the waveguide array (cf. Figure 1.1), and ψ_m has one boundstate eigenvalue λ_0. We then substitute equation (1.4.26) into equation (1.4.25), multiply the result by $\psi_n^*\exp(i\lambda_0 z)$, integrate over x (x ranges from $-\infty < x < \infty$), and assume that $\psi_m(x)$ satisfies

$$\partial_{xx}\psi_m + (f_m(x) - \lambda_0^2)\psi_m = 0, \qquad (1.4.27)$$

where $f_m(x) = \tilde{f}(x - md)$ is a localized function. By carrying out these calculations, we obtain

$$\sum_{m=-\infty}^{+\infty} \int_{-\infty}^{\infty} dx \left[-2i\delta\lambda_0 \psi_m \frac{\partial E_m}{\partial Z} + \left(\frac{\partial^2 \psi_m}{\partial x^2} + (f(x) - \lambda_0^2)\psi_m \right) E_m \right.$$
$$\left. + \delta \sum_{m',m''=-\infty}^{\infty} E_m E_{m'} E_{m''}^* \psi_m \psi_{m'} \psi_{m''}^* \right] \psi_n^* = 0. \qquad (1.4.28)$$

Using (1.4.27), we find that

$$\sum_{m=-\infty}^{\infty} \left[-2i\delta\lambda_0 \frac{\partial E_m}{\partial Z} \int_{-\infty}^{\infty} \psi_m \psi_n^* dx\, E_m \int_{\infty}^{\infty} \Delta f_m \psi_m \psi_n^* dx \right.$$
$$\left. + \delta \sum_{m',m''} E_m E_{m'} E_{m''}^* \int_{-\infty}^{\infty} \psi_m \psi_{m'} \psi_{m''}^* \psi_n^* dx \right] = 0, \qquad (1.4.29)$$

where $\Delta f_m = f(x) - f_m(x)$. The eigenfunctions are assumed to be localized corresponding to waveguides that are well separated (see Figure 1.1). Maximal balance is achieved when the relevant integrals are defined by

$$\int dx(f(x) - f_m(x))|\psi_m|^2 = \delta c_0$$

$$\int dx(f(x) - f_{m\pm1}(x))\psi_m\psi_{m\pm1}^* = \delta c_1$$

$$\int dx|\psi_m|^4 = \delta g_0,$$

where c_0, c_1, g_0 are $O(1)$ constants and higher order terms in δ are dropped. The resulting equation turns out to be the discrete NLS (DNLS) equation in the form

$$i\partial_Z E_n + c_1(E_{n+1} + E_{n-1}) + c_0 E_n + g_0|E_n|^2 E_n = 0. \qquad (1.4.30)$$

By transforming variables, one can put the equation in the form

$$i\partial_Z\phi_n + (\phi_{n+1} + \phi_{n-1} - 2\phi_n)/h^2 + 2|\phi_n|^2\phi_n = 0. \qquad (1.4.31)$$

The DNLS equation was first derived in the context of nonlinear optics by Christodoulides and Joseph [54]. It was studied earlier by Davydov [60] in molecular biology and by Su, Schieffer, and Heeger in condensed matter physics [151]; there are numerous other applications as well (cf. [148], [149], [123]). Various authors studied the interaction and collision processes associated with the localized solitary wave solutions of the DNLS equation (cf. [22], [23], [113]).

Experimentally, the solitary waves of the DNLS equation were observed in a nonlinear optical array by Eisenberg et al. [70] and Morandotti et al. [137]. Linear diffraction management of the optical array system was subsequently studied by Eisenberg et al. [71] and nonlinear focusing and defocusing by Morandotti et al. [138]. Theoretically speaking, diffraction-managed solitons, whose width and peak amplitude vary periodically and which are the discrete analog of dispersion-managed solitons, have been obtained [14]. More recently, DNLS equations have been proposed for Bose–Einstein condensation [166]. Vector extensions of the DNLS equation have also been derived and studied (cf. [15], [16], [57]). While these DNLS equations are not transformable to the integrable discrete systems studied in this book, the integrable systems nevertheless provide useful insight into discrete equations and solitary wave phenomena. For more information about DNLS equations, related problems, and references, the reader may wish to consult the review article [105].

1.5 Outline of the work

Chapter 2 is dedicated to the scalar NLS equation (1.3.1). Chapter 3 is concerned with the IDNLS equation (1.3.8), which is the integrable discretization of the NLS equation. Chapter 4 describes the IST for the matrix nonlinear Schrödinger equation (MNLS) and, in particular, for the VNLS system (1.3.3), and Chapter 5 describes the integrable discrete matrix NLS (IDMNLS), that is, the system (1.3.11a)–(1.3.11b) with special attention to the case where the symmetry (1.3.12)–(1.3.13) holds, including the reduction to the integrable discrete vector NLS equation (1.3.10).

For pedagogical reasons we study separately the scalar and the vector cases. We believe that understanding the scalar case is helpful in the comprehension of the vector systems. We have followed a similar outline in each chapter. For each system, we give explicit conditions on the solution of the evolution equation such that the eigenfunctions are analytic, and we show how to obtain the scattering data from the analytic eigenfunctions. Moreover, for each system, we determine symmetries in the scattering data that are induced by symmetries in the eigenfunctions, and, in the formulation of the inverse problem, we include these symmetries in the scattering data. For the inverse problem, we first formulate a generalized Riemann–Hilbert problem. Then we obtain a system of linear algebraic–integral equations from the Riemann–Hilbert problem (see also [6], where this approach is described for the KdV equation). These algebraic–integral equations determine the eigenfunctions in terms of the scattering data, and from these equations we also show how to obtain Gel'fand–Levitan–Marchenko (GLM) integral equations. Moreover, we prove that the GLM equations do not have any homogeneous solutions. As a special case of the inverse problem, the respective soliton solutions are constructed for each of the nonlinear Schrödinger systems. The evolution of the solitons of each system follows from the evolution of the scattering data, as determined by the associated linear time-dependence operators. Then soliton interactions are analyzed, and, for the vector systems, the description of the soliton interactions includes a determination of the polarization shift induced by soliton collisions. Finally, we show how the IST machinery can be used to generate conserved quantities for each of these systems. Beyond the basic outline, the parallel structure of these chapters illustrates the generality of the IST approach for this class of systems. From chapter to chapter, one can compare the IST for each of these systems. Both discretization and extension from scalar to vector systems introduce some additional complications in what is essentially the same IST scheme. We note that an additional symmetry in the discrete vector system introduces a further difficulty in the IST and in the solutions. A careful handling of this symmetry is required for the correct formulation of the IST in this case.

In the appendices we have included, among other items, a detailed discussion of the direct and inverse scattering associated with the linear discrete Schrödinger equation. Importantly, this equation is connected to the IDNLS equation. In fact, an appropriate linearization of the IDNLS leads to this linear scattering problem. Moreover, the linear Schrödinger scattering problem forms the basis of the IST for both the Toda lattice and the nonlinear ladder network. Hence, for completeness we have included a discussion of the IST for these two cases. Moreover, we have also briefly described the IST associated with NLS systems with an additional potential term. Finally, we discuss the "extreme" limit of large amplitude for these systems and show that they are integrable.

Chapter 2
Nonlinear Schrödinger equation (NLS)

2.1 Overview

In this chapter we develop the inverse scattering transform for the NLS equation,

$$iq_t = q_{xx} \pm 2|q|^2 q, \qquad (2.1.1)$$

on the infinite line. In particular, we formulate the IST procedure for the somewhat more general system (cf. [10])

$$iq_t = q_{xx} - 2rq^2 \qquad (2.1.2a)$$
$$-ir_t = r_{xx} - 2qr^2 \qquad (2.1.2b)$$

and then consider the reductions $r = \mp q^*$ as a special case.

The IST can be broken into three parts: (i) The direct problem – constructing x-independent scattering data from the "potentials" q, r (see (2.2.3)). As far as nomenclature is concerned, the term "potential" is used in reference to the solution of a scattering problem such as (2.2.3). The functions q, r, supplemented with suitable time-dependence, as will be described later, are the solution to the nonlinear evolution equations (2.1.2). (ii) The inverse problem – reconstructing the potentials from the scattering data. (iii) Time evolution – determining the evolution of the scattering data by making use of the time-dependence operator (see (2.2.4)). The IST procedure for solving the initial-value problem proceeds by first constructing the scattering data at $t = t_i$ from the initial data $q(t_i)$, $r(t_i)$ – that is, step (i) – then computing the evolution of the scattering data from t_i to $t \neq t_i$ – step (iii) – and, finally, recovering $q(t), r(t)$ by solving the inverse problem – step (ii).

The treatment of the direct problem given here follows [6], [21] while the inverse problem is formulated as a Riemann–Hilbert boundary-value problem, following [6]. The Gel'fand–Levitan–Marchenko integral equation formulation of the inverse problem follows from the Riemann–Hilbert formulation.

As is typical for soliton equations, the soliton solutions of the NLS correspond to the case in which the inverse problem reduces from a system of linear algebraic–integral equations to a system of linear algebraic equations and can be solved explicitly. By solving the algebraic system, we obtain an analytic expression for a single soliton and an explicit formula for the phase shifts that result from soliton interactions.

Finally, with the machinery of the inverse scattering transform, one can derive infinitely many conserved quantities for the NLS. Among these is the Hamiltonian of the NLS. We give an account of the formulation of the NLS as a Hamiltonian system.

2.2 The inverse scattering transform for NLS

2.2.1 Operator pair

As noted previously, an essential prerequisite to the development of the IST for a nonlinear evolution equation is the association of the evolution equation with a pair of linear equations. The association of the scalar NLS equation (2.1.1) with the linear equations

$$v_x = \begin{pmatrix} -ik & q \\ r & ik \end{pmatrix} v, \tag{2.2.3}$$

sometimes referred to in the literature as Zakharov–Shabat or ZS/AKNS or AKNS spectral problem, and

$$v_t = \begin{pmatrix} 2ik^2 + iqr & -2kq - iq_x \\ -2kr + ir_x & -2ik^2 - iqr \end{pmatrix} v, \tag{2.2.4}$$

where v is a two-component vector, $v(x, t) = \left(v^{(1)}(x, t), v^{(2)}(x, t)\right)^T$, is well known (cf. [21], [140]). A straightforward calculation shows that the equality of the mixed derivatives of v, that is, $v_{xt} = v_{tx}$, is equivalent to the statement that q and r satisfy the evolution equations (2.1.2a)–(2.1.2b), if k, the scattering parameter, is independent of x and t. For this reason, the evolution equations (2.1.2a)–(2.1.2b) are sometimes referred to as the *compatibility condition* of equations (2.2.3)–(2.2.4). We refer to the equation with the x derivative, equation (2.2.3), as the *scattering problem* and the equation with the t derivative, equation (2.2.4), as the *time-dependence*.

2.2.2 Direct scattering problem

Jost functions and integral equations

We refer to solutions of the scattering problem (2.2.3) as eigenfunctions with respect to the parameter k. When the potentials $q, r \to 0$ rapidly as $x \to \pm\infty$, the eigenfunctions are asymptotic to the solutions of

$$v_x = \begin{pmatrix} -ik & 0 \\ 0 & ik \end{pmatrix} v$$

when $|x|$ is large. Therefore it is natural to introduce the eigenfunctions defined by the following boundary conditions:

$$\phi(x, k) \sim \begin{pmatrix} 1 \\ 0 \end{pmatrix} e^{-ikx}, \qquad \bar{\phi}(x, k) \sim \begin{pmatrix} 0 \\ 1 \end{pmatrix} e^{ikx} \qquad \text{as } x \to -\infty \quad (2.2.5)$$

$$\psi(x, k) \sim \begin{pmatrix} 0 \\ 1 \end{pmatrix} e^{ikx}, \qquad \bar{\psi}(x, k) \sim \begin{pmatrix} 1 \\ 0 \end{pmatrix} e^{-ikx} \qquad \text{as } x \to +\infty. \quad (2.2.6)$$

In the following analysis, it is convenient to consider functions with constant boundary conditions. Hence, we define the Jost functions as follows:

$$M(x, k) = e^{ikx}\phi(x, k), \qquad \bar{M}(x, k) = e^{-ikx}\bar{\phi}(x, k), \qquad (2.2.7a)$$

$$N(x, k) = e^{-ikx}\psi(x, k), \qquad \bar{N}(x, k) = e^{ikx}\bar{\psi}(x, k). \qquad (2.2.7b)$$

If the scattering problem (2.2.3) is rewritten as

$$v_x = (ik\mathbf{J} + \mathbf{Q})\, v, \qquad (2.2.8)$$

where

$$\mathbf{J} = \begin{pmatrix} -1 & 0 \\ 0 & 1 \end{pmatrix}, \qquad \mathbf{Q} = \begin{pmatrix} 0 & q \\ r & 0 \end{pmatrix}, \qquad (2.2.9)$$

and $\mathbf{I}$ denotes the 2×2 identity matrix, then the Jost functions $M(x, k)$ and $\bar{N}(x, k)$ are solutions of the differential equation

$$\chi_x(x, k) = ik(\mathbf{J} + \mathbf{I})\chi(x, k) + (\mathbf{Q}\chi)(x, k), \qquad (2.2.10)$$

while $N(x, k)$ and $\bar{M}(x, k)$ satisfy

$$\tilde{\chi}_x(x, k) = ik(\mathbf{J} - \mathbf{I})\tilde{\chi}(x, k) + (\mathbf{Q}\tilde{\chi})(x, k) \qquad (2.2.11)$$

with the constant boundary conditions

$$M(x, k) \to \begin{pmatrix} 1 \\ 0 \end{pmatrix}, \qquad \bar{M}(x, k) \to \begin{pmatrix} 0 \\ 1 \end{pmatrix} \qquad \text{as } x \to -\infty \quad (2.2.12a)$$

$$N(x, k) \to \begin{pmatrix} 0 \\ 1 \end{pmatrix}, \qquad \bar{N}(x, k) \to \begin{pmatrix} 1 \\ 0 \end{pmatrix} \qquad \text{as } x \to +\infty. \quad (2.2.12b)$$

Solutions of the differential equations (2.2.10)–(2.2.11) can be represented by means of the following integral equations:

$$\chi(x, k) = w + \int_{-\infty}^{+\infty} \mathbf{G}(x - x', k)(\mathbf{Q}\chi)(x', k)dx'$$

$$\tilde{\chi}(x, k) = \tilde{w} + \int_{-\infty}^{+\infty} \tilde{\mathbf{G}}(x - x', k)(\mathbf{Q}\tilde{\chi})(x', k)dx'$$

or, in component form, for $j = 1, 2,$

$$\chi^{(j)}(x, k) = w^{(j)} + \int_{-\infty}^{+\infty} \sum_{\ell=1}^{2} G^{(j\ell)}(x - x', k)(\mathbf{Q}\chi)^{(\ell)}(x', k)dx'$$

$$\tilde{\chi}^{(j)}(x, k) = \tilde{w}^{(j)} + \int_{-\infty}^{+\infty} \sum_{\ell=1}^{2} \tilde{G}^{(j\ell)}(x - x', k)(\mathbf{Q}\tilde{\chi})^{(\ell)}(x', k)dx',$$

where $w = (w^{(1)}, 0)^T$, $\tilde{w} = (0, \tilde{w}^{(2)})^T$ and the (matrix) Green's functions $\mathbf{G}(x, k) = (G^{(j\ell)}(x, k))_{j,\ell=1,2}$ and $\tilde{\mathbf{G}}(x, k) = (\tilde{G}^{(j\ell)}(x, k))_{j,\ell=1,2}$ satisfy the differential equations

$$\mathcal{L}_0 \mathbf{G}(x, k) = \delta(x)\mathbf{I}, \qquad \tilde{\mathcal{L}}_0 \tilde{\mathbf{G}}(x, k) = \delta(x)\mathbf{I} \qquad (2.2.13)$$

$$\mathcal{L}_0 = \mathbf{I}\,\partial_x - ik(\mathbf{J} + \mathbf{I}), \qquad \tilde{\mathcal{L}}_0 = \mathbf{I}\,\partial_x - ik(\mathbf{J} - \mathbf{I}). \qquad (2.2.14)$$

The Green's functions are not unique, and, as we will show later, the choice of the Green's function and the choice of the inhomogeneous term together uniquely determine the Jost function and its analytic properties.

By using the Fourier transform method, it is easy to find

$$\mathbf{G}(x, k) = \frac{1}{2\pi i} \int_C \begin{pmatrix} p^{-1} & 0 \\ 0 & (p - 2k)^{-1} \end{pmatrix} e^{ipx} dp$$

$$\tilde{\mathbf{G}}(x, k) = \frac{1}{2\pi i} \int_{\tilde{C}} \begin{pmatrix} (p + 2k)^{-1} & 0 \\ 0 & p^{-1} \end{pmatrix} e^{ipx} dp,$$

where C and $\tilde{C}$ are appropriate contours. It is natural to consider $\mathbf{G}_\pm(x, k)$ and $\tilde{\mathbf{G}}_\pm(x, k)$ defined by

$$\mathbf{G}_\pm(x, k) = \frac{1}{2\pi i} \int_{C_\pm} \begin{pmatrix} p^{-1} & 0 \\ 0 & (p - 2k)^{-1} \end{pmatrix} e^{ipx} dp$$

$$\tilde{\mathbf{G}}_\pm(x, k) = \frac{1}{2\pi i} \int_{C_\pm} \begin{pmatrix} (p + 2k)^{-1} & 0 \\ 0 & p^{-1} \end{pmatrix} e^{ipx} dp,$$

where $C_\pm$ are the contours from $-\infty$ to $+\infty$ that, respectively, pass below and above both the singularities at $p = 0$ and $p = 2k$ (see Figure 2.1).

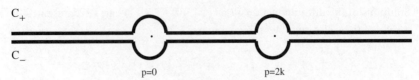

p=0   p=2k

Figure 2.1: The contours C_+ and C_-

Therefore,

$$\mathbf{G}_{\pm}(x, k) = \pm\theta(\pm x)\begin{pmatrix} 1 & 0 \\ 0 & e^{2ikx} \end{pmatrix}, \qquad \tilde{\mathbf{G}}_{\pm}(x, k) = \mp\theta(\mp x)\begin{pmatrix} e^{-2ikx} & 0 \\ 0 & 1 \end{pmatrix},$$
(2.2.15)

where $\theta(x)$ is the Heaviside function ($\theta(x) = 1$ if $x > 0$ and $\theta(x) = 0$ if $x < 0$).
The " $+$ " functions are analytic in the upper half-plane of k and the " $-$ " functions are analytic in the lower half-plane. By taking into account the boundary
conditions (2.2.12a)–(2.2.12b), we obtain the following integral equations for
the Jost solutions:

$$M(x, k) = \begin{pmatrix} 1 \\ 0 \end{pmatrix} + \int_{-\infty}^{+\infty} \mathbf{G}_+(x - x', k)(\mathbf{Q}M)(x', k)dx' \qquad (2.2.16a)$$

$$N(x, k) = \begin{pmatrix} 0 \\ 1 \end{pmatrix} + \int_{-\infty}^{+\infty} \tilde{\mathbf{G}}_+(x - x', k)(\mathbf{Q}N)(x', k)dx' \qquad (2.2.16b)$$

$$\bar{M}(x, k) = \begin{pmatrix} 0 \\ 1 \end{pmatrix} + \int_{-\infty}^{+\infty} \tilde{\mathbf{G}}_-(x - x', k)(\mathbf{Q}\bar{M})(x', k)dx' \qquad (2.2.16c)$$

$$\bar{N}(x, k) = \begin{pmatrix} 1 \\ 0 \end{pmatrix} + \int_{-\infty}^{+\infty} \mathbf{G}_-(x - x', k)(\mathbf{Q}\bar{N})(x', k)dx'. \qquad (2.2.16d)$$

Equations (2.2.16a)–(2.2.16d) are Volterra integral equations. We show in the
following Lemma 2.1 that if $q, r \in L^1(\mathbb{R})$, the Neumann series of the integral
equations for M and N converge absolutely and uniformly (in x and k) in the
upper k-plane, while the Neumann series of the integral equations for $\bar{M}$ and $\bar{N}$
converge absolutely and uniformly (in x and k) in the lower k-plane. These facts
immediately imply that the Jost functions $M(x, k)$ and $N(x, k)$ are analytic functions of k for Im $k > 0$ and continuous for Im $k \geq 0$, while $\bar{M}(x, k)$, and $\bar{N}(x, k)$
are analytic functions of k for Im $k < 0$ and continuous for Im $k \leq 0$.

Lemma 2.1 *If* $q, r \in L^1(\mathbb{R})$, *then* $M(x, k)$, $N(x, k)$ *defined by (2.2.16a)–
(2.2.16b) are analytic functions of k for* Im $k > 0$ *and continuous for* Im $k \geq 0$,
while $\bar{M}(x, k)$, $\bar{N}(x, k)$ *defined by (2.2.16c)–(2.2.16d) are analytic functions of
k for* Im $k < 0$ *and continuous for* Im $k \leq 0$. *Moreover, the solutions are unique
in the space of continuous functions.*

Proof We prove the result for $M(x, k)$. The proofs are analogous for the remaining eigenfunctions.

The Neumann series

$$M(x, k) = \sum_{j=0}^{\infty} C_j(x, k), \qquad (2.2.17)$$

where

$$C_{j+1}(x, k) = \int_{-\infty}^{+\infty} \mathbf{G}_+(x - x', k) \left(\mathbf{Q}C_j\right)(x', k)dx' \qquad (2.2.18a)$$

$$C_0(x, k) = \begin{pmatrix} 1 \\ 0 \end{pmatrix}, \qquad (2.2.18b)$$

is, formally, a solution of the integral equation (2.2.16a). In component form,

$$C_{j+1}^{(1)}(x, k) = \int_{-\infty}^{x} q(x')C_j^{(2)}(x', k)dx'$$

$$C_{j+1}^{(2)}(x, k) = \int_{-\infty}^{x} e^{2ik(x-x')}r(x')C_j^{(1)}(x', k)dx'.$$

Because $C_0^{(2)} = 0$, we have for any integer $j \geq 0$

$$C_{2j+1}^{(1)} = 0, \qquad C_{2j}^{(2)} = 0.$$

Using the identities

$$\frac{1}{j!} \int_{-\infty}^{x} |f(\xi)| \left[\int_{-\infty}^{\xi} |f(\xi')| d\xi'\right]^{j} d\xi$$

$$= \frac{1}{(j+1)!} \int_{-\infty}^{x} \frac{d}{d\xi} \left[\int_{-\infty}^{\xi} |f(\xi')| d\xi'\right]^{j+1} d\xi$$

$$= \frac{1}{(j+1)!} \left[\int_{-\infty}^{x} |f(\xi)| d\xi\right]^{j+1}, \qquad (2.2.19)$$

where $f \in L^1(\mathbb{R})$, one can show by induction that for $\operatorname{Im} k \geq 0$

$$\left|C_{2j+1}^{(2)}(x, k)\right| \leq \frac{\left(\int_{-\infty}^{x} |q(x')| dx'\right)^{j}}{j!} \frac{\left(\int_{-\infty}^{x} |r(x')| dx'\right)^{j+1}}{(j+1)!}$$

$$\left|C_{2j}^{(1)}(x, k)\right| \leq \frac{\left(\int_{-\infty}^{x} |q(x')| dx'\right)^{j}}{j!} \frac{\left(\int_{-\infty}^{x} |r(x')| dx'\right)^{j}}{j!}.$$

Therefore, if $q, r \in L^1(\mathbb{R})$, the series (2.2.17) is majorized in norm by a uniformly convergent power series. Hence, the Neumann series (2.2.17) is itself uniformly convergent for $\operatorname{Im} k \geq 0$. It follows that $M(x, k)$ is analytic for

$\text{Im} \, k > 0$ because a uniformly convergent series of analytic functions converges to an analytic function (cf. [7]). Similarly, $M(x, k)$ is continuous for $\text{Im} \, k \geq 0$.

A proof of the uniqueness of the solution for the integral equation (2.2.16a) can be obtained as follows. Assume $M_1(x, k)$ and $M_2(x, k)$ both solve the integral equation (2.2.16a). Then their difference, $m(x, k) = M_1(x, k) - M_2(x, k)$, is such that

$$m(x, k) = \int_{-\infty}^{+\infty} \mathbf{G}_+(x - \xi, k) \, (\mathbf{Q}m) \, (\xi, k) d\xi$$

or, explicitly,

$$m^{(1)}(x, k) = \int_{-\infty}^{x} d\xi \, q(\xi) \int_{-\infty}^{\xi} d\xi' \, e^{2ik(\xi - \xi')} r(\xi') m^{(1)}(\xi', k)$$

$$m^{(2)}(x, k) = \int_{-\infty}^{x} e^{2ik(x - \xi)} r(\xi) m^{(1)}(\xi, k) d\xi.$$

The first equation yields the bound

$$\left| m^{(1)}(x, k) \right| \leq \int_{-\infty}^{x} d\xi \, |q(\xi)| \int_{-\infty}^{\xi} d\xi' \, |r(\xi')| \left| m^{(1)}(\xi', k) \right|$$

and iterating once

$$\left| m^{(1)}(x, k) \right| \leq \int_{-\infty}^{x} d\xi_1 \, |q(\xi_1)| \int_{-\infty}^{\xi_1} d\xi_1' \, |r(\xi_1')|$$
$$\times \int_{-\infty}^{\xi_1'} d\xi_2 \, |q(\xi_2)| \int_{-\infty}^{\xi_2} d\xi_2' \, |r(\xi_2')| \left| m^{(1)}(\xi_2', k) \right|.$$

Assuming that $m^{(1)}(x, k)$ is bounded, that is, $\left| m^{(1)}(x, k) \right| \leq C$, and using the identity (2.2.19), we obtain the bound

$$\left| m^{(1)}(x, k) \right| \leq C \frac{\left(\int_{-\infty}^{+\infty} d\xi \, |q(\xi)| \right)^2}{2} \frac{\left(\int_{-\infty}^{+\infty} d\xi \, |r(\xi)| \right)^2}{2}.$$

Iterating n times yields

$$\left| m^{(1)}(x, k) \right| \leq C \frac{\left(\int_{-\infty}^{+\infty} d\xi \, |q(\xi)| \right)^n}{n!} \frac{\left(\int_{-\infty}^{+\infty} d\xi \, |r(\xi)| \right)^n}{n!}.$$

Since the right-hand side tends to 0 as $n \to \infty$, $m(x, k)$ is identically zero and the solution of the integral equation (2.2.16a) is unique in the space of continuous functions.

Simply requiring $q, r \in L^1(\mathbb{R})$ does not yield analyticity on the real axis; more stringent conditions must be imposed. For instance, using the ideas in

Lemma 2.1, one can show that if

$$|r(x)| \le Ce^{-2K|x|}, \qquad |q(x)| \le Ce^{-2K|x|},$$

where C and K are some positive constants, then M and N are analytic for all k with Im $k \ge -K$ and $\bar{M}, \bar{N}$ are analytic for all k with Im $k < K$. Having r, q vanishing faster than any exponential as $|x| \to \infty$ implies that all four eigenfunctions are entire functions of k. We recall that (Volterra equations on a finite interval always have absolutely convergent Neumann series solutions.)

From the integral equations (2.2.16a)–(2.2.16d) we can compute the asymptotic expansion for large k of the Jost functions. Integration by parts yields

$$M(x, k) = \begin{pmatrix} 1 - \frac{1}{2ik} \int_{-\infty}^{x} q(x')r(x')dx' \\ -\frac{1}{2ik}r(x) \end{pmatrix} + O(k^{-2}) \qquad (2.2.20a)$$

$$\bar{N}(x, k) = \begin{pmatrix} 1 + \frac{1}{2ik} \int_{x}^{+\infty} q(x')r(x')dx' \\ -\frac{1}{2ik}r(x) \end{pmatrix} + O(k^{-2}) \qquad (2.2.20b)$$

$$N(x, k) = \begin{pmatrix} \frac{1}{2ik}q(x) \\ 1 - \frac{1}{2ik} \int_{x}^{+\infty} q(x')r(x')dx' \end{pmatrix} + O(k^{-2}) \qquad (2.2.20c)$$

$$\bar{M}(x, k) = \begin{pmatrix} \frac{1}{2ik}q(x) \\ 1 + \frac{1}{2ik} \int_{-\infty}^{x} q(x')r(x')dx' \end{pmatrix} + O(k^{-2}). \qquad (2.2.20d)$$

Scattering data

The two eigenfunctions with fixed boundary conditions as $x \to -\infty$ are linearly independent, as are the two eigenfunctions with fixed boundary conditions as $x \to +\infty$. Indeed, if $u(x, k) = \left(u^{(1)}(x, k), u^{(2)}(x, k)\right)^T$ and $v(x, k) = (v^{(1)}(x, k), v^{(2)}(x, k))^T$ are any two solutions of (2.2.3), we have

$$\frac{d}{dx}W(u, v) = 0, \qquad (2.2.21)$$

where the Wronskian of u and v, $W(u, v)$, is given by

$$W(u, v) = u^{(1)}v^{(2)} - u^{(2)}v^{(1)}. \qquad (2.2.22)$$

From the asymptotics (2.2.5)–(2.2.6) it follows that

$$W\left(\phi, \bar{\phi}\right) = \lim_{x \to -\infty} W\left(\phi(x, k), \bar{\phi}(x, k)\right) = 1 \qquad (2.2.23a)$$

$$W\left(\psi, \bar{\psi}\right) = \lim_{x \to +\infty} W\left(\psi(x, k), \bar{\psi}(x, k)\right) = -1, \qquad (2.2.23b)$$

which proves that the functions ϕ and $\bar{\phi}$ are linearly independent, as are ψ and $\bar{\psi}$. Therefore, we can write $\phi(x, k)$ and $\bar{\phi}(x, k)$ as linear combinations of $\psi(x, k)$

and $\bar{\psi}(x, k)$, or vice versa. The coefficients of these linear combinations depend on k. Hence, the relations

$$\phi(x, k) = b(k)\psi(x, k) + a(k)\bar{\psi}(x, k) \tag{2.2.24a}$$
$$\bar{\phi}(x, k) = \bar{a}(k)\psi(x, k) + \bar{b}(k)\bar{\psi}(x, k) \tag{2.2.24b}$$

hold for any k such that all four eigenfunctions exist. In particular, (2.2.24a)–(2.2.24b) hold for Im $k = 0$ and define the scattering coefficients $a(k)$, $\bar{a}(k)$, $b(k)$, and $\bar{b}(k)$. Comparing the asymptotics of $W\left(\phi, \bar{\phi}\right)$ as $x \to \pm\infty$ with equations (2.2.24a)–(2.2.24b) shows that the scattering data satisfy the following characterization equation:

$$a(k)\bar{a}(k) - b(k)\bar{b}(k) = 1. \tag{2.2.25}$$

The scattering coefficients can be represented as Wronskians of the Jost functions. Indeed, from equations (2.2.24a)–(2.2.24b) it follows that

$$a(k) = W(\phi, \psi), \qquad \bar{a}(k) = -W(\bar{\phi}, \bar{\psi}) \tag{2.2.26a}$$
$$b(k) = -W\left(\phi, \bar{\psi}\right), \qquad \bar{b}(k) = W\left(\bar{\phi}, \psi\right). \tag{2.2.26b}$$

Therefore, as long as $q, r \in L^1(\mathbb{R})$, Lemma 2.1 and equations (2.2.26a) immediately imply that $a(k)$ is analytic in the upper k-plane while $\bar{a}(k)$ is analytic in the lower k-plane. In general, $b(k)$ and $\bar{b}(k)$ cannot be extended off the real k-axis.

Alternatively, one can derive the following integral relationships for the scattering coefficients:

$$a(k) = 1 + \int_{-\infty}^{+\infty} q(x')M^{(2)}(x', k)dx' \tag{2.2.27a}$$

$$b(k) = \int_{-\infty}^{+\infty} e^{-2ikx'}r(x')M^{(1)}(x', k)dx' \tag{2.2.27b}$$

$$\bar{a}(k) = 1 + \int_{-\infty}^{+\infty} r(x')\bar{M}^{(1)}(x', k)dx' \tag{2.2.27c}$$

$$\bar{b}(k) = \int_{-\infty}^{+\infty} e^{2ikx'}q(x')\bar{M}^{(2)}(x', k)dx', \tag{2.2.27d}$$

where $M^{(j)}$, $\bar{M}^{(j)}$ for $j = 1, 2$ denote the j-th component of vectors M and $\bar{M}$, respectively.

Indeed, let us introduce

$$\Delta(x, k) = M(x, k) - a(k)\bar{N}(x, k).$$

Using the integral equations (2.2.16a), (2.2.16d) for M and $\bar{N}$ and the relation

$$\mathbf{G}_+(x, k) - \mathbf{G}_-(x, k) = \begin{pmatrix} 1 & 0 \\ 0 & e^{2ikx} \end{pmatrix},$$

we can write

$$\Delta(x,k) - \int_{-\infty}^{+\infty} \mathbf{G}_-(x-x',k)\,(\mathbf{Q}\Delta)\,(x',k)dx'$$

$$= \begin{pmatrix} 1 - a(k) \\ 0 \end{pmatrix} - \int_{-\infty}^{+\infty} \begin{pmatrix} q(x')M^{(2)}(x',k) \\ e^{2ik(x-x')}r(x')M^{(1)}(x',k) \end{pmatrix} dx'. \quad (2.2.28)$$

From the other side, the scattering equation (2.2.24a) yields

$$\Delta(x,k) = b(k)e^{2ikx}N(x,k),$$

and then

$$\Delta(x,k) - \int_{-\infty}^{+\infty} \mathbf{G}_-(x-x',k)\,(\mathbf{Q}\Delta)\,(x',k)\,dx' = \begin{pmatrix} 0 \\ b(k)e^{2ikx} \end{pmatrix}, \quad (2.2.29)$$

where we used the integral equation (2.2.16b) for $N(x,k)$ as well as the identity

$$\mathbf{G}_-(x,k) = e^{2ikx}\tilde{\mathbf{G}}_+(x,k).$$

By comparing (2.2.28) and (2.2.29) we get the integral representations (2.2.27a) and (2.2.27b). Equations (2.2.27c) and (2.2.27d) are derived analogously, by considering $\bar{\Delta}(x,k) = \bar{M}(x,k) - \bar{a}(k)N(x,k)$.

From the integral representations (2.2.27a) and (2.2.27c) and the asymptotics (2.2.20a), (2.2.20d), it also follows that

$$a(k) = 1 - \frac{1}{2ik}\int_{-\infty}^{+\infty} q(x')r(x')dx' + O(k^{-2}) \qquad \text{Im } k > 0 \quad (2.2.30a)$$

$$\bar{a}(k) = 1 + \frac{1}{2ik}\int_{-\infty}^{+\infty} q(x')r(x')dx' + O(k^{-2}) \qquad \text{Im } k < 0. \quad (2.2.30b)$$

Note that equations (2.2.24a) and (2.2.24b) can be written as

$$\mu(x,k) = \bar{N}(x,k) + \rho(k)e^{2ikx}N(x,k) \qquad (2.2.31a)$$
$$\bar{\mu}(x,k) = N(x,k) + \bar{\rho}(k)e^{-2ikx}\bar{N}(x,k), \qquad (2.2.31b)$$

where we introduced

$$\mu(x,k) = M(x,k)a^{-1}(k), \qquad \bar{\mu}(x,k) = \bar{M}(x,k)\bar{a}^{-1}(k) \qquad (2.2.32)$$

and the reflection coefficients

$$\rho(k) = b(k)a^{-1}(k), \qquad \bar{\rho}(k) = \bar{b}(k)\bar{a}^{-1}(k). \qquad (2.2.33)$$

Proper eigenvalues and norming constants

A proper eigenvalue of the scattering problem (2.2.3) is a complex value of k (Im $k \neq 0$) corresponding to a bounded solution v such that $v \to 0$ as $x \to \pm\infty$.

Suppose $a(k_j) = 0$ for some $k_j = \xi_j + i\eta_j$, $\eta_j > 0$. Then from (2.2.26a) it follows that $W(\phi(x, k_j), \psi(x, k_j)) = 0$, and therefore $\phi_j(x) = \phi(x, k_j)$ and $\psi_j(x) = \psi(x, k_j)$ are linearly dependent, that is, there exists a complex constant c_j such that

$$\phi_j(x) = c_j \psi_j(x). \tag{2.2.34}$$

Hence, by (2.2.5) and (2.2.6) it follows that

$$\phi_j(x) \sim \begin{pmatrix} 1 \\ 0 \end{pmatrix} e^{\eta_j x - i\xi_j x} \qquad \text{as } x \to -\infty$$

$$\phi_j(x) = c_j \psi_j(x) \sim c_j \begin{pmatrix} 0 \\ 1 \end{pmatrix} e^{-\eta_j x + i\xi_j x} \qquad \text{as } x \to +\infty,$$

and therefore k_j is a proper eigenvalue. On the other hand, if $a(k) \neq 0$, then any solution of the scattering problem blows up in one or both directions. Hence, the proper eigenvalues in the region Im $k > 0$ are the zeros of $a(k)$.

Similarly, the eigenvalues in the region Im $k < 0$ are the zeros of $\bar{a}(k)$, and these zeros $\bar{k}_j = \bar{\xi}_j + i\bar{\eta}_j$, $\bar{\eta}_j < 0$ for $j = 1, \dots, \bar{J}$, are such that

$$\bar{\phi}_j(x) = \bar{c}_j \bar{\psi}_j(x) \tag{2.2.35}$$

for some complex constant $\bar{c}_j$, where, as before, $\bar{\phi}_j(x) = \bar{\phi}(x, \bar{k}_j)$ and $\bar{\psi}_j(x) = \bar{\psi}(x, \bar{k}_j)$. The coefficients $\{c_j\}_{j=1}^J$ and $\{\bar{c}_j\}_{j=1}^{\bar{J}}$ are called *norming constants*. In terms of the eigenfunctions with fixed boundary conditions, the norming constants are defined by

$$M_j(x) = e^{2ik_j x} c_j N_j(x), \qquad \bar{M}_j(x) = e^{-2i\bar{k}_j x} \bar{c}_j \bar{N}_j(x). \tag{2.2.36}$$

In the following we assume that all the eigenvalues are simple zeros of $a(k)$ and $\bar{a}(k)$. If the eigenvalues are not simple zeros, one can study the situation by the coalescence of simple poles (see [192]).

Note that if the potentials are rapidly decaying such that equations (2.2.24a)–(2.2.24b) can be analytically extended off the real axis, then

$$c_j = b(k_j), \qquad j = 1, \dots, J \qquad \bar{c}_l = \bar{b}(\bar{k}_l), \qquad l = 1, \dots, \bar{J}.$$

Symmetry reductions

The evolution equation (2.1.1) is a special case of the system (2.1.2a)–(2.1.2b) under the symmetry reduction

$$r = \mp q^*. \tag{2.2.37}$$

This symmetry in the potential induces a symmetry between the Jost functions analytic in the upper k-plane and the Jost functions analytic in the lower k-plane.

In turn, this symmetry of the Jost functions induces a symmetry in the scattering data.

Indeed, if $v(x, k) = (v^{(1)}(x, k), v^{(2)}(x, k))^T$ satisfies equation (2.2.3) and symmetry (2.2.37) holds, then $\hat{v}(x, k) = (v^{(2)}(x, k^*), \mp v^{(1)}(x, k^*))^H$ also satisfies equation (2.2.3). Taking into account the boundary conditions (2.2.5)–(2.2.6), we get

$$\bar{\psi}(x, k) = \begin{pmatrix} \psi^{(2)}(x, k^*) \\ \mp \psi^{(1)}(x, k^*) \end{pmatrix}^*, \qquad \bar{\phi}(x, k) = \begin{pmatrix} \mp \phi^{(2)}(x, k^*) \\ \phi^{(1)}(x, k^*) \end{pmatrix}^* \qquad (2.2.38a)$$

$$\bar{N}(x, k) = \begin{pmatrix} N^{(2)}(x, k^*) \\ \mp N^{(1)}(x, k^*) \end{pmatrix}^*, \qquad \bar{M}(x, k) = \begin{pmatrix} \mp M^{(2)}(x, k^*) \\ M^{(1)}(x, k^*) \end{pmatrix}^*.$$

$$(2.2.38b)$$

Then, from the integral representations (2.2.27a)–(2.2.27d) for the scattering data, it follows that

$$\bar{a}(k) = a^*(k^*) \qquad (2.2.39a)$$
$$\bar{b}(k) = \mp b^*(k^*) \qquad (2.2.39b)$$

and consequently

$$\bar{\rho}(k) = \mp \rho^*(k) \qquad \text{Im } k = 0. \qquad (2.2.40)$$

From (2.2.39a) it follows that k_j is a zero of $a(k)$ in the upper k-plane iff k_j^* is a zero for $\bar{a}(k)$ in the lower k-plane. Therefore, due to (2.2.36), $J = \bar{J}$ and

$$\bar{k}_j = k_j^*, \qquad \bar{c}_j = \mp c_j^* \qquad j = 1, \dots, J. \qquad (2.2.41)$$

Note that when $r = q^*$, the operator (2.2.3) is Hermitian. In this case, the spectrum lies on the real axis, and from the characterization equation (2.2.25) and (2.2.39a)–(2.2.39b) it follows that

$$|a(\xi)|^2 - |b(\xi)|^2 = 1, \qquad \xi \in \mathbb{R}, \qquad (2.2.42)$$

so that $|a(\xi)| > 0$. Moreover, the problem is self-adjoint; hence $|a(k)| > 0$ for any k in the upper k-plane. This implies that, for $q \to 0$ sufficiently rapidly as $|x| \to \infty$, there are no eigenvalues with Im $k \neq 0$.

Trace formula

We assume that $a(k)$ and $\bar{a}(k)$ have the simple zeros $\{k_j : \text{Im } k_j > 0\}_{j=1}^J$ and $\{\bar{k}_j : \text{Im } \bar{k}_j < 0\}_{j=1}^J$, respectively. Moreover, we define

$$\alpha(k) = \prod_{m=1}^J \frac{k - k_m^*}{k - k_m} a(k), \qquad \bar{\alpha}(k) = \prod_{m=1}^J \frac{k - \bar{k}_m^*}{k - \bar{k}_m} \bar{a}(k). \qquad (2.2.43)$$

$\alpha(k)$ is analytic in the upper k-plane, where it has no zeros, while $\bar{\alpha}(k)$ is analytic in the lower k-plane, where it has no zeros; moreover, due to (2.2.30a), $\alpha(k), \bar{\alpha}(k) \to 1$ as $|k| \to \infty$ in the respective half-planes. Therefore we have

$$\log \alpha(k) = \frac{1}{2\pi i} \int_{-\infty}^{+\infty} \frac{\log \alpha(\xi)}{\xi - k} d\xi \qquad \text{for Im } k > 0,$$

$$\frac{1}{2\pi i} \int_{-\infty}^{+\infty} \frac{\log \bar{\alpha}(\xi)}{\xi - k} d\xi = 0 \qquad \text{for Im } k > 0,$$

$$\log \bar{\alpha}(k) = -\frac{1}{2\pi i} \int_{-\infty}^{+\infty} \frac{\log \bar{\alpha}(\xi)}{\xi - k} d\xi \qquad \text{for Im } k < 0,$$

$$\frac{1}{2\pi i} \int_{-\infty}^{+\infty} \frac{\log \alpha(\xi)}{\xi - k} d\xi = 0 \qquad \text{for Im } k < 0.$$

Subtracting the equations from one another and using (2.2.43), we obtain

$$\log a(k) = \sum_{m=1}^{J} \log \left(\frac{k - k_m}{k - k_m^*} \right)$$

$$+ \frac{1}{2\pi i} \int_{-\infty}^{+\infty} \frac{\log (\alpha(\xi)\bar{\alpha}(\xi))}{\xi - k} d\xi, \qquad \text{Im } k > 0 \quad (2.2.44a)$$

$$\log \bar{a}(k) = \sum_{m=1}^{\bar{J}} \log \left(\frac{k - \bar{k}_m}{k - \bar{k}_m^*} \right)$$

$$- \frac{1}{2\pi i} \int_{-\infty}^{+\infty} \frac{\log (\alpha(\xi)\bar{\alpha}(\xi))}{\xi - k} d\xi, \qquad \text{Im } k < 0. \quad (2.2.44b)$$

This allows one to recover $a(k)$, $\bar{a}(k)$ from knowledge of $\left\{ k_j, \text{ Im } k_j > 0 \right\}_{j=1}^{J}$, $\left\{ \bar{k}_j, \text{ Im } \bar{k}_j > 0 \right\}_{j=1}^{\bar{J}}$, and $a(\xi)\bar{a}(\xi) = (1 + \rho(\xi)\bar{\rho}(\xi))^{-1}$. In particular, if $r = -q^*$, then from (2.2.39a)–(2.2.41) it follows that $\alpha(\xi)\bar{\alpha}(\xi) = a(\xi)a^*(\xi)$, and consequently (2.2.44a) can be written as

$$\log a(k) = \sum_{m=1}^{J} \log \left(\frac{k - k_m}{k - k_m^*} \right)$$

$$- \frac{1}{2\pi i} \int_{-\infty}^{+\infty} \frac{\log \left(1 + |\rho(\xi)|^2 \right)}{\xi - k} d\xi, \qquad \text{Im } k > 0. \quad (2.2.45)$$

Note that, by writing

$$\log a(k) = \sum_{n=0}^{+\infty} \frac{\gamma_n}{(2ik)^{n+1}} \qquad (2.2.46)$$

and using the well-known expansion for $\log(1 - x) = -\sum_{n=0}^{+\infty} \frac{x^{n+1}}{n+1}$, one can obtain from (2.2.44a) and (2.2.46) the relation

$$\gamma_n = \sum_{m=1}^{J} \left[\frac{(2ik_m^*)^{n+1} - (2ik_m)^{n+1}}{n+1} \right] - \frac{1}{\pi} \int_{-\infty}^{+\infty} (2i\xi)^n \log(\alpha(\xi)\bar{\alpha}(\xi))d\xi,$$

(2.2.47)

which is usually referred to as the trace formula.

2.2.3 Inverse scattering problem

The inverse problem consists of constructing a map from the scattering data, that is, (i) the reflection coefficients $\rho(k)$ and $\bar{\rho}(k)$ on the real axis, (ii) the eigenvalues $\{k_j\}_{j=1}^{J}$ and $\{\bar{k}_j\}_{j=1}^{\bar{J}}$, and (iii) the norming constants $\{c_j\}_{j=1}^{J}$ and $\{\bar{c}_j\}_{j=1}^{\bar{J}}$, back to the potentials. First we use these data to recover the Jost functions. Then we recover the potentials in terms of these Jost functions.

In the previous section, we showed that the functions $N(x, k)$ and $\bar{N}(x, k)$ exist and are analytic in the regions Im $k > 0$ and Im $k < 0$, respectively, if $q, r \in L^1(\mathbb{R})$. Similarly, under the same conditions on the potentials, the functions $\mu(x, k)$ and $\bar{\mu}(x, k)$ defined by (2.2.32) are meromorphic in the regions Im $k > 0$ and Im $k < 0$, respectively. Therefore, in the inverse problem we assume that the unknown functions are sectionally meromorphic. With this assumption, equations (2.2.31a)–(2.2.31b) can be considered to be the jump conditions of a Riemann–Hilbert problem. To recover the sectionally meromorphic functions from the scattering data, we convert the Riemann–Hilbert problem to a system of linear integral equations with the use of the Plemelej formula [7].

Case of no poles

We begin by solving the Riemann–Hilbert problem in the case where μ and $\bar{\mu}$ have no poles. Introducing the 2×2 matrices

$$\mathbf{m}_+(x, k) = (\mu(x, k), N(x, k)), \qquad \mathbf{m}_-(x, k) = \left(\bar{N}(x, k), \bar{\mu}(x, k)\right),$$

(2.2.48)

we can write the "jump" conditions (2.2.31a)–(2.2.31b) as

$$\mathbf{m}_+(x, k) - \mathbf{m}_-(x, k) = \mathbf{m}_-(x, k)\mathbf{V}(x, k),$$

(2.2.49)

where

$$\mathbf{V}(x, k) = \begin{pmatrix} -\rho(k)\bar{\rho}(k) & -\bar{\rho}(k)e^{-2ikx} \\ \rho(k)e^{2ikx} & 0 \end{pmatrix}$$

(2.2.50)

and

$$\mathbf{m}_\pm(x, k) \to \mathbf{I}$$

as $|k| \to \infty$ in the proper half-plane. We consider the projection operators

$$P^{\pm}(f)(k) = \frac{1}{2\pi i} \int_{-\infty}^{+\infty} \frac{f(\zeta)}{\zeta - (k \pm i0)} d\zeta. \tag{2.2.51}$$

If $f_{\pm}(k)$ is analytic in the upper/lower k-plane and $f_{\pm}(k) \to 0$ as $|k| \to \infty$ for $\operatorname{Im} k \gtrless 0$, then

$$P^{\pm}(f_{\mp})(k) = 0, \qquad P^{\pm}(f_{\pm})(k) = \pm f_{\pm}(k).$$

Applying P^- to (2.2.49) yields

$$\mathbf{m}_-(x, k) = \mathbf{I} + \frac{1}{2\pi i} \int_{-\infty}^{+\infty} \frac{\mathbf{m}_-(x, \xi)\mathbf{V}(x, \xi)}{\xi - (k - i0)} d\xi, \tag{2.2.52}$$

which allows one, in principle, to find $\mathbf{m}_-(x, k)$. Note that, as $|k| \to \infty$,

$$\mathbf{m}_-(x, k) = \mathbf{I} - \frac{1}{2\pi i k} \int_{-\infty}^{+\infty} \mathbf{m}_-(x, \xi)\mathbf{V}(x, \xi)d\xi + O(k^{-2}). \tag{2.2.53}$$

Taking into account the asymptotics (2.2.20a)–(2.2.20d), the definitions (2.2.49)–(2.2.50), and comparing with equation (2.2.53) we obtain the potentials in terms of the scattering data, that is,

$$r(x) = \frac{1}{\pi} \int_{-\infty}^{+\infty} \rho(k)e^{2ikx} N^{(2)}(x, k)dk \tag{2.2.54a}$$

$$q(x) = \frac{1}{\pi} \int_{-\infty}^{+\infty} \bar{\rho}(k)e^{-2ikx} \bar{N}^{(1)}(x, k)dk. \tag{2.2.54b}$$

Case of poles

Suppose now that the potential is such that $a(k)$ and $\bar{a}(k)$ have a finite number of simple zeros in the regions $\operatorname{Im} k > 0$ and $\operatorname{Im} k < 0$, respectively, which we denote as $\{k_j, \operatorname{Im} k_j > 0\}_{j=1}^J$ and $\{\bar{k}_j, \operatorname{Im} \bar{k}_j < 0\}_{j=1}^{\bar{J}}$. We shall also assume that $a(\xi) \neq 0$, $\bar{a}(\xi) \neq 0$ for any $\xi \in \mathbb{R}$. As before, we apply P^- to both sides of (2.2.31a) and P^+ to both sides of (2.2.31b). Taking into account the analytic properties of N, $\bar{N}$, μ, and $\bar{\mu}$ and the asymptotics (2.2.20a)–(2.2.20d) and using (2.2.36), we obtain

$$\bar{N}(x, k) = \begin{pmatrix} 1 \\ 0 \end{pmatrix} + \sum_{j=1}^{J} \frac{C_j e^{2ik_j x} N_j(x)}{(k - k_j)}$$

$$+ \frac{1}{2\pi i} \int_{-\infty}^{+\infty} \frac{\rho(\xi)e^{2i\xi x} N(x, \xi)}{\xi - (k - i0)} d\xi \tag{2.2.55a}$$

$$N(x, k) = \begin{pmatrix} 0 \\ 1 \end{pmatrix} + \sum_{j=1}^{\bar{J}} \frac{\bar{C}_j e^{-2i\bar{k}_j x} \bar{N}_j(x)}{(k - \bar{k}_j)}$$

$$- \frac{1}{2\pi i} \int_{-\infty}^{+\infty} \frac{\bar{\rho}(\xi) e^{-2i\xi x} \bar{N}(x, \xi)}{\xi - (k + i0)} d\xi, \qquad (2.2.55b)$$

where $N_j(x) = N(x, k_j)$, $\bar{N}_j(x) = \bar{N}(x, \bar{k}_j)$, and we introduced

$$C_j = \frac{c_j}{a'(k_j)}, \qquad \bar{C}_j = \frac{\bar{c}_j}{\bar{a}'(\bar{k}_j)}, \qquad (2.2.56)$$

with $'$ denoting the derivative with respect to the spectral parameter k. Note that the equations defining the inverse problem for $N(x, k)$ and $\bar{N}(x, k)$ now depend on the extra terms $\{N_j(x)\}_{j=1}^{J}$ and $\{\bar{N}_l(x)\}_{l=1}^{\bar{J}}$. In order to close the system, we evaluate equation (2.2.55a) at $k = \bar{k}_l$ for any $l = 1, \ldots, \bar{J}$ and (2.2.55b) at $k = k_j$ for any $j = 1, \ldots, J$, thus obtaining

$$\bar{N}_l(x) = \begin{pmatrix} 1 \\ 0 \end{pmatrix} + \sum_{j=1}^{J} \frac{C_j e^{2ik_j x} N_j(x)}{(\bar{k}_l - k_j)}$$

$$+ \frac{1}{2\pi i} \int_{-\infty}^{+\infty} \frac{\rho(\xi) e^{2i\xi x} N(x, \xi)}{\xi - \bar{k}_l} d\xi \qquad (2.2.57a)$$

$$N_j(x) = \begin{pmatrix} 0 \\ 1 \end{pmatrix} + \sum_{m=1}^{\bar{J}} \frac{\bar{C}_m e^{-2i\bar{k}_m x} \bar{N}_m(x)}{(k_j - \bar{k}_m)}$$

$$- \frac{1}{2\pi i} \int_{-\infty}^{+\infty} \frac{\bar{\rho}(\xi) e^{-2i\xi x} \bar{N}(x, \xi)}{\xi - k_m} d\xi. \qquad (2.2.57b)$$

Equations (2.2.55a)–(2.2.55b) and (2.2.57a)–(2.2.57b) together constitute a linear algebraic–integral system of equations which, in principle, solve the inverse problem for the eigenfunctions $N(x, k)$ and $\bar{N}(x, k)$.

By comparing the asymptotic expansions at large k of the right-hand sides of (2.2.55a) and (2.2.55b) to the expansions (2.2.20b) and (2.2.20c), respectively, we obtain

$$r(x) = -2i \sum_{j=1}^{J} e^{2ik_j x} C_j N_j^{(2)}(x) + \frac{1}{\pi} \int_{-\infty}^{+\infty} \rho(\xi) e^{2i\xi x} N^{(2)}(x, \xi) d\xi \qquad (2.2.58a)$$

$$q(x) = 2i \sum_{j=1}^{\bar{J}} e^{-2i\bar{k}_j x} \bar{C}_j \bar{N}_j^{(1)}(x) + \frac{1}{\pi} \int_{-\infty}^{+\infty} \bar{\rho}(\xi) e^{-2i\xi x} \bar{N}^{(1)}(x, \xi) d\xi,$$

$$(2.2.58b)$$

which reconstruct the potentials and thus complete the formulation of the inverse problem (as before, the superscript ℓ denotes the ℓ-th component of the corresponding vector).

If the potentials decay rapidly enough at infinity, so that $\rho(k)$ can be analytically continued above all poles $\{k_j,\ \text{Im } k_j > 0\}_{j=1}^{J}$ and $\bar{\rho}(k)$ can be analytically continued below all poles $\{\bar{k}_j,\ \text{Im } \bar{k}_j < 0\}_{j=1}^{J}$, then the system of equations (2.2.55a)–(2.2.55b) and (2.2.57a)–(2.2.57b) can be simplified as follows:

$$\bar{N}(x,k) = \begin{pmatrix} 1 \\ 0 \end{pmatrix} + \frac{1}{2\pi i} \int_{C_0} \frac{\rho(\xi) e^{2i\xi x} N(x,\xi)}{\xi - k} d\xi \qquad (2.2.59a)$$

$$N(x,k) = \begin{pmatrix} 0 \\ 1 \end{pmatrix} - \frac{1}{2\pi i} \int_{\bar{C}_0} \frac{\bar{\rho}(\xi) e^{-2i\xi x} \bar{N}(x,\xi)}{\xi - k} d\xi, \qquad (2.2.59b)$$

where C_0 is a contour from $-\infty$ to $+\infty$ that passes above all zeros of $a(k)$ and $\bar{C}_0$ is a contour from $-\infty$ to $+\infty$ that passes below all zeros of $\bar{a}(k)$. With the same hypothesis, equations (2.2.58a)–(2.2.58b) can be written as

$$r(x) = \frac{1}{\pi} \int_{C_0} \rho(\xi) e^{2i\xi x} N^{(2)}(x,\xi) d\xi \qquad (2.2.60a)$$

$$q(x) = \frac{1}{\pi} \int_{\bar{C}_0} \bar{\rho}(\xi) e^{-2i\xi x} \bar{N}^{(1)}(x,\xi) d\xi. \qquad (2.2.60b)$$

Gel'fand–Levitan–Marchenko equations

We can also provide a reconstruction for the potentials by developing the Gel'fand–Levitan–Marchenko integral equations, instead of using the projection operators (cf. [21], [192]). Indeed, let us represent the eigenfunctions in terms of triangular kernels,

$$N(x,k) = \begin{pmatrix} 0 \\ 1 \end{pmatrix} + \int_x^{+\infty} K(x,s) e^{-ik(x-s)} ds$$

$$s > x, \quad \text{Im } k > 0 \qquad (2.2.61a)$$

$$\bar{N}(x,k) = \begin{pmatrix} 1 \\ 0 \end{pmatrix} + \int_x^{+\infty} \bar{K}(x,s) e^{ik(x-s)} ds$$

$$s > x, \quad \text{Im } k < 0. \qquad (2.2.61b)$$

Applying the operator $\frac{1}{2\pi} \int_{-\infty}^{+\infty} dk\ e^{-ik(x-y)}$ for $y > x$ to the equation (2.2.59a), we obtain

$$\bar{K}(x,y) + \begin{pmatrix} 0 \\ 1 \end{pmatrix} F(x+y) + \int_x^{+\infty} K(x,s) F(s+y) ds = 0, \qquad (2.2.62)$$

where

$$F(x) = \frac{1}{2\pi} \int_{C_0} \rho(\xi)e^{i\xi x}d\xi = \frac{1}{2\pi} \int_{-\infty}^{+\infty} \rho(\xi)e^{i\xi x}d\xi - i \sum_{j=1}^{J} C_j e^{ik_j x}.$$

(2.2.63)

Analogously, operating on equation (2.2.59b) with $\frac{1}{2\pi}\int_{-\infty}^{+\infty}dk\,e^{ik(x-y)}$ for $y > x$ gives

$$K(x, y) + \begin{pmatrix} 1 \\ 0 \end{pmatrix}\bar{F}(x+y) + \int_{x}^{+\infty}\bar{K}(x, s)\bar{F}(s+y)ds = 0, \qquad (2.2.64)$$

where

$$\bar{F}(x) = \frac{1}{2\pi}\int_{\bar{C}_0}\bar{\rho}(\xi)e^{-i\xi x}d\xi = \frac{1}{2\pi}\int_{-\infty}^{+\infty}\bar{\rho}(\xi)e^{-i\xi x}d\xi + i\sum_{j=1}^{\bar{J}}\bar{C}_j e^{-i\bar{k}_j x}.$$

(2.2.65)

Equations (2.2.62) and (2.2.64) constitute the Gel'fand–Levitan–Marchenko (GLM) equations.

Inserting the representations (2.2.61a)–(2.2.61b) for the eigenfunctions into equations (2.2.60a)–(2.2.60b), we obtain the reconstruction of the potentials in terms of the kernels of GLM equations, that is,

$$q(x) = -2K^{(1)}(x, x), \qquad r(x) = -2\bar{K}^{(2)}(x, x), \qquad (2.2.66)$$

where, as usual, $K^{(j)}$ and $\bar{K}^{(j)}$ for $j = 1, 2$ denote the j-th component of the vectors K and $\bar{K}$, respectively.

If the symmetry $r = \mp q^*$ holds, then, taking into account (2.2.39a)–(2.2.41), it is easy to check that

$$\bar{F}(x) = \mp F^*(x) \qquad (2.2.67)$$

and consequently

$$\bar{K}(x, y) = \begin{pmatrix} K^{(2)}(x, y) \\ \mp K^{(1)}(x, y) \end{pmatrix}^*. \qquad (2.2.68)$$

In this case, equations (2.2.62)–(2.2.64) solving the inverse problem reduce to

$$K^{(1)}(x, y) = \pm F^*(x+y) \mp \int_{x}^{+\infty}ds\int_{x}^{+\infty}ds'K^{(1)}(x, s')F(s+s')F^*(y+s),$$

(2.2.69)

and the potentials are reconstructed by means of the first of (2.2.66).

Existence and uniqueness of solutions

The question of existence and uniqueness of solutions of linear integral equations is usually examined by the use of the Fredholm alternative. For example, under suitable assumptions on the decay of the potentials, the restriction $r = \mp q^*$ is sufficient to guarantee that the solutions of (2.2.62) and (2.2.64) exist and are unique. To show this, consider the homogeneous equations corresponding to (2.2.62) and (2.2.64) ($y > x$):

$$h_1(y) + \int_x^{+\infty} h_2(s)F(s+y)ds = 0 \tag{2.2.70}$$

$$h_2(y) + \int_x^{+\infty} h_1(s)\bar{F}(s+y)ds = 0. \tag{2.2.71}$$

Suppose $h(y) = (h_1(y), h_2(y))$ is a solution of (2.2.70)–(2.2.71) that vanishes identically for $y < x$. Multiply (2.2.70)–(2.2.71) by (h_1^*, h_2^*), integrate in y, and use

$$\int_x^\infty |h_j(y)|^2 dy = \int_{-\infty}^\infty |h_j(y)|^2 dy.$$

One obtains

$$\int_{-\infty}^\infty \left\{ |h_1(y)|^2 + |h_2(y)|^2 + \int_{-\infty}^\infty \left[h_2(s)h_1^*(y)F(s+y) \right. \right.$$
$$\left. \left. + h_1(s)h_2^*(y)\bar{F}(s+y) \right] ds \right\} dy = 0. \tag{2.2.72}$$

If $r = -q^*$, then the symmetry condition (2.2.67) allows the latter equation to be written as

$$\int_{-\infty}^\infty \left\{ |h_1(y)|^2 + |h_2(y)|^2 + 2i \operatorname{Im} \int_{-\infty}^\infty h_2(s)h_1^*(y)F(s+y)ds \right\} dy = 0.$$

The real and imaginary parts must both vanish, from which it follows that

$$h(y) \equiv 0.$$

Second, if $r(x) = q^*(x)$, using (2.2.67), equation (2.2.72) becomes

$$\int_{-\infty}^\infty \left\{ |h_1(y)|^2 + |h_2(y)|^2 + 2\operatorname{Re} \int_{-\infty}^\infty h_2(s)h_1^*(y)F(s+y)ds \right\} dy = 0. \tag{2.2.73}$$

Moreover, in this case, the scattering problem is formally self-adjoint and there are no discrete eigenvalues; therefore

$$F(x) = \frac{1}{2\pi} \int_{-\infty}^{+\infty} \rho(\xi)e^{i\xi x}d\xi. \tag{2.2.74}$$

The Fourier transform of $h_j(y)$ is

$$\hat{h}_j(\xi) = \int_{-\infty}^\infty h_j(y)e^{-i\xi y}dy,$$

which satisfies Parseval's identity

$$\int_{-\infty}^{\infty} |h_j(y)|^2 dy = \frac{1}{2\pi} \int_{-\infty}^{\infty} |\hat{h}_j(\xi)|^2 d\xi. \tag{2.2.75}$$

Substituting these results into (2.2.73) and reversing the order of integration yields

$$\int_{-\infty}^{\infty} \left\{ |\hat{h}_1(-\xi)|^2 + |\hat{h}_2^*(\xi)|^2 + 2\,\mathrm{Re}\left[\frac{b}{a}(\xi)\hat{h}_1(-\xi)\hat{h}_2^*(\xi)\right] \right\} d\xi = 0. \tag{2.2.76}$$

Since $|(b/a)(\xi)| < 1$, we have

$$\left| 2\,\mathrm{Re}\left[\frac{b}{a}(\xi)\hat{h}_1(-\xi)\hat{h}_2^*(\xi)\right]\right| < 2|\hat{h}_1(-\xi)||\hat{h}_2^*(\xi)| \le |\hat{h}_1(-\xi)|^2 + |\hat{h}_2(\xi)|^2,$$

and hence

$$h(y) \equiv 0.$$

We conclude that, when $r = \mp q^*$, the integral equations (2.2.62), (2.2.64) admit no homogenous solutions except the trivial one. The complete study of the inverse scattering problem is outside the scope of this book. However, we can say that when the kernel $F(x + y)$ is compact, as it would be when the class of potentials is suitably restricted, then the Gel'fand–Levitan–Marchenko equations are Fredholm. In this case, the Fredholm alternative applies, and the latter equation implies that the solution of (2.2.62), (2.2.64) exists and is unique.

Note that existence and uniqueness issues regarding the inverse problem have been considered in a more general context in [27], [28].

2.2.4 Time evolution

We now show how the time evolution of solutions of the NLS can be obtained. We do this by calculating the evolution of the scattering data. Then, by the method of the previous section, one can reconstruct the evolution of the solution.

The operator (2.2.4) determines the evolution of the Jost functions, which can be written as

$$\partial_t v = \begin{pmatrix} A & B \\ C & -A \end{pmatrix} v, \tag{2.2.77}$$

where $B, C \to 0$ as $x \to \pm\infty$ (since we have assumed that $q, r \to 0$ as $x \to \pm\infty$). Then the time-dependence must asymptotically satisfy

$$\partial_t v = \begin{pmatrix} A_\infty & 0 \\ 0 & -A_\infty \end{pmatrix} v \qquad \text{as } x \to \pm\infty, \tag{2.2.78}$$

where

$$A_\infty = \lim_{|x| \to \infty} A(x, k) = 2ik^2. \qquad (2.2.79)$$

The system (2.2.78) has solutions that are linear combinations of

$$v^+ = \begin{pmatrix} e^{A_\infty t} \\ 0 \end{pmatrix}, \qquad v^- = \begin{pmatrix} 0 \\ e^{-A_\infty t} \end{pmatrix}.$$

However, such solutions are not compatible with the fixed boundary conditions of the Jost functions (2.2.5)–(2.2.6). Therefore we define time-dependent functions

$$\Phi(x, t) = e^{A_\infty t} \phi(x, t), \qquad \bar\Phi(x, t) = e^{-A_\infty t} \bar\phi(x, t) \qquad (2.2.80a)$$

$$\Psi(x, t) = e^{-A_\infty t} \psi(x, t), \qquad \bar\Psi(x, t) = e^{A_\infty t} \bar\psi(x, t) \qquad (2.2.80b)$$

to be solutions of the time-differential equation (2.2.77). Then the evolution equations for ϕ and $\bar\phi$ become

$$\partial_t \phi = \begin{pmatrix} A - A_\infty & B \\ C & -A - A_\infty \end{pmatrix} \phi, \qquad \partial_t \bar\phi = \begin{pmatrix} A + A_\infty & B \\ C & -A + A_\infty \end{pmatrix} \bar\phi,$$

$$(2.2.81)$$

so that, by taking into account equations (2.2.24a)–(2.2.24b) and evaluating (2.2.81) as $x \to +\infty$, one obtains

$$\partial_t a(k) = 0, \qquad \partial_t \bar a(k) = 0$$

$$\partial_t b(k) = -2A_\infty b(k), \qquad \partial_t \bar b(k) = 2A_\infty \bar b(k)$$

or, explicitly,

$$a(k, t) = a(k, 0), \qquad \bar a(k, t) = \bar a(k, 0) \qquad (2.2.82a)$$

$$b(k, t) = e^{-4ik^2 t} b(k, 0), \qquad \bar b(k, t) = e^{4ik^2 t} \bar b(k, 0). \qquad (2.2.82b)$$

From (2.2.82a) it is clear that the eigenvalues (i.e., the zeros of $a(k)$ and $\bar a(k)$) are constant as the solution evolves. Not only the number of eigenvalues, but also their locations, are fixed. Thus, the eigenvalues are time-independent discrete states of the evolution. In fact, this time invariance is the underlying mechanism of the elastic soliton interaction in the NLS (2.1.1) and in integrable soliton equations in general. On the other hand, the evolution of the reflection coefficients is given by

$$\rho(k, t) = \rho(k, 0)e^{-4ik^2 t}, \qquad \bar\rho(k, t) = \bar\rho(k, 0)e^{4ik^2 t}, \qquad (2.2.83)$$

and this also gives the evolution of the norming constants:

$$C_j(t) = C_j(0)e^{-4ik_j^2 t}, \qquad \bar C_j(t) = \bar C_j(0)e^{4i\bar k_j^2 t}. \qquad (2.2.84)$$

The expressions for the evolution of the scattering data allow one to solve the initial-value problem for the NLS. The procedure is the following: (i) The scattering data are calculated from the initial time the (e.g., at $t = 0$) according to the procedure illustrated in Section 2.1.1; (ii) the scattering data at later time (say, $t = t_1$) is determined by formulas (2.2.82a)–(2.2.84); and (iii) the solution at $t = t_1$ is recovered from the scattering data using, for instance, the reconstruction formulas (2.2.55a)–(2.2.58b).

2.3 Soliton solutions

In the case where the scattering data comprise proper eigenvalues, but $\rho(\xi) = \bar{\rho}(\xi) = 0$ for all $\xi \in \mathbb{R}$, the algebraic–integral system (2.2.55a)–(2.2.55b) and (2.2.57a)–(2.2.57b) reduces to the linear algebraic system

$$\bar{N}_l(x) = \begin{pmatrix} 1 \\ 0 \end{pmatrix} + \sum_{j=1}^{J} \frac{C_j e^{2ik_j x} N_j(x)}{(\bar{k}_l - k_j)} \tag{2.3.85a}$$

$$N_j(x) = \begin{pmatrix} 0 \\ 1 \end{pmatrix} + \sum_{m=1}^{\bar{J}} \frac{\bar{C}_m e^{-2i\bar{k}_m x} \bar{N}_m(x)}{(k_j - \bar{k}_m)}. \tag{2.3.85b}$$

The one-soliton solution is obtained for $J = \bar{J} = 1$. In the relevant physical case, when the symmetry $r = -q^*$ holds, by taking into account (2.2.38b) and (2.2.41), we get

$$N_1^{(1)}(x) = -\frac{C_1^*}{k_1 - k_1^*} e^{-2ik_1^* x} \left[1 - \frac{|C_1|^2 e^{2i(k_1 - k_1^*)x}}{\left(k_1 - k_1^*\right)^2} \right]^{-1}$$

$$N_1^{(2)}(x) = \left[1 - \frac{|C_1|^2 e^{2i(k_1 - k_1^*)x}}{\left(k_1 - k_1^*\right)^2} \right]^{-1},$$

where, as usual, $N_1(x) = (N_1^{(1)}(x), N_1^{(2)}(x))^T$. Then from (2.2.58b) it follows that

$$q(x) = -2i\eta \frac{C_1^*}{|C_1|} e^{-2i\xi x} \operatorname{sech}(2\eta x - 2\delta), \tag{2.3.86}$$

where

$$k_1 = \xi + i\eta, \qquad e^{2\delta} = \frac{|C_1|}{2\eta} \tag{2.3.87}$$

and, taking into account the time dependence of C_1 as given by (2.2.84), one finally gets the well-known one-soliton solution of the scalar NLS equation,

$$q(x, t) = 2\eta e^{-2i\xi x + 4i(\xi^2 - \eta^2)t - i(\psi_0 + \pi/2)} \operatorname{sech}(2\eta x - 8\xi\eta t - 2\delta_0) \tag{2.3.88}$$

with

$$e^{2\delta_0} = \frac{|C_1(0)|}{2\eta}, \qquad \psi_0 = \arg C_1(0) \tag{2.3.89}$$

or, equivalently,

$$C_1(0) = 2\eta e^{2\delta_0 + i\psi_0}. \tag{2.3.90}$$

Note that the velocity of this solution is 4ξ and its amplitude is 2η. Therefore, unlike the KdV solitons, in the NLS equation the height and the speed of a soliton may each be specified independently. As a consequence, it is also possible to construct a solution having two (or more) peaks with different amplitudes each traveling at the same speed. In this case, the peaks will oscillate periodically in amplitude and the separation between peaks will not increase in the long-time limit [192]. Such a solution is considered a special two-soliton solution and not a "breather" solution because the corresponding scattering data do not consist of a minimal set of eigenvalues (i.e., there are four eigenvalues, not two).

We now consider the solution of the NLS ($r = -q^*$) corresponding to the scattering data

$$\left\{ k_j = \xi_j + i\eta_j, \ \eta_j > 0, \ C_j \right\}_{j=1}^{J} \cup \{\rho(\xi) = 0 \text{ for } \xi \in \mathbb{R}\},$$

which is sometimes referred to as a reflectionless state. The reflectionless state with $2J$ eigenvalues is a pure J-soliton solution of NLS. In order to get the pure J-soliton solution, one can in principle solve the linear system (2.3.85a)–(2.3.85b).

The problem of a J-soliton collision can be investigated by looking at the asymptotic states as $t \to \pm\infty$ and proceeding in a similar way as in [192] or [120]. If $\xi_j \neq \xi_l$ for $j \neq l$, then for $t \to \pm\infty$ the potential breaks up into individual solitons of the form (2.3.88), that is,

$$q^{\pm}(x,t) \sim \sum_{j=1}^{J} q_j^{\pm}(x,t) \qquad t \to \pm\infty \tag{2.3.91}$$

with

$$q_j^{\pm}(x,t) = 2\eta_j e^{-2i\xi_j x + 4i(\xi_j^2 - \eta_j^2)t - i(\psi_j^{\pm} + \pi/2)} \operatorname{sech}(2\eta_j x - 8\xi_j \eta_j t - 2\delta_j^{\pm}). \tag{2.3.92}$$

Let us fix the values of the soliton parameters such that $\xi_1 < \xi_2 < \cdots < \xi_J$. Then, as $t \to -\infty$, the solitons are distributed along the x-axis in the order corresponding to $\xi_J, \xi_{J-1}, \ldots, \xi_1$; the order of the soliton sequence is reversed as $t \to +\infty$. In order to determine the result of the interaction between solitons, we trace the passage of the eigenfunctions through the asymptotic states. We denote the "soliton coordinates" (i.e., the center of the soliton) at the instant of time t by $x_j(t)$ ($|t|$ is assumed large enough so that one can talk about individual solitons). If $t \to -\infty$, then $x_J \ll x_{J-1} \ll \cdots \ll x_1$. The function $\phi(x, k_j)$ has the form $\phi(x, k_j) \sim e^{-ik_j x} (1, \ 0)^T$ when $x \ll x_J$. After passing through the J-th soliton, it will be of the form $\phi(x, k_j) \sim a_J(k_j)e^{-ik_j x} (1, 0)^T$, where $a_J(k)$ is the transmission coefficient relative to the J-th soliton. By repeating the argument, we find

$$\phi(x, k_j) \sim \prod_{l=j+1}^{J} a_l(k_j) \begin{pmatrix} e^{-ik_j x} \\ 0 \end{pmatrix} \qquad x_{j+1} \ll x \ll x_j.$$

Upon passing through the j-th soliton, since the corresponding state is a bound state, we get

$$\phi(x, k_j) \sim S_j \prod_{l=j+1}^{J} a_l(k_j) \begin{pmatrix} 0 \\ e^{ik_j x} \end{pmatrix} \qquad x_j \ll x \ll x_{j-1}. \tag{2.3.93}$$

On the other hand, starting from $x \gg x_1$ and proceeding in a similar way, we find for the eigenfunction ψ the following asymptotic behavior:

$$\psi(x, k_j) \sim \prod_{l=1}^{j-1} a_l(k_j) \begin{pmatrix} 0 \\ e^{ik_j x} \end{pmatrix} \qquad x_j \ll x \ll x_{j-1}, \tag{2.3.94}$$

where we have used (2.2.24a)–(2.2.24b) and solved for ψ, $\bar{\psi}$ in terms of ϕ, $\bar{\phi}$. Then, comparing (2.3.93) and (2.3.94) and recalling (2.2.34) and (2.2.56), we get

$$S_j \prod_{l=j+1}^{J} a_l(k_j) = a'(k_j)C_j \prod_{l=1}^{j-1} a_l(k_j). \tag{2.3.95}$$

From (2.3.90) we deduce that

$$S_j(t) = 2\eta_j e^{-4ik_j^2 t + 2\delta_j + i\psi_j},$$

where δ_j and ψ_j are real functions of time. Equation (2.3.95) yields

$$C_j(t) \sim 2\eta_j e^{-4ik_j^2 t + 2\delta_j^- + i\psi_j^-} \frac{1}{a'(k_j)} \prod_{l=j+1}^{J} a_l(k_j) \prod_{m=1}^{j-1} a_m(k_j)^{-1} \qquad t \to -\infty,$$

where δ_j^-, ψ_j^- denote the asymptotics of the functions δ_j, ψ_j as $t \to -\infty$.

By proceeding in a similar fashion as $t \to +\infty$ and taking into account that the order of solitons is reversed, we get

$$C_j(t) \sim 2\eta_j e^{-4ik_j^2 t + 2\delta_j^+ + i\psi_j^+} \frac{1}{a'(k_j)} \prod_{l=1}^{j-1} a_l(k_j) \prod_{m=j+1}^{J} a_m(k_j)^{-1} \qquad t \to +\infty;$$

therefore we conclude that

$$e^{2(\delta_j^+ - \delta_j^-) + i(\psi_j^+ - \psi_j^-)} = \prod_{l=1}^{j-1} a_l(k_j)^2 \prod_{m=j+1}^{J} a_m(k_j)^{-2}.$$

Using formula (2.2.45) for the pure one-soliton transmission coefficient, we obtain

$$e^{2(\delta_j^+ - \delta_j^-) + i(\psi_j^+ - \psi_j^-)} = \prod_{l=1}^{j-1} \left(\frac{k_j - k_l}{k_j - k_l^*} \right)^2 \prod_{m=j+1}^{J} \left(\frac{k_j - k_m^*}{k_j - k_m} \right)^2. \tag{2.3.96}$$

According to (2.3.89), (2.3.92), this last formula provides the phase shift of the j-th soliton on the transition between the asymptotic states $t \to \pm\infty$ due to the interaction with the other solitons.

For instance, in the two-soliton case we obtain

$$\delta_1^+ - \delta_1^- = -\left(\delta_2^+ - \delta_2^- \right) = \log \left| \frac{k_1 - k_2^*}{k_1 - k_2} \right| \tag{2.3.97a}$$

$$\psi_1^+ - \psi_1^- = \arg \left(\frac{k_1 - k_2^*}{k_1 - k_2} \right)^2 \qquad \psi_2^+ - \psi_2^- = \arg \left(\frac{k_1 - k_2}{k_1^* - k_2} \right)^2. \tag{2.3.97b}$$

2.4 Conserved quantities and Hamiltonian structure

Equation (2.2.82a) shows that $a(k, t)$ and $\bar{a}(k, t)$ are conserved in time. Recalling (2.2.6) and (2.2.24a), we have

$$a(k) = \lim_{x \to +\infty} \phi^{(1)}(x, k) e^{ikx}, \tag{2.4.98}$$

where $\phi = \left(\phi^{(1)}, \phi^{(2)} \right)^T$ satisfies the scattering problem (2.2.3) with boundary condition (2.2.5). Eliminating $\phi^{(2)}$ from the scattering problem (2.2.3) and substituting

$$\phi^{(1)}(x, k) = e^{-ikx + \hat{\phi}} \tag{2.4.99}$$

results in a Riccati equation for $\gamma = \hat{\phi}_x$:

$$2ik\gamma = \gamma^2 - qr + q \left(\frac{\gamma}{q} \right)_x. \tag{2.4.100}$$

Because $\hat{\phi}$ vanishes as $|k| \to \infty$ (Im $k > 0$), we may expand

$$\gamma(x, t) = \sum_{n=1}^{+\infty} \frac{\gamma_n(x, t)}{(2ik)^n}. \tag{2.4.101}$$

Substituting this into (2.4.100) and matching the corresponding power of $2ik$ yields

$$\gamma_1 = -qr, \qquad \gamma_2 = -qr_x$$

$$\gamma_{n+1} = q \left(\frac{\gamma_n}{q} \right)_x + \sum_{k=1}^{n-1} \gamma_k \gamma_{n-k}.$$

From (2.2.46) and the fact that $\hat{\phi}$ vanishes as $x \to -\infty$, it follows that

$$\log a(k) = \sum_{n=1}^{+\infty} \frac{\Gamma_n}{(2ik)^n}, \tag{2.4.102}$$

where

$$\Gamma_n = \int_{-\infty}^{+\infty} \gamma_n(x) dx. \tag{2.4.103}$$

But $\log a(k)$ is time-independent (for all k with $\mathrm{Im}\, k > 0$), so Γ_n must be time-independent as well. Thus we get an infinite set of (global) constants of the motion, and the first few are

$$\Gamma_1 = -\int q(x)r(x)dx, \qquad \Gamma_2 = -\int q(x)r_x(x)dx, \tag{2.4.104a}$$

$$\Gamma_3 = \int \left(q_x(x)r_x(x) + (q(x)r(x))^2 \right) dx, \tag{2.4.104b}$$

and so on. With the reductions $r = \mp q^*$, these constants of the motion can be written as

$$\Gamma_1 = \pm \int |q(x)|^2\, dx, \qquad \Gamma_2 = \pm \int q(x)q_x^*(x)dx, \tag{2.4.105a}$$

$$\Gamma_3 = \int \left(\mp q_x(x)q_x^*(x) + |q(x)|^4 \right) dx, \tag{2.4.105b}$$

and so on.

It is known that the equations solvable by the IST are completely integrable Hamiltonian systems and IST amounts to a canonical transformation from physical variables to (an infinite set of) action-angle variables. The phase space $\mathcal{M}_0$ is an infinite-dimensional real linear space with complex coordinates defined by pairs of functions $q(x)$, $r(x)$. By analogy with finite-dimensional coordinates labeled by a discrete parameter, the variable x may be thought of as a coordinate label.

On the algebra of smooth functionals on the phase space $\mathcal{M}_0$ one can define (see, for instance, [74]) a Poisson structure by the following Poisson brackets:

$$\{F, G\} = i \int_{-\infty}^{+\infty} \left(\frac{\delta F}{\delta q(x)} \frac{\delta G}{\delta r(x)} - \frac{\delta F}{\delta r(x)} \frac{\delta G}{\delta q(x)} \right) dx, \tag{2.4.106}$$

where the variational derivative is defined according to

$$\delta F(q, r) = F(q + \delta q, r + \delta r) - F(q, r)$$

$$= \int_{-\infty}^{+\infty} \left(\frac{\delta F}{\delta q(x)} \delta q(x) + \frac{\delta F}{\delta r(x)} \delta r(x) \right) dx.$$

Obviously, the bracket (2.4.106) possesses the basic properties of Poisson brackets, namely, it is (i) skew-symmetric:

$$\{F, G\} = -\{G, F\}, \tag{2.4.107}$$

(ii) linear, that is, for any constants a, b:

$$\{aF + bG, H\} = a\{F, H\} + b\{G, H\}, \tag{2.4.108}$$

and (iii) satisfies the Jacobi identity:

$$\{F, \{G, H\}\} + \{H, \{F, G\}\} + \{G, \{H, F\}\} = 0. \tag{2.4.109}$$

The bracket (2.4.106) is the infinite-dimensional generalization of the usual Poisson bracket in the phase space $\mathbb{R}^{2n}$ with real coordinates p_k, q_k for $k = 1, \ldots, n$:

$$\{f, g\} = \sum_{k=1}^{n} \left(\frac{\partial f}{\partial q_k} \frac{\partial g}{\partial p_k} - \frac{\partial f}{\partial p_k} \frac{\partial g}{\partial q_k} \right).$$

The coordinates $q(x), r(x)$ themselves may be considered as functionals on $\mathcal{M}_0$. Their variational derivatives are generalized functions,

$$\frac{\delta q(x)}{\delta q(y)} = \delta(x - y), \qquad \frac{\delta r(x)}{\delta q(y)} = \delta(x - y) \tag{2.4.110}$$

$$\frac{\delta q(x)}{\delta r(y)} = \frac{\delta r(x)}{\delta q(y)} = 0, \tag{2.4.111}$$

where $\delta(x - y)$ is the Dirac δ-function. Substituting (2.4.110) into (2.4.106) gives

$$\{q(x), q(y)\} = \{r(x), r(y)\} = 0$$

$$\{q(x), r(y)\} = i\delta(x - y).$$

These formulas also yield

$$\frac{\delta F}{\delta q(x)} = -i\{F, r(x)\}, \qquad \frac{\delta F}{\delta r(x)} = -i\{F, q(x)\}.$$

A dynamical system is said to be Hamiltonian if it is possible to identify generalized coordinates q, momenta p, and a Hamiltonian $H(p, q, t)$ such that the equations of motion of the system can be written as

$$\frac{\partial q}{\partial t} = \{q, H\} \qquad (2.4.112a)$$

$$\frac{\partial p}{\partial t} = \{p, H\}, \qquad (2.4.112b)$$

where $\{,\}$ denotes the Poisson brackets. Equations (2.4.112a)–(2.4.112b) are called Hamilton's equations of motion, and the variables (p, q) are called conjugates.

The dynamical system (2.1.2a)–(2.1.2b) is Hamiltonian, as may be seen from the identifications:

$$\text{coordinates } (q): \qquad q(x, t) \qquad (2.4.113a)$$

$$\text{momenta } (p): \qquad r(x, t) \qquad (2.4.113b)$$

$$\text{Hamiltonian } (H): \qquad i \int_{-\infty}^{+\infty} \left(q_x r_x + (qr)^2 \right) dx \qquad (2.4.113c)$$

with the canonical brackets

$$\{q(x, t), r(y, t)\} = i\delta(x - y) \qquad (2.4.114a)$$

$$\{q(x, t), q(y, t)\} = \{r(x, t), r(y, t)\} = 0. \qquad (2.4.114b)$$

Note that the Hamiltonian is given by the conserved quantity (2.4.104b).

In the case when the initial data satisfy the additional symmetry $r = \mp q^*$, we identify coordinates $q(x, t)$, momenta $q^*(x, t)$. The corresponding Hamiltonian and brackets are given by

$$H = i \int_{-\infty}^{+\infty} \left(\mp |q_x|^2 + |q|^4 \right) dx \qquad (2.4.115)$$

$$\left\{ q(x, t), q^*(y, t) \right\} = i\delta(x - y) \qquad (2.4.116)$$

$$\{q(x, t), q(y, t)\} = \left\{ q^*(x, t), q^*(y, t) \right\} = 0. \qquad (2.4.117)$$

Chapter 3

Integrable discrete nonlinear Schrödinger equation (IDNLS)

3.1 Overview

In general, any given discretization of an integrable PDE is most likely to be nonintegrable. Even though the integrable PDE is the compatibility condition of a linear operator pair, there is, in general, no pair of linear equations corresponding to a generic discretization. Furthermore, given a discretization that appears to be integrable – for example, one for which numerical simulations suggest the elastic interaction of solitons – there is no algorithmic method to construct the associated linear pair from the discrete system. However, the IDNLS,

$$i\frac{d}{dt}q_n = \frac{1}{h^2}(q_{n+1} - 2q_n + q_{n-1}) \pm |q_n|^2 (q_{n+1} + q_{n-1}), \qquad (3.1.1)$$

is a straightforward discretization of the NLS (2.1.1), for which there is such an associated operator pair (see equations (3.2.4)–(3.2.5)). With this operator pair, the initial-value problem for the IDNLS can be solved via the inverse scattering transform. In this chapter, we formulate the IST, on the doubly infinite lattice (i.e., $-\infty < n < +\infty$), for the somewhat more general system

$$i\frac{d}{d\tau}Q_n = Q_{n+1} - 2Q_n + Q_{n-1} - Q_n R_n (Q_{n+1} + Q_{n-1}) \qquad (3.1.2a)$$

$$-i\frac{d}{d\tau}R_n = R_{n+1} - 2R_n + R_{n-1} - Q_n R_n (R_{n+1} + R_{n-1}) \qquad (3.1.2b)$$

and include the IDNLS (3.1.1) as a special case.

The essential outline of the IST for the IDNLS (3.1.1) matches step-by-step the IST for the NLS as described in Chapter 2. Moreover, the continuum limits of the key equations in the IST for the IDNLS have as their continuum (i.e., $h \to 0$) limit the corresponding equations in the IST for the NLS. While there are important differences in the details, the steps described in Chapter 2 provide a useful guide for this formulation of the IST for the IDNLS. By

46

comparing the corresponding sections of Chapters 2 and 3, the reader can identify the modifications required to adapt the IST to the discrete equation.

The treatment of the direct problem given here closely mirrors that in [11] (see also [21]). However, here we allow the potential to have infinite support on the lattice, whereas [11] formulates the direct problem for potentials with finite support. Indeed, we show that the direct problem is solvable if Q_n and R_n are such that $\|Q_n\|_1, \|R_n\|_1 < \infty$, where $\|Q_n\|_1 = \sum_{n=-\infty}^{+\infty} |Q_n|$ is the (discrete) ℓ^1-norm. This extension to potentials with infinite support is significant in light of the fact that the soliton solutions, while decaying (exponentially) as $n \to \pm\infty$, have infinite support.

We formulate the inverse problem as a Riemann–Hilbert boundary-value problem, in analogy with the inverse problem for the differential scattering problem associated with the NLS, as described in Chapter 2 [6]. As in the inverse problem for the NLS, the Gel'fand–Levitan–Marchenko integral equations follow from the Riemann–Hilbert problem. The Riemann–Hilbert approach was partially developed in [41], but no solutions were explicitly calculated. Here, we show how to obtain solutions concretely, including the soliton solutions, from the Riemann–Hilbert approach to the inverse problem.

Just as in the IST for the NLS, the soliton solutions for the system (3.1.2a)–(3.1.2b) and the IDNLS (3.1.1) correspond to the case in which the inverse problem reduces to a linear–algebraic system of equations that can be solved explicitly. Hence, we solve this system to obtain the soliton solutions. In particular, we show that nonsingular soliton solutions can exist even when $R_n \neq -Q_n^*$ (i.e., when (3.1.2a)–(3.1.2b) does not reduce to IDNLS). Instead, we identify conditions on the scattering data that are sufficient for the existence of a nonsingular soliton solution. Soliton solutions for which $R_n \neq -Q_n^*$ are a generalization of the usual soliton solutions of the IDNLS. Moreover, for the case $R_n = -Q_n^*$, we obtain a formula for the soliton phase shifts that result from soliton interactions.

The IDNLS has infinitely many conserved quantities and a Hamiltonian structure. As in the case of the NLS, the conserved quantities are derived from the IST machinery. We describe these in the final part of this chapter.

We encourage the reader to also consult some of the references (e.g., Ablowitz et al. [1, 13]; Black et al. [32]; Bobenko et al. [33, 34]; Boiti et al. [35–37]; Bruschi et al. [42, 43, 45]; Case et al. [48, 49]; Chiu and Ladik [53]; Common [56]; Doliwa and Santini [66, 67]; Gerdjikov et al. [86]; Herbst and Weideman [93]; Hirota [94–96]; Kac and van Moerbeke [98, 99]; Konotop et al. [109, 111]; Nihoff and Capel [139]; Orfanidis [141]; Ramani et al. [143]; Suris [156–160]; Taha and Ablowitz [161–164]; Tsuchida et al. [168–172]; Vakhnenenko et al. [173]; Vekslerchik [175, 176] ; Veselov and Shabat [177] and references therein) for related studies involving discrete integrable systems from various perspectives.

3.2 The inverse scattering transform for IDNLS

3.2.1 Operator pair

One approach to the construction of integrable semi-discrete nonlinear evolution equations is to compute the compatibility condition of a spatially discrete scattering problem and a time-dependence equation. In particular, in order to construct an integrable spatial discretization of an integrable PDE, one can first discretize the scattering problem associated with the PDE. Then, by suitably expanding the time-dependence matrix in powers of the scattering parameter, one can derive compatible systems of ODEs. This method was used successfully in [11] to construct the integrable discretization of NLS, that is, the IDNLS (3.1.1).

A natural discretization of the scattering problem (2.2.3) is

$$\frac{v_{n+1} - v_n}{h} = \begin{pmatrix} -ik & q_n \\ r_n & ik \end{pmatrix} v_n + O(h^2), \tag{3.2.3}$$

where $v_n = v(nh) = \left(v_n^{(1)}, v_n^{(2)}\right)^T$, $q_n = q(nh)$, and $r_n = r(nh)$. We rewrite this finite difference as

$$v_{n+1} = \begin{pmatrix} z & Q_n \\ R_n & z^{-1} \end{pmatrix} v_n, \tag{3.2.4}$$

where

$$Q_n = hq_n, \qquad R_n = hr_n, \qquad z = e^{-ikh} = 1 - ikh + O(h^2),$$
$$z^{-1} = e^{ikh} = 1 + ikh + O(h^2)$$

and then drop terms of $O(h^2)$ and higher. The scattering problem (3.2.4) is sometimes referred to as the Ablowitz–Ladik scattering problem. Given the scattering problem (3.2.4) and the time-dependence equation (cf. [21])

$$\frac{d}{d\tau} v_n = \begin{pmatrix} i Q_n R_{n-1} - \frac{i}{2}(z - z^{-1})^2 & -i(zQ_n - z^{-1}Q_{n-1}) \\ i(z^{-1}R_n - zR_{n-1}) & -i R_n Q_{n-1} + \frac{i}{2}(z - z^{-1})^2 \end{pmatrix} v_n, \tag{3.2.5}$$

the discrete compatibility condition $\frac{d}{d\tau} v_{n+1} = \left(\frac{d}{d\tau} v_m\right)_{m=n+1}$ is equivalent to the evolution equations (3.1.2a)–(3.1.2b).

In order to obtain the IDNLS (3.1.1) from the system (3.1.2a)–(3.1.2b), we change variables as follows:

$$Q_n \to hq_n, \qquad R_n \to hr_n, \qquad \tau \to h^{-2}t. \tag{3.2.6}$$

Then the system (3.1.2a)–(3.1.2b) becomes

$$i\frac{d}{dt}q_n = \frac{1}{h^2}(q_{n+1} - 2q_n + q_{n-1}) - q_n r_n (q_{n+1} + q_{n-1}) \quad (3.2.7a)$$

$$-i\frac{d}{dt}r_n = \frac{1}{h^2}(r_{n+1} - 2r_n + r_{n-1}) - q_n r_n (r_{n+1} + r_{n-1}), \quad (3.2.7b)$$

which corresponds to the IDNLS (3.1.1) under the reduction $r_n = \mp q_n^*$.

In the limit $h \to 0, nh \to x$, the system (3.2.7a)–(3.2.7b) becomes (2.1.2a)–(2.1.2b). Moreover, in this limit the time-dependence equation (3.2.5) becomes (2.2.4), that is, the time-dependence equation for the NLS. It is noteworthy that the usual form of the time-dependence matrix does not have a convergent continuum limit (cf., e.g., [21]), but here we have chosen the gauge so that the continuum limit exists. Also, the discrete scattering problem (3.2.4) becomes, in the continuum limit, the scattering problem (2.2.3) associated with the NLS.

3.2.2 Direct scattering problem

Jost functions and summation equations

We refer to solutions of the discrete scattering problem (3.2.4) as eigenfunctions with respect to the parameter z. When the potentials $|Q_n|, |R_n| \to 0$ as $n \to \pm\infty$, the eigenfunctions are asymptotic to the solutions of

$$v_{n+1} = \begin{pmatrix} z & 0 \\ 0 & z^{-1} \end{pmatrix} v_n.$$

Therefore, it is natural to introduce the eigenfunctions defined by the following boundary conditions:

$$\phi_n(z) \sim z^n \begin{pmatrix} 1 \\ 0 \end{pmatrix}, \qquad \bar{\phi}_n(z) \sim z^{-n} \begin{pmatrix} 0 \\ 1 \end{pmatrix} \qquad \text{as } n \to -\infty \quad (3.2.8a)$$

$$\psi_n(z) \sim z^{-n} \begin{pmatrix} 0 \\ 1 \end{pmatrix}, \qquad \bar{\psi}_n(z) \sim z^n \begin{pmatrix} 1 \\ 0 \end{pmatrix} \qquad \text{as } n \to +\infty. \quad (3.2.8b)$$

In the following analysis, it is convenient to consider functions with constant boundary conditions. Hence, we define the Jost functions as follows:

$$M_n(z) = z^{-n}\phi_n(z), \qquad \bar{M}_n(z) = z^n \bar{\phi}_n(z), \qquad (3.2.9a)$$

$$N_n(z) = z^n \psi_n(z), \qquad \bar{N}_n(z) = z^{-n} \bar{\psi}_n(z). \qquad (3.2.9b)$$

If the scattering problem (3.2.4) is rewritten as

$$v_{n+1} - \mathbf{Z}v_n = \tilde{\mathbf{Q}}_n v_n,$$

where

$$\mathbf{Z} = \begin{pmatrix} z & 0 \\ 0 & z^{-1} \end{pmatrix}, \qquad \tilde{\mathbf{Q}}_n = \begin{pmatrix} 0 & Q_n \\ R_n & 0 \end{pmatrix}, \qquad (3.2.10)$$

then the Jost functions are solutions of the difference equations

$$M_{n+1}(z) - z^{-1}\mathbf{Z}M_n(z) = z^{-1}\tilde{\mathbf{Q}}_n M_n(z) \qquad (3.2.11a)$$

$$\bar{M}_{n+1}(z) - z\mathbf{Z}\bar{M}_n(z) = z\tilde{\mathbf{Q}}_n \bar{M}_n(z) \qquad (3.2.11b)$$

$$N_{n+1}(z) - z\mathbf{Z}N_n(z) = z\tilde{\mathbf{Q}}_n N_n(z) \qquad (3.2.11c)$$

$$\bar{N}_{n+1}(z) - z^{-1}\mathbf{Z}\bar{N}_n(z) = z^{-1}\tilde{\mathbf{Q}}_n \bar{N}_n(z) \qquad (3.2.11d)$$

with the constant boundary conditions

$$M_n(z) \to \begin{pmatrix} 1 \\ 0 \end{pmatrix}, \qquad \bar{M}_n(z) \to \begin{pmatrix} 0 \\ 1 \end{pmatrix} \qquad \text{as } n \to -\infty \quad (3.2.12a)$$

$$N_n(z) \to \begin{pmatrix} 0 \\ 1 \end{pmatrix}, \qquad \bar{N}_n(z) \to \begin{pmatrix} 1 \\ 0 \end{pmatrix} \qquad \text{as } n \to +\infty. \quad (3.2.12b)$$

In analogy with the continuous case, we use the method of Green's functions to construct a set of summation equations whose solutions satisfy, respectively, the difference equations (3.2.11a)–(3.2.11b) with the appropriate boundary conditions (3.2.12a)–(3.2.12b).

The Green's function corresponding to (3.2.11a), or, equivalently, (3.2.11d), is a solution of the summation equation

$$\mathbf{G}_{n+1} - z^{-1}\mathbf{Z}\mathbf{G}_n = z^{-1}\delta_{0,n}\mathbf{I}, \qquad (3.2.13)$$

where

$$\delta_{0,n} = \begin{cases} 0 & n \neq 0 \\ 1 & n = 0. \end{cases}$$

Now, if v_n satisfies the summation equation

$$v_n = w + \sum_{k=-\infty}^{+\infty} \mathbf{G}_{n-k}\tilde{\mathbf{Q}}_k v_k, \qquad (3.2.14)$$

where $\mathbf{G}_n$ is a solution of (3.2.13) and w satisfies

$$w - z^{-1}\mathbf{Z}w = \mathbf{0}, \qquad (3.2.15)$$

then v_n is a solution of the difference equation (3.2.11a) or, equivalently, (3.2.11d). The Green's function is not unique, and, as we will show in the following, the choice of the Green's function and the choice of the inhomogeneous term w together determine the Jost function and its analytical properties.

To find the Green's function explicitly, we first note that the equations for the components of (3.2.13) are uncoupled. Hence, the off-diagonal terms can be set to zero and

$$\mathbf{G}_n = \begin{pmatrix} g_n^{(1)} & 0 \\ 0 & g_n^{(2)} \end{pmatrix}, \tag{3.2.16}$$

where, according to (3.2.13), $g_n^{(j)}$ must satisfy

$$g_{n+1}^{(j)} - b^{(j)} g_n^{(j)} = z^{-1} \delta_{0,n} \qquad j = 1, 2 \tag{3.2.17}$$

with

$$b^{(1)} = 1, \qquad b^{(2)} = z^{-2}. \tag{3.2.18}$$

Next, we represent $g_n^{(j)}$ and $\delta_{0,n}$ as Fourier integrals,

$$g_n^{(j)} = \frac{1}{2\pi i} \oint_{|p|=1} p^{n-1} \hat{g}^{(j)}(p) dp, \qquad \delta_{0,n} = \frac{1}{2\pi i} \oint_{|p|=1} p^{n-1} dp.$$

Substituting these integrals into the difference equations (3.2.17), we obtain

$$\hat{g}^{(j)}(p) = z^{-1} \frac{1}{p - b^{(j)}}$$

and

$$g_n^{(j)} = z^{-1} \frac{1}{2\pi i} \oint_{|p|=1} \frac{1}{p - b^{(j)}} p^{n-1} dp. \tag{3.2.19}$$

The integral in (3.2.19) depends only on whether the pole $b^{(j)}$ is located inside or outside the contour of integration. However, when $|z| = 1$ we have $|b^{(1)}| = |b^{(2)}| = 1$, that is, the poles are on the contour $|p| = 1$. In analogy with the approach used for the continuous scattering problem, we consider contours that are perturbed away from $|p| = 1$ to avoid the singularities. Let C^{out} be a contour enclosing $p = 0$ and $p = b^{(j)}$, and let C^{in} be a contour enclosing $p = 0$ but not $p = b^{(j)}$ (see Figure 3.1).

Consequently, we get

$$g_n^{(j),\text{out}} = z^{-1} \frac{1}{2\pi i} \oint_{C^{\text{out}}} \frac{1}{p - b^{(j)}} p^{n-1} dp = z^{-1} \begin{cases} \left(b^{(j)}\right)^{n-1} & n \geq 1 \\ 0 & n \leq 0 \end{cases} \tag{3.2.20}$$

and

$$g_n^{(j),\text{in}} = z^{-1} \frac{1}{2\pi i} \oint_{C^{\text{in}}} \frac{1}{p - b^{(j)}} p^{n-1} dp = z^{-1} \begin{cases} 0 & n \geq 1 \\ -\left(b^{(j)}\right)^{n-1} & n \leq 0. \end{cases} \tag{3.2.21}$$

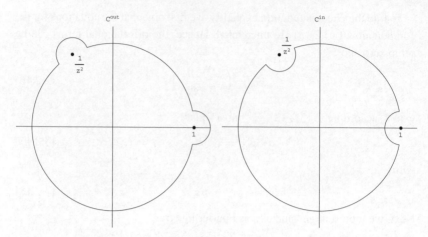

Figure 3.1: The contours C^{out} and C^{in} for the integrals in (3.2.19) that avoid singularities at $p = 0$ and $p = z^{-2}$.

By substituting one or the other of (3.2.20) or (3.2.21) into (3.2.16), with $b^{(j)}$ given by (3.2.18), we obtain two Green's functions satisfying (3.2.13), that is,

$$\mathbf{G}_n^\ell(z) = z^{-1}\theta(n-1)\begin{pmatrix} 1 & 0 \\ 0 & z^{-2(n-1)} \end{pmatrix} \qquad (3.2.22a)$$

$$\bar{\mathbf{G}}_n^r(z) = -z^{-1}\theta(-n)\begin{pmatrix} 1 & 0 \\ 0 & z^{-2(n-1)} \end{pmatrix}, \qquad (3.2.22b)$$

where $\theta(n)$ is the discrete version of the Heaviside function, that is,

$$\theta(n) = \sum_{k=-\infty}^{n} \delta_{0,k} = \begin{cases} 1 & n \geq 0 \\ 0 & n < 0. \end{cases} \qquad (3.2.23)$$

Taking into account the boundary conditions (3.2.12a)–(3.2.12b) and the relation (3.2.15) for the inhomogeneous term in (3.2.14), if Q_n, $R_n \to 0$ as $n \to \pm\infty$, we obtain the following summation equations for $M_n(z)$ and $\bar{N}_n(z)$:

$$M_n(z) = \begin{pmatrix} 1 \\ 0 \end{pmatrix} + \sum_{k=-\infty}^{+\infty} \mathbf{G}_{n-k}^\ell(z)\tilde{\mathbf{Q}}_k M_k(z) \qquad (3.2.24a)$$

$$\bar{N}_n(z) = \begin{pmatrix} 1 \\ 0 \end{pmatrix} + \sum_{k=-\infty}^{+\infty} \bar{\mathbf{G}}_{n-k}^r(z)\tilde{\mathbf{Q}}_k \bar{N}_k(z). \qquad (3.2.24b)$$

A similar approach yields summation equations for $\bar{M}_n(z)$ and $N_n(z)$:

$$\bar{M}_n(z) = \begin{pmatrix} 0 \\ 1 \end{pmatrix} + \sum_{k=-\infty}^{+\infty} \bar{\mathbf{G}}_{n-k}^\ell(z)\tilde{\mathbf{Q}}_k \bar{M}_k(z) \qquad (3.2.24c)$$

$$N_n(z) = \begin{pmatrix} 0 \\ 1 \end{pmatrix} + \sum_{k=-\infty}^{+\infty} \mathbf{G}_{n-k}^r(z)\tilde{\mathbf{Q}}_k N_k(z), \qquad (3.2.24d)$$

where

$$\bar{\mathbf{G}}_n^\ell(z) = z\theta(n-1)\begin{pmatrix} z^{2(n-1)} & 0 \\ 0 & 1 \end{pmatrix} \qquad (3.2.25a)$$

$$\mathbf{G}_n^r(z) = -z\theta(-n)\begin{pmatrix} z^{2(n-1)} & 0 \\ 0 & 1 \end{pmatrix}. \qquad (3.2.25b)$$

Existence and analyticity of the Jost functions

We now prove the existence and the sectional analyticity of the Jost functions. We will make use of Lemmas A1 and A2 proved in Appendix A.

Lemma 3.1 *If* $\|Q\|_1 = \sum_{-\infty}^{+\infty} |Q_n| < \infty$ *and* $\|R\|_1 = \sum_{-\infty}^{+\infty} |R_n| < \infty$, *then* $M_n(z)$, $N_n(z)$ *defined by (3.2.24a), (3.2.24d) are analytic functions of z for $|z| > 1$ and continuous for $|z| \geq 1$, and $\bar{M}_n(z)$, $\bar{N}_n(z)$ defined by (3.2.24b), (3.2.24c) are analytic functions of z for $|z| < 1$ and continuous for $|z| \leq 1$. Moreover, the solution of the summation equations (3.2.24a)–(3.2.24d) is unique in the space of bounded functions.*

Proof First we consider the Jost function $M_n(z)$. The Neumann series

$$M_n(z) = \sum_{j=0}^{\infty} C_n^j(z), \qquad (3.2.26)$$

where

$$C_n^0(z) = \begin{pmatrix} 1 \\ 0 \end{pmatrix} \qquad (3.2.27)$$

$$C_n^{j+1}(z) = \sum_{k=-\infty}^{+\infty} \mathbf{G}_{n-k}^\ell(z)\tilde{\mathbf{Q}}_k C_k^j(z), \qquad j \geq 0, \qquad (3.2.28)$$

is, formally, a solution of the summation equation (3.2.24a). To make this rigorous, we establish a bound on the C_n^j such that the series representation (3.2.26) converges absolutely and uniformly in n and uniformly in z in the region $|z| \geq 1$.

The componentwise summation equation for $M_n(z)$ is

$$M_n^{(1)}(z) = 1 + z^{-1}\sum_{k=-\infty}^{n-1} Q_k M_k^{(2)}(z) \qquad (3.2.29a)$$

$$M_n^{(2)}(z) = \sum_{k=-\infty}^{n-1} z^{-2(n-k-1)-1} R_k M_k^{(1)}(z), \qquad (3.2.29b)$$

and equation (3.2.28) in component form is

$$C_n^{j+1,(1)}(z) = z^{-1} \sum_{k=-\infty}^{n-1} Q_k C_k^{j,(2)}(z),$$

$$C_n^{j+1,(2)}(z) = z^{-1} \sum_{k=-\infty}^{n-1} z^{-2(n-1-k)} R_k C_k^{j,(1)}(z). \tag{3.2.30}$$

Since $C_n^{0,(2)} = 0$, it follows also that $C_n^{2j+1,(1)} = 0$, $C_n^{2j,(2)} = 0$ for any $j \geq 0$, and we only need to find bounds for $C_n^{2j,(1)}$ and $C_n^{2j+1,(2)}$. Then we prove by induction on j that, for $|z| \geq 1$,

$$\left| C_n^{2j+1,(2)}(z) \right| \leq \frac{\left(\sum_{k=-\infty}^{n-1} |R_k| \right)^{j+1}}{(j+1)!} \frac{\left(\sum_{k=-\infty}^{n-1} |Q_k| \right)^{j}}{j!} \tag{3.2.31a}$$

$$\left| C_n^{2(j+1),(1)}(z) \right| \leq \frac{\left(\sum_{k=-\infty}^{n-1} |R_k| \right)^{j+1}}{(j+1)!} \frac{\left(\sum_{k=-\infty}^{n-1} |Q_k| \right)^{j+1}}{(j+1)!}. \tag{3.2.31b}$$

The inductive step is the application of (3.2.30) twice. One iteration yields

$$\left| C_n^{2j+1,(2)}(z) \right| \leq \sum_{k=-\infty}^{n-1} |R_k| \frac{\left(\sum_{\ell=-\infty}^{k-1} |R_\ell| \right)^{j}}{j!} \frac{\left(\sum_{\ell=-\infty}^{k-1} |Q_\ell| \right)^{j}}{j!}$$

$$\leq \frac{\left(\sum_{k=-\infty}^{n-1} |Q_k| \right)^{j}}{j!} \sum_{k=-\infty}^{n-1} |R_k| \frac{\left(\sum_{\ell=-\infty}^{k-1} |R_\ell| \right)^{j}}{j!};$$

then, applying Lemma A.2 completes the induction for (3.2.31a).

The next iteration of (3.2.30) yields for $|z| \geq 1$

$$\left| C_n^{2j+2,(1)}(z) \right| \leq \sum_{k=-\infty}^{n-1} |Q_k| \left| C_k^{2j+1,(2)} \right|$$

$$\leq \sum_{k=-\infty}^{n-1} |Q_k| \frac{\left(\sum_{\ell=-\infty}^{k-1} |Q_\ell| \right)^{j}}{j!} \frac{\left(\sum_{\ell=-\infty}^{k-1} |R_\ell| \right)^{j+1}}{(j+1)!}$$

$$\leq \frac{\left(\sum_{k=-\infty}^{n-1} |R_k| \right)^{j+1}}{(j+1)!} \sum_{k=-\infty}^{n-1} |Q_k| \frac{\left(\sum_{\ell=-\infty}^{k-1} |Q_\ell| \right)^{j}}{j!},$$

and we again use Lemma A2 to complete the induction.

The bounds (3.2.31a)–(3.2.31b) are absolutely and uniformly (in n) summable if $\|Q\|_1, \|R\|_1 < \infty$, where $\|\cdot\|_1$ is the ℓ^1-norm. Moreover, in this case, the Neumann series (3.2.26) converges uniformly for all $|z| \geq 1$. Hence, since we have a uniformly convergent series of functions analytic in $|z| > 1$ and continuous in $|z| \geq 1$, the function $M_n(z)$ exists and is analytic in the region $|z| > 1$ and continuous for $|z| \geq 1$. Note that we originally constructed $M_n(z)$ for $|z| = 1$, but the summation equation allows us to extend $M_n(z)$ to the region $|z| \geq 1$.

Because the Neumann series (3.2.26) converges absolutely, this yields a convergent Laurent series in powers of z^{-1} for the solution $M_n(z)$. Thus, the bounds (3.2.31a)–(3.2.31b) establish that $M_n(z)$ is analytic in the region $|z| > 1$ when $Q_n, R_n \in \ell^1$.

To prove the existence and analyticity of the Jost function $\bar{N}_n(z)$, it is convenient to rewrite the difference equation (3.2.11d) as

$$(1 - Q_n R_n)\bar{N}_n - z\mathbf{Z}^{-1}\bar{N}_{n+1} = -z\tilde{\mathbf{Q}}_n\bar{N}_{n+1},$$

with $\mathbf{Z}$ and $\tilde{\mathbf{Q}}_n$ given in (3.2.10), and to define the modified Jost function as

$$\hat{N}_n(z) = c_n\bar{N}_n(z), \tag{3.2.32}$$

where

$$c_n = \prod_{k=n}^{+\infty}(1 - Q_k R_k). \tag{3.2.33}$$

Note that the product c_n converges absolutely if $\|Q\|_1, \|R\|_1 < \infty$. The modified Jost function $\hat{N}_n(z)$ must satisfy the difference equation

$$\hat{N}_n - z\mathbf{Z}^{-1}\hat{N}_{n+1} = -z\tilde{\mathbf{Q}}_n\hat{N}_{n+1} \tag{3.2.34}$$

with the boundary condition

$$\hat{N}_n(z) \to \begin{pmatrix} 1 \\ 0 \end{pmatrix} \qquad \text{as } n \to +\infty. \tag{3.2.35}$$

By using the method of Green's functions, we obtain the summation equation

$$\hat{N}_n = \begin{pmatrix} 1 \\ 0 \end{pmatrix} + \sum_{k=-\infty}^{+\infty} \hat{\mathbf{G}}^r_{n-k}\tilde{\mathbf{Q}}_k\hat{N}_{k+1}, \tag{3.2.36}$$

where

$$\hat{\mathbf{G}}^r_n(z) = -z\theta(-n)\begin{pmatrix} 1 & 0 \\ 0 & z^{-2n} \end{pmatrix}$$

for the solution of the difference equation (3.2.34) with the boundary condition (3.2.35).

Now, to prove the existence of the Jost function $\bar{N}_n(z)$, we first prove the existence of the modified Jost function $\hat{N}_n(z)$. As before, we construct the formal Neumann series,

$$\hat{N}_n(z) = \sum_{j=0}^{+\infty} \hat{C}_n^j(z),$$ (3.2.37)

where

$$\hat{C}_n^0(z) = \begin{pmatrix} 1 \\ 0 \end{pmatrix}$$

$$\hat{C}_n^{j+1}(z) = \sum_{k=-\infty}^{+\infty} \hat{\mathbf{G}}_{n-k}^r \tilde{\mathbf{Q}}_k \hat{C}_{k+1}^j(z), \qquad j \geq 0.$$ (3.2.38)

In the same way as before, one can show by induction that $\hat{C}_n^{2j+1,(1)} = 0$, $\hat{C}_n^{2j,(2)} = 0$ for any $j \geq 0$ and that for $|z| \leq 1$

$$\left| \hat{C}_n^{2j+1,(2)}(z) \right| \leq \frac{\left(\sum_{k=n}^{\infty} |R_k| \right)^{j+1} \left(\sum_{k=n+1}^{\infty} |Q_k| \right)^j}{(j+1)! \, j!}$$ (3.2.39)

$$\left| \hat{C}_n^{2(j+1),(1)}(z) \right| \leq \frac{\left(\sum_{k=n+1}^{\infty} |R_k| \right)^{j+1} \left(\sum_{k=n}^{\infty} |Q_k| \right)^{j+1}}{(j+1)! \, (j+1)!}.$$ (3.2.40)

The bounds (3.2.39)–(3.2.40) guarantee that, if $\|Q\|_1, \|R\|_1 < \infty$, then the modified Jost function $\hat{N}_n(z)$ – that is, the solution of the summation equation (3.2.34) and the difference equation (3.2.34) – exists and is continuous in the region $|z| \leq 1$. Moreover, $\hat{N}_n(z)$ is analytic in the region $|z| < 1$. Then, from (3.2.32) it follows that

$$\bar{N}_n(z) = c_n^{-1} \hat{N}_n(z),$$ (3.2.41)

where $\hat{N}_n(z)$ is the solution of the summation equation (3.2.36) and c_n is given by (3.2.33), provided $c_n \neq 0$ for all n. Note that, if $R_n = -Q_n^*$, then $R_n Q_n \leq 0$, and, therefore, $c_n > 0$ for all n. Alternatively, if $|R_n Q_n| < 1$ for all n, then, also, $c_n > 0$ for all n.

If, as we already assumed, the potentials are such that $\|Q\|_1, \|R\|_1 < \infty$, then the condition $c_n \neq 0 \, \forall n \in \mathbb{Z}$ is equivalent to the condition

$$c_{-\infty} = \prod_{k=-\infty}^{+\infty} (1 - Q_k R_k) \neq 0$$ (3.2.42)

because all partial products $\prod_{k=m}^{n} (1 - Q_k R_k)$ are well defined and finite. Furthermore, $c_n \neq 0$ for all n if, and only if, $(1 - Q_n R_n) \neq 0$ for all n. Moreover, we will show that the quantity $c_{-\infty}$ is time-independent when the potentials evolve according to (3.1.2a)–(3.1.2b) (cf. [11]). We conclude that, under the assumptions $\|Q\|_1, \|R\|_1 < \infty$ and $(1 - Q_n R_n) \neq 0 \, \forall n \in \mathbb{Z}$, $\bar{N}_n(z)$ as defined by (3.2.41) satisfies the difference equation (3.2.11d) with the boundary condition (3.2.12b) and the summation equation (3.2.24b). Finally, we note that $\bar{N}_n(z)$ is continuous in the region $|z| \leq 1$ and analytic in $|z| < 1$ since the series is uniformly convergent and c_n is independent of z.

Similar calculations show that $N_n(z)$ exists and is continuous in the region $|z| \geq 1$, and that $\bar{M}_n(z)$ exists and is continuous in the region $|z| \leq 1$ when $\|Q\|_1, \|R\|_1 < \infty$. Moreover, under the same condition, $N_n(z)$ is analytic in the region $|z| > 1$ and $\bar{M}_n(z)$ is analytic in the region $|z| < 1$.

We remark that, as in the IST for the continuous NLS, the sectional analyticity of the Jost functions is the key property in the formulation of the inverse problem. Finally we note that M_n, $\bar{M}_n$, N_n, $\bar{N}_n$ are unique in the space of bounded functions. The proof of uniqueness described previously can be applied to achieve this result (cf. Chapter 2).

Asymptotic behavior of the Jost functions in the complex z-plane

Because the Jost functions $M_n(z)$ and $N_n(z)$ are analytic in the region $|z| > 1$, they have a convergent Laurent series expansion about the point $z = \infty$. Similarly, because $\bar{M}_n(z)$ and $\bar{N}_n(z)$ are analytic in the region $|z| < 1$, they have a convergent power series expansion about $z = 0$.

If we write the leading terms of the Laurent expansions of the components of $M_n(z)$ as

$$M_n^{(1)} = M_n^{(1),0} + z^{-1} M_n^{(1),-1} + O(z^{-2})$$
$$M_n^{(2)} = M_n^{(2),0} + z^{-1} M_n^{(2),-1} + O(z^{-2}),$$

then substituting these expansions into the componentwise summation equations (3.2.29a)–(3.2.29b) and matching the powers of z^{-1} yields

$$\begin{pmatrix} M_n^{(1),0} \\ M_n^{(2),0} \end{pmatrix} = \begin{pmatrix} 1 \\ 0 \end{pmatrix}, \tag{3.2.43}$$

$$\begin{pmatrix} M_n^{(1),-1} \\ M_n^{(2),-1} \end{pmatrix} = \begin{pmatrix} \sum_{k=-\infty}^{n-1} Q_k M_k^{(2),0} \\ R_{n-1} M_{n-1}^{(1),0} \end{pmatrix} = \begin{pmatrix} 0 \\ R_{n-1} \end{pmatrix}, \tag{3.2.44}$$

$$\begin{pmatrix} M_n^{(1),-2} \\ M_n^{(2),-2} \end{pmatrix} = \begin{pmatrix} \sum_{k=-\infty}^{n-1} Q_k M_k^{(2),-1} \\ R_{n-1} M_{n-1}^{(1),-1} \end{pmatrix} = \begin{pmatrix} \sum_{k=-\infty}^{n-1} Q_k R_{k-1} \\ 0 \end{pmatrix}, \tag{3.2.45}$$

and so on. Moreover, one can show by induction that

$$M_n^{(1),-2j-1} = 0, \qquad M_n^{(2),-2j} = 0 \qquad \forall j \geq 0 \tag{3.2.46}$$

and that for any $j \geq 1$

$$M_n^{(2),-2j+1} = \sum_{k=n-j}^{n-1} R_k M_k^{(1),-2(j+k-n)} \tag{3.2.47}$$

$$M_n^{(1),-2j} = \sum_{k=-\infty}^{n-1} Q_k M_k^{(2),-2j+1}, \tag{3.2.48}$$

so that the Laurent expansion of $M_n(z)$ as $|z| \to \infty$ is of the form

$$M_n(z) = \begin{pmatrix} 1 + O(z^{-2}, \text{even}) \\ z^{-1} R_{n-1} + O(z^{-3}, \text{odd}) \end{pmatrix}, \tag{3.2.49}$$

where "even" indicates that the higher order terms are even powers of z^{-1} while "odd" indicates that the higher order terms are odd powers.

Analogously, one obtains for the coefficients of the power series expansion of $\bar{M}_n(z)$ about $z = 0$, that is,

$$\bar{M}_n^{(1)}(z) = \sum_{j=0}^{+\infty} z^j \bar{M}_n^{(1),j}, \qquad \bar{M}_n^{(2)}(z) = \sum_{j=0}^{+\infty} z^j \bar{M}_n^{(2),j}, \tag{3.2.50}$$

the following relations:

$$\bar{M}_n^{(1),0} = 0, \qquad \bar{M}_n^{(2),0} = 1, \tag{3.2.51}$$

while for any integer $j \geq 0$

$$\bar{M}_n^{(1),2j} = 0, \qquad \bar{M}_n^{(2),2j+1} = 0$$

and for $j \geq 1$

$$\bar{M}_n^{(1),2j-1} = \sum_{k=n-j}^{n-1} Q_k \bar{M}_k^{(2),2(j-n+k)} \tag{3.2.52}$$

$$\bar{M}_n^{(2),2j} = \sum_{k=-\infty}^{n-1} R_k \bar{M}_k^{(1),2j-1}. \tag{3.2.53}$$

Therefore

$$\bar{M}_n(z) = \begin{pmatrix} z Q_{n-1} + O(z^3, \text{odd}) \\ 1 + O(z^2, \text{even}) \end{pmatrix} \qquad z \to 0. \tag{3.2.54}$$

A similar procedure gives the leading terms of the power series expansion of $\bar{N}_n(z)$ as well as the structure of the higher order terms. However, instead of deriving the power series expansion of $\bar{N}_n(z)$ directly, we derive the power series expansion of the modified Jost function $\hat{N}_n(z)$ (3.2.32). In the same manner as before, one can show that

$$\hat{N}_n(z) = \begin{pmatrix} 1 + O(z^2, \text{even}) \\ -zR_n + O(z^3, \text{odd}) \end{pmatrix}$$

around $z = 0$. Recalling that $\bar{N}_n(z) = c_n^{-1}\hat{N}_n(z)$, where c_n, given by (3.2.33), is independent of z and by assumption $c_n \neq 0$, we also have

$$\bar{N}_n(z) = \begin{pmatrix} c_n^{-1} + O(z^2, \text{even}) \\ -zc_n^{-1}R_n + O(z^3, \text{odd}) \end{pmatrix} \qquad (3.2.55)$$

around $z = 0$.

Similar calculations show that

$$N_n(z) = \begin{pmatrix} -z^{-1}c_n^{-1}Q_n + O(z^{-3}, \text{odd}) \\ c_n^{-1} + O(z^{-2}, \text{even}) \end{pmatrix} \qquad |z| > 1. \qquad (3.2.56)$$

Scattering data

The two eigenfunctions with fixed boundary conditions as $n \to -\infty$ are linearly independent, as are the two eigenfunctions with fixed boundary conditions as $n \to +\infty$. We show this by calculating the Wronskian of these solutions. Let

$$W(v, w) = \det |v, w| = v^{(1)}w^{(2)} - v^{(2)}w^{(1)}$$

for any two vectors $v = (v^{(1)}, v^{(2)})^T$ and $w = (w^{(1)}, w^{(2)})^T$. The vector-valued sequences v_n and w_n are linearly independent if $W(v_n, w_n) \neq 0$ for all n.

In particular, if v_n and w_n are any two solutions of the scattering problem (3.2.4), their Wronskian satisfies the recursive relation

$$W(v_{n+1}, w_{n+1}) = (1 - R_n Q_n) W(v_n, w_n);$$

therefore, for any integer $s \geq 1$,

$$W\left(\phi_n(z), \bar{\phi}_n(z)\right) = \left\{ \prod_{k=n-s}^{n-1} (1 - R_k Q_k) \right\} W\left(\phi_{n-s}(z), \bar{\phi}_{n-s}(z)\right)$$

$$= \left\{ \prod_{k=n-s}^{n-1} (1 - R_k Q_k) \right\} W\left(M_{n-s}(z), \bar{M}_{n-s}(z)\right),$$

and in the limit as $s \to +\infty$ we get

$$W\left(\phi_n(z), \bar{\phi}_n(z)\right) = \prod_{k=-\infty}^{n-1} (1 - R_k Q_k). \qquad (3.2.57)$$

Recall that we have assumed that Q_n and R_n are such that $1 - Q_n R_n \neq 0$ for all n, and therefore (3.2.57) proves that ϕ_n, $\bar{\phi}_n$ are linearly independent. Similarly, for any integer $s \geq 1$,

$$W\left(\bar{\psi}_n(z), \psi_n(z)\right) = \left\{\prod_{k=n}^{n+s-1} (1 - R_k Q_k)^{-1}\right\} W\left(\bar{\psi}_{n+s}(z), \psi_{n+s}(z)\right)$$

$$= \left\{\prod_{k=n}^{n+s-1} (1 - R_k Q_k)^{-1}\right\} W\left(\bar{N}_{n+s}(z), N_{n+s}(z)\right),$$

and the limit as $s \to +\infty$, due to the boundary conditions (3.2.12b), yields

$$W\left(\bar{\psi}_n(z), \psi_n(z)\right) = \prod_{k=n}^{+\infty}(1 - R_k Q_k)^{-1}. \tag{3.2.58}$$

The discrete scattering problem (3.2.4) is a second-order difference equation. Therefore, there are at most two linearly independent solutions for any fixed value of z. Any other solution can be expressed as a linear combination of the first two. Since both the "left" eigenfunctions, $\phi_n(z)$ and $\bar{\phi}_n(z)$, and the "right" eigenfunctions, $\psi_n(z)$ and $\bar{\psi}_n(z)$, are pairs of linearly independent solutions, we can write $\phi_n(z)$ and $\bar{\phi}_n(z)$ as linear combinations of $\psi_n(z)$ and $\bar{\psi}_n(z)$, or vice versa. The coefficients of these linear combinations depend on z. Hence, the relations

$$\phi_n(z) = b(z)\psi_n(z) + a(z)\bar{\psi}_n(z) \tag{3.2.59a}$$

$$\bar{\phi}_n(z) = \bar{a}(z)\psi_n(z) + \bar{b}(z)\bar{\psi}_n(z) \tag{3.2.59b}$$

hold for any z such that all four eigenfunctions $\phi_n(z)$, $\bar{\phi}_n(z)$, $\psi_n(z)$, and $\bar{\psi}_n(z)$ exist. In particular, (3.2.59a)–(3.2.59b) hold on $|z| = 1$ and define the scattering coefficients $a(z)$, $\bar{a}(z)$, $b(z)$, and $\bar{b}(z)$.

Calculating $W\left(\phi_n(z), \bar{\phi}_n(z)\right)$ using (3.2.59a)–(3.2.59b) yields

$$W\left(\phi_n(z), \bar{\phi}_n(z)\right) = \left[a(z)\bar{a}(z) - b(z)\bar{b}(z)\right] W\left(\bar{\psi}_n(z), \psi_n(z)\right).$$

Using equations (3.2.57)–(3.2.58) to evaluate this as $n \to +\infty$, we obtain

$$a(z)\bar{a}(z) - b(z)\bar{b}(z) = c_{-\infty} \tag{3.2.60}$$

$$c_{-\infty} = \lim_{n \to -\infty} c_n = \prod_{k=-\infty}^{+\infty} (1 - R_k Q_k). \tag{3.2.61}$$

This is a significant difference with respect to the continuous problem, where the characterization equation (2.2.25) does not depend on the potentials.

Note that we can also introduce the "right" scattering data

$$\psi_n(z) = d(z)\phi_n(z) + c(z)\bar{\phi}_n(z) \tag{3.2.62a}$$
$$\bar{\psi}_n(z) = \bar{c}(z)\phi_n(z) + \bar{d}(z)\bar{\phi}_n(z), \tag{3.2.62b}$$

which, according to (3.2.59a)–(3.2.59b) and (3.2.60), are related to the "left" data through

$$c(z) = \prod_{k=-\infty}^{+\infty} (1 - R_k Q_k)^{-1} a(z) \qquad \bar{c}(z) = \prod_{k=-\infty}^{+\infty} (1 - R_k Q_k)^{-1} \bar{a}(z)$$

$$\tag{3.2.63a}$$

$$d(z) = - \prod_{k=-\infty}^{+\infty} (1 - R_k Q_k)^{-1} \bar{b}(z) \qquad \bar{d}(z) = - \prod_{k=-\infty}^{+\infty} (1 - R_k Q_k)^{-1} b(z).$$

$$\tag{3.2.63b}$$

The scattering coefficients can be represented as Wronskians of the Jost functions. Indeed, using (3.2.59a)–(3.2.59b), we obtain

$$b(z) = c_n W \left(\bar{\psi}_n(z), \phi_n(z) \right) = z^{2n} c_n W \left(\bar{N}_n(z), M_n(z) \right) \tag{3.2.64a}$$
$$\bar{b}(z) = c_n W \left(\bar{\phi}_n(z), \psi_n(z) \right) = z^{-2n} c_n W \left(\bar{M}_n(z), N_n(z) \right) \tag{3.2.64b}$$

$$a(z) = c_n W \left(\phi_n(z), \psi_n(z) \right) = c_n W \left(M_n(z), N_n(z) \right) \tag{3.2.64c}$$
$$\bar{a}(z) = c_n W \left(\bar{\psi}_n(z), \bar{\phi}_n(z) \right) = c_n W \left(\bar{N}_n(z), \bar{M}_n(z) \right), \tag{3.2.64d}$$

where c_n is given by (3.2.33). Recalling the analytic properties of the Jost functions, the expressions (3.2.64c)–(3.2.64d) show that $a(z)$ has an analytic extension in the region $|z| > 1$ while $\bar{a}(z)$ has an analytic extension in the region $|z| < 1$. Because the Jost functions are continuous up to $|z| = 1$, the functions $a(z)$ and $\bar{a}(z)$ are also continuous up to $|z| = 1$. Moreover, substituting the z- expansions of the Jost functions (3.2.49) and (3.2.54)–(3.2.56) into (3.2.64c)–(3.2.64d) immediately yields

$$a(z) = 1 + O(z^{-2}, \text{even}) \qquad \text{as } |z| \to \infty \tag{3.2.65a}$$
$$\bar{a}(z) = 1 + O(z^{2}, \text{even}) \qquad \text{as } |z| \to 0. \tag{3.2.65b}$$

The scattering coefficients can also be written as explicit sums of the Jost functions. The formulas are derived as follows. First, we obtain the relation

$$M_n(z) - \bar{N}_n(z)a(z)$$
$$= \begin{pmatrix} 1 - a(z) \\ 0 \end{pmatrix} + \sum_{k=-\infty}^{+\infty} \left\{ \mathbf{G}_{n-k}^{\ell}(z)\tilde{\mathbf{Q}}_k M_k(z) - \bar{\mathbf{G}}_{n-k}^{r}(z)\tilde{\mathbf{Q}}_k \bar{N}_k(z)a(z) \right\}$$

by substituting the right-hand sides of the summation equations (3.2.24a) and (3.2.24b) for $M_n(z)$ and $\bar{N}_n(z)$, respectively. Then, we use the identity

$$\mathbf{G}_n^{\ell}(z) = \bar{\mathbf{G}}_n^{r}(z) + z^{-1} \begin{pmatrix} 1 & 0 \\ 0 & z^{-2(n-1)} \end{pmatrix}$$

to get

$$M_n(z) - \bar{N}_n(z)a(z) = \begin{pmatrix} 1 - a(z) \\ 0 \end{pmatrix} + \sum_{k=-\infty}^{+\infty} \bar{\mathbf{G}}_{n-k}^{r}(z)\tilde{\mathbf{Q}}_k \left\{ M_k(z) - N_k(z)a(z) \right\}$$
$$+ \sum_{k=-\infty}^{+\infty} \begin{pmatrix} z^{-1} & 0 \\ 0 & z^{-2(n-k)+1} \end{pmatrix} \tilde{\mathbf{Q}}_k M_k(z).$$

Now, by the relation (3.2.59a), we can replace $M_n(z) - \bar{N}_n(z)a(z)$ with $z^{-2n} N_n(z)b(z)$ on both sides and use the identity $\bar{\mathbf{G}}_n^{r}(z) = z^{-2n} \mathbf{G}_n^{r}(z)$ to obtain

$$z^{-2n} \left\{ N_n(z) - \sum_{k=-\infty}^{+\infty} \mathbf{G}_{n-k}^{r}(z)\tilde{\mathbf{Q}}_k N_k(z) \right\} b(z)$$
$$= \begin{pmatrix} 1 - a(z) \\ 0 \end{pmatrix} + \sum_{k=-\infty}^{+\infty} \begin{pmatrix} z^{-1} & 0 \\ 0 & z^{-2(n-k)+1} \end{pmatrix} \tilde{\mathbf{Q}}_k M_k(z).$$

If we assume that the summation equation (3.2.24d) for $N_n(z)$ has a unique solution, the term in curly braces is $(0, 1)^T$, and then

$$a(z) = 1 + \sum_{k=-\infty}^{+\infty} z^{-1} Q_k M_k^{(2)}(z) \tag{3.2.66a}$$

$$b(z) = \sum_{k=-\infty}^{+\infty} z^{2k+1} R_k M_k^{(1)}(z). \tag{3.2.66b}$$

The same approach works for $\bar{a}(z)$ and $\bar{b}(z)$, and the corresponding expressions are

$$\bar{a}(z) = 1 + \sum_{k=-\infty}^{+\infty} z R_k \bar{M}_k^{(1)}(z) \tag{3.2.66c}$$

$$\bar{b}(z) = \sum_{k=-\infty}^{+\infty} z^{-2k-1} Q_k \bar{M}_k^{(2)}(z). \tag{3.2.66d}$$

In the formulation of the inverse problem, we replace the Jost functions $M_n(z)$ and $\bar{M}_n(z)$ with the functions

$$\mu_n(z) = \frac{M_n(z)}{a(z)} = \begin{pmatrix} 1 + O(z^{-2}) \\ z^{-1}R_{n-1} + O(z^{-3}) \end{pmatrix} \tag{3.2.67}$$

$$\bar{\mu}_n(z) = \frac{\bar{M}_n(z)}{\bar{a}(z)} = \begin{pmatrix} zQ_{n-1} + O(z^3) \\ 1 + O(z^2) \end{pmatrix}. \tag{3.2.68}$$

Observe that $\mu_n(z)$ is meromorphic in the region $|z| > 1$ with poles corresponding to the zeros of $a(z)$, while $\bar{\mu}_n(z)$ is meromorphic in the region $|z| < 1$ with poles at the zeros of $\bar{a}(z)$. Also, we define the reflection coefficients

$$\rho(z) = \frac{b(z)}{a(z)}, \qquad \bar{\rho}(z) = \frac{\bar{b}(z)}{\bar{a}(z)} \tag{3.2.69}$$

so that the conditions (3.2.59a)–(3.2.59b) are equivalent to

$$\mu_n(z) - \bar{N}_n(z) = z^{-2n}\rho(z)N_n(z) \tag{3.2.70a}$$

$$\bar{\mu}_n(z) - N_n(z) = z^{2n}\bar{\rho}(z)\bar{N}_n(z). \tag{3.2.70b}$$

Note that (3.2.69), as well as the relations (3.2.70a)–(3.2.70b), are defined, in the general case, only for $|z| = 1$.

Proper eigenvalues and norming constants

The discrete scattering problem (3.2.4) can possess discrete eigenvalues (bound states). These occur whenever $a(z_j) = 0$ for some z_j such that $|z_j| > 1$ or $\bar{a}(\bar{z}_\ell) = 0$ for $\bar{z}_\ell$ with $|\bar{z}_\ell| < 1$. Indeed, for such values of the spectral parameter $W(\phi_n(z_j), \psi_n(z_j)) = 0$ and $W(\bar{\phi}_n(\bar{z}_\ell), \bar{\psi}_n(\bar{z}_\ell)) = 0$, and therefore $\phi_n(z_j), \psi_n(z_j)$ and $\bar{\phi}_n(\bar{z}_\ell), \bar{\psi}_n(\bar{z}_\ell)$ are linearly dependent, that is,

$$\phi_n(z_j) = b_j\psi_n(z_j), \qquad \bar{\phi}_n(\bar{z}_\ell) = \bar{b}_\ell\bar{\psi}_n(\bar{z}_\ell) \tag{3.2.71}$$

for some complex constants $b_j, \bar{b}_\ell$. In terms of the Jost functions, (3.2.71) can be written as

$$M_n(z_j) = b_j z_j^{-2n} N_n(z_j), \qquad \bar{M}_n(\bar{z}_\ell) = \bar{b}_\ell \bar{z}_\ell^{2n} \bar{N}_n(\bar{z}_\ell), \tag{3.2.72}$$

which hold if, and only if, z_j and $\bar{z}_\ell$ are eigenvalues. Note that the boundary conditions (3.2.8a) and (3.2.8b), together with (3.2.71), imply that $\phi_n(z_j), \bar{\phi}_n(\bar{z}_\ell) \to 0$ for $|n| \to \infty$.

We will assume that neither $a(z)$ nor $\bar{a}(z)$ vanishes on the unit circle. Since the eigenvalues correspond to zeros of the sectionally analytic functions $a(z)$

and $\bar{a}(z)$, there are no accumulation points of eigenvalues in the regions of analyticity, that is, $|z| > 1$ and $|z| < 1$, respectively. Moreover, because $a(z) \to 1$ as $|z| \to \infty$ and $\bar{a}(z) \to 1$ as $z \to 0$, there is only a finite number of eigenvalues in the regions $|z| \leq A^{-1}$ and $|z| \geq A$ for any $A > 1$. If there were an infinite number of zeros of either $a(z)$ or of $\bar{a}(z)$ (i.e., an infinite number of eigenvalues), these zeros would necessarily have an accumulation point on $|z| = 1$. Recall, however, that $a(z)$ and $\bar{a}(z)$ are continuous, respectively, in the regions $|z| \geq 1$ and $|z| \leq 1$. Hence, the accumulation of zeros at a point on $|z| = 1$ implies that there is a zero of $a(z)$ or of $\bar{a}(z)$ on $|z| = 1$, which contradicts our previous assumption. We conclude that, under the generic assumption $a(z), \bar{a}(z) \neq 0$ on $|z| = 1$, there is a finite number of eigenvalues.

We further assume that the eigenvalues are simple zeros of $a(z)$ and $\bar{a}(z)$. If the eigenvalues are not simple zeros, one can study the situation by the coalescence of simple poles, in analogy with the continuous case [192].

Let us assume $a(z)$ has J simple zeros $\{z_j : |z_j| > 1\}_{j=1}^{J}$ and $\bar{a}(z)$ has $\bar{J}$ simple zeros at the points $\{\bar{z}_j : |\bar{z}_j| < 1\}_{j=1}^{\bar{J}}$. Then, by equations (3.2.67)–(3.2.68) and (3.2.72),

$$\mathrm{Res}(\mu_n; z_j) = \frac{M_n(z_j)}{a'(z_j)} = \frac{b_j}{a'(z_j)} z_j^{-2n} N_n(z_j) = z_j^{-2n} C_j N_n(z_j) \qquad (3.2.73a)$$

$$\mathrm{Res}(\bar{\mu}_n; \bar{z}_\ell) = \frac{\bar{M}_n(\bar{z}_\ell)}{\bar{a}'(\bar{z}_\ell)} = \frac{\bar{b}_\ell}{\bar{a}'(\bar{z}_\ell)} \bar{z}_\ell^{2n} \bar{N}_n(\bar{z}_\ell) = \bar{z}_\ell^{2n} \bar{C}_\ell \bar{N}_n(\bar{z}_\ell), \qquad (3.2.73b)$$

where $'$ denotes the derivative with respect to the spectral parameter. We refer to C_j and $\bar{C}_\ell$ as the norming constants associated with the eigenvalues z_j and $\bar{z}_\ell$, respectively. If the potentials decay rapidly enough as $|n| \to \infty$ such that $b(z)$ and $\bar{b}(z)$ can be extended off the unit circle in correspondence of the discrete eigenvalues z_j and $\bar{z}_\ell$, respectively, then the norming constants are simply

$$C_j = \frac{b(z_j)}{a'(z_j)}, \qquad \bar{C}_\ell = \frac{\bar{b}(z_\ell)}{\bar{a}'(\bar{z}_\ell)}.$$

Symmetries and symmetry reductions

First we note that the expansions of $a(z)$ and $\bar{a}(z)$ in (3.2.65a)–(3.2.65b) contain only even powers of z^{-1} and z, respectively. Hence, if z_j is a zero of $a(z)$, so is $-z_j$, and the same holds for $\bar{a}(z)$. Therefore, the eigenvalues appear in pairs $\pm z_j, \pm \bar{z}_\ell$. Let us denote the norming constants associated with the paired poles $\pm z_j$ as $C_j^{\pm}$. Similarly, we label the constants $b_j^{\pm}$ in (3.2.72) that are associated with $\pm z_j$. Given the expansions (3.2.49) and (3.2.56), we conclude that $b_j^- = -b_j^+$. On the other hand, $a(z)$ is even, so $a'(z)$ is an odd function of z. Thus,

$$C_j^- = \frac{b_j^-}{a'(-z_j)} = \frac{-b_j^+}{-a'(z_j)} = \frac{b_j^+}{a'(z_j)} = C_j^+. \qquad (3.2.74)$$

Since the two norming constants associated with $\pm z_j$ are equal, we will drop the superscript $\pm$ on the norming constant and refer to both norming constants as C_j. Similarly, one can show that

$$\bar{C}_\ell^- = \bar{C}_\ell^+ = \bar{C}_\ell \qquad (3.2.75)$$

for the norming constants $\bar{C}_\ell^\pm$ associated with $\pm \bar{z}_\ell$. Moreover, we also have

$$\rho(-z) = -\rho(z), \qquad \bar{\rho}(-z) = -\bar{\rho}(z). \qquad (3.2.76)$$

The eigenvalues $\{\pm z_j\}_{j=1}^{J}$, $\{\pm \bar{z}_j\}_{j=1}^{\bar{J}}$ and the associated norming constants $\{C_j\}_{j=1}^{J}$, $\{\bar{C}_j\}_{j=1}^{\bar{J}}$, together with the reflection coefficients (3.2.69), constitute the set of the scattering data $S(z)$.

The evolution equation (3.1.1) is a special case of the compatibility condition (3.2.7) where $r_n = \mp q_n^*$. This symmetry in the potentials induces a symmetry between the Jost functions defined for $|z| \leq 1$, analytic in the region $|z| < 1$, and the Jost functions defined for $|z| \geq 1$, analytic in the region $|z| > 1$. In turn, this symmetry of the Jost functions induces a symmetry in the scattering data. Indeed, a direct computation shows that the Green's functions (3.2.22a)–(3.2.22b) and (3.2.25a)–(3.2.25b) satisfy the identities

$$\hat{\mathbf{P}}_\mp \left(\mathbf{G}_n^\ell \left(1/z^*\right)\right)^* \hat{\mathbf{P}}_\mp^{-1} = \bar{\mathbf{G}}_n^\ell(z), \qquad \hat{\mathbf{P}}_\mp \left(\mathbf{G}_n^r \left(1/z^*\right)\right)^* \hat{\mathbf{P}}_\mp^{-1} = \bar{\mathbf{G}}_n^r(z), \quad (3.2.77)$$

where

$$\hat{\mathbf{P}}_\mp = \begin{pmatrix} 0 & \mp 1 \\ 1 & 0 \end{pmatrix}.$$

Moreover, under the symmetry reductions $R_n = \mp Q_n^*$, the matrix potential $\tilde{\mathbf{Q}}_n$ in (3.2.10) is such that

$$\hat{\mathbf{P}}_\mp \tilde{\mathbf{Q}}_n^* \hat{\mathbf{P}}_\mp^{-1} = \tilde{\mathbf{Q}}_n, \qquad (3.2.78)$$

and from (3.2.77)–(3.2.78) it follows that $\hat{\mathbf{P}}_\mp M_n^*(1/z^*)$ and $\hat{\mathbf{P}}_\mp N_n^*(1/z^*)$ satisfy, respectively, the same summation equations as $\bar{M}_n(z)$ and $\mp \bar{N}_n(z)$, that is, equations (3.2.24c) and (3.2.24b). Therefore, assuming that such equations have unique solutions, we obtain the symmetries

$$\bar{M}_n(z) = \hat{\mathbf{P}}_\mp M_n^*(1/z^*), \qquad \bar{N}_n(z) = \mp \hat{\mathbf{P}}_\mp N_n^*(1/z^*) \qquad (3.2.79)$$

or, equivalently,

$$\bar{\phi}_n(z) = \hat{\mathbf{P}}_{\mp}\phi_n^*(1/z^*), \qquad \bar{\psi}_n(z) = \mp\hat{\mathbf{P}}_{\mp}\psi_n^*(1/z^*). \qquad (3.2.80)$$

The symmetry of the scattering coefficients $a(z)$, $\bar{a}(z)$ and $b(z)$, $\bar{b}(z)$ follows from the symmetry of the Jost functions and the Wronskian formulas (3.2.64a)–(3.2.64b), that is,

$$\bar{a}(z) = a^*(1/z^*) \qquad (3.2.81a)$$

$$\bar{b}(z) = \mp b^*(1/z^*) \qquad (3.2.81b)$$

when $R_n = \mp Q_n^*$. From (3.2.81a)–(3.2.81b) one obtains the symmetry between the reflection coefficients, namely

$$\bar{\rho}(z) = \mp \rho^*(1/z^*). \qquad (3.2.82)$$

Note that (3.2.81b) and (3.2.82) hold, in general, only for $|z| = 1$, that is, for $z = 1/z^*$.

The symmetry (3.2.81a) implies that the eigenvalues are paired; that is, for each eigenvalue z_j such that $|z_j| > 1$, there is an eigenvalue $\bar{z}_j = 1/z_j^*$ with $|\bar{z}_j| < 1$, and vice versa. Consequently, it also follows that $J = \bar{J}$ (i.e., the number of eigenvalues outside the unit circle equals that of the eigenvalues inside).

We now compute the symmetry of the norming constants. With the pairing of the eigenvalues, equations (3.2.72) give

$$\bar{b}_j = \mp b_j^*. \qquad (3.2.83)$$

Moreover, from (3.2.81a) it also follows that

$$\bar{a}'(\bar{z}_j) = -\left(z_j^2 a'(z_j)\right)^*;$$

hence

$$\bar{C}_j = \frac{\bar{b}_j}{\bar{a}'(\bar{z}_j)} = \frac{\mp b_j^*}{-(z_j^2 a'(z_j))^*} = \pm(z_j^*)^{-2} C_j^*, \qquad (3.2.84)$$

when $R_n = \mp Q_n^*$ and the zeros of $a(z)$ and $\bar{a}(z)$ are simple.

Note that when $R_n = Q_n^*$ there are no discrete eigenvalues with $|z_j| \neq 1$. Indeed, as in the continuous case, the scattering problem with this reduction is formally self-adjoint. Moreover, from (3.2.60) and the symmetry relations (3.2.81a)–(3.2.81b) it follows that, on the unit circle, $|z| = 1$, $|a(z)|^2 - |b(z)|^2 = c_{-\infty}$, where, according to (3.2.61), $c_{-\infty} = \prod_{k=-\infty}^{+\infty}(1 - |Q_k|)^2$ under the symmetry reduction we are considering. Therefore, assuming for the potential the small norm condition $|Q_n| < 1$ for any $n \in \mathbb{Z}$, $|a(z)|^2 > 0$ on the unit circle. Since, in addition, the scattering problem is self-adjoint under this reduction, $|a(z)| > 0$ for any z such that $|z| \geq 1$.

To summarize, we have shown that:

Symmetry 3.1 All the eigenvalues of the scattering problem (3.2.4) appear in pairs $\pm z_j$ with $|z_j| > 1$, that is, outside the unit circle, or $\pm \bar{z}_\ell$, with $|\bar{z}_\ell| < 1$, and the norming constant associated with $-z_j$ (resp. $-\bar{z}_\ell$) is equal to the norming constant associated with $+z_j$ (resp. $+\bar{z}_\ell$). Correspondingly, the norming constant associated with the pair of eigenvalues $\pm z_j$ outside the unit circle will be denoted by C_j, and the norming constant associated with the pair of eigenvalues $\pm \bar{z}_\ell$ inside the unit circle will be denoted by $\bar{C}_\ell$.

Symmetry 3.2 If the potentials satisfy the symmetry $R_n = \mp Q_n^*$, then the scattering coefficients satisfy the symmetry

$$\bar{a}(z) = a^*(1/z^*), \qquad \bar{\rho}(z) = \mp \rho^*(1/z^*). \qquad (3.2.85)$$

It follows that $\bar{z}_j = 1/z_j^*$ is an eigenvalue such that $|\bar{z}_j| < 1$ if, and only if, z_j is an eigenvalue such that $|z_j| > 1$. Therefore $J = \bar{J}$, that is, the number of eigenvalues inside the unit circle equals the number of eigenvalues outside, and, taking also into account Symmetry 3.1, the eigenvalues come in quartets,

$$\left\{ \pm z_j, \pm 1/z_j^* \right\}_{j=1}^J . \qquad (3.2.86)$$

Moreover, according to (3.2.84), the norming constants associated with these paired eigenvalues satisfy the symmetry

$$\bar{C}_j = \pm (z_j^*)^{-2} C_j^* . \qquad (3.2.87)$$

Trace formula

We assume that $a(z)$ and $\bar{a}(z)$ have the simple zeros $\left\{ \pm z_j : |z_j| > 1 \right\}_{j=1}^J$ and $\left\{ \pm \bar{z}_j : |\bar{z}_j| < 1 \right\}_{j=1}^{\bar{J}}$, respectively. Moreover, we define

$$\alpha(z) = \prod_{m=1}^J \frac{z^2 - (z_m^*)^{-2}}{z^2 - z_m^2} \, a(z), \qquad \bar{\alpha}(z) = \prod_{m=1}^{\bar{J}} \frac{z^2 - (\bar{z}_m^*)^{-2}}{z^2 - \bar{z}_m^2} \, \bar{a}(z). \qquad (3.2.88)$$

According to these definitions, the function $\alpha(z)$ is analytic outside the unit circle, where it has no zeros, while $\bar{\alpha}(z)$ is analytic inside the unit circle, and it has no zeros; moreover, due to (3.2.65a), $\alpha(z) \to 1$ as $|z| \to \infty$. Therefore, taking into account that both a and $\bar{a}$ are even functions of z, we have

$$\log \alpha(z) = -\frac{1}{2\pi i} \oint_{|w|=1} \frac{w \log \alpha(w)}{w^2 - z^2} dw,$$

$$\frac{1}{2\pi i} \oint_{|w|=1} \frac{w \log \bar{\alpha}(w)}{w^2 - z^2} dw = 0 \qquad |z| > 1$$

$$\log \bar{\alpha}(z) = \frac{1}{2\pi i} \oint_{|w|=1} \frac{w \log \bar{\alpha}(w)}{w^2 - z^2} dw,$$

$$\frac{1}{2\pi i} \oint_{|w|=1} \frac{w \log \alpha(w)}{w^2 - z^2} dw = 0 \qquad |z| < 1.$$

Subtracting the equations from one another and using (3.2.88) yields

$$\log a(z) = \sum_{m=1}^{J} \log \left(\frac{z^2 - z_m^2}{z^2 - (z_m^*)^{-2}} \right)$$
$$- \frac{1}{2\pi i} \oint_{|w|=1} \frac{w \log (\alpha(w)\bar{\alpha}(w))}{w^2 - z^2} dw, \qquad |z| > 1 \quad (3.2.89a)$$

$$\log \bar{a}(z) = \sum_{m=1}^{J} \log \left(\frac{z^2 - \bar{z}_m^2}{z^2 - (\bar{z}_m^*)^{-2}} \right)$$
$$+ \frac{1}{2\pi i} \oint_{|w|=1} \frac{w \log (\alpha(w)\bar{\alpha}(w))}{w^2 - z^2} dw \qquad |z| < 1. \quad (3.2.89b)$$

This allows one to recover $a(z)$, $\bar{a}(z)$ from knowledge of $\left\{ \pm z_j : \left| z_j \right| > 1 \right\}_{j=1}^{J}$, $\left\{ \pm \bar{z}_j : \left| \bar{z}_j \right| < 1 \right\}_{j=1}^{J}$ and $a(w)\bar{a}(w) = c_{-\infty} (1 - \rho(w)\bar{\rho}(w))^{-1}$ for $|w| = 1$.

Note that if $R_n = -Q_n^*$, then, from (3.2.81a)–(3.2.83) and (3.2.60), it follows that $\alpha(w)\bar{\alpha}(w) = c_{-\infty}(1 + |\rho(w)|^2)^{-1}$ for $|w| = 1$, and consequently (3.2.89a) can be written as

$$\log a(z) = \sum_{m=1}^{J} \log \left(\frac{z^2 - z_m^2}{z^2 - (z_m^*)^{-2}} \right)$$
$$+ \frac{1}{2\pi i} \oint_{|w|=1} \frac{w \log \left(1 + |\rho(w)|^2\right)}{w^2 - z^2} dw \qquad |z| > 1. \quad (3.2.90)$$

3.2.3 Inverse scattering problem

The inverse problem consists of reconstructing the potentials in terms of the scattering data $\{\rho(w), \bar{\rho}(w)$ for $|w| = 1\} \cup \left\{ \pm z_j, C_j \right\}_{j=1}^{J} \cup \left\{ \pm \bar{z}_j, \bar{C}_j \right\}_{j=1}^{J}$. In the previous section, we showed that the functions $N_n(z)$ and $\bar{N}_n(z)$ exist and are analytic in the regions $|z| > 1$ and $|z| < 1$, respectively, if $\|Q\|_1, \|R\|_1 < 1$. Similarly, under the same conditions on the potentials, the functions $\mu_n(z)$ and

$\bar{\mu}_n(z)$ defined by (3.2.67)–(3.2.68) are meromorphic in the regions $|z| > 1$ and $|z| < 1$, respectively. Therefore, in the inverse problem we assume these analytic properties for the unknown functions. With this assumption, equations (3.2.70a)–(3.2.70b) can be considered to be the jump conditions of a Riemann–Hilbert problem. As in the continuous case, to recover the sectionally meromorphic functions from the scattering data, we convert the Riemann–Hilbert problem to a system of linear integral equations with the use of the Plemelj formula (cf. [7], [41]).

Because the functions $\mu_n(z)$ and $\bar{\mu}_n(z)$ can be meromorphic, the jump conditions (3.2.70a)–(3.2.70b) are insufficient to fix the solution of the Riemann–Hilbert problem. We also need information about the residues of the poles. Furthermore, to fix the solution, it is also necessary to specify the boundary conditions.

Boundary conditions and residues

The functions $\mu_n(z)$ and $N_n(z)$ are meromorphic in the region $|z| > 1$ and have the limits

$$N_n(z) \to \begin{pmatrix} 1 \\ c_n^{-1} \end{pmatrix}, \qquad \mu_n(z) \to \begin{pmatrix} 1 \\ 0 \end{pmatrix} \qquad \text{as } |z| \to \infty$$

(cf. (3.2.49), (3.2.56), and (3.2.65a)). Here, as before, $c_n = \prod_{k=n}^{\infty}(1 - Q_k R_k)$. The boundary condition for $N_n(z)$ depends on Q_k and R_k for all $k \geq n$. However, Q_n and R_n are unknowns in the inverse problem. To remove this dependence, we introduce the functions

$$N_n' = \begin{pmatrix} 1 & 0 \\ 0 & c_n \end{pmatrix} N_n = \begin{pmatrix} -z^{-1}c_n^{-1}Q_n \\ 1 \end{pmatrix} + O(z^{-2}) \qquad \text{as } |z| \to \infty \quad (3.2.91a)$$

$$\mu_n' = \begin{pmatrix} 1 & 0 \\ 0 & c_n \end{pmatrix} \mu_n = \begin{pmatrix} 1 \\ z^{-1}c_n R_{n-1} \end{pmatrix} + O(z^{-2}) \qquad \text{as } |z| \to \infty \quad (3.2.91b)$$

$$\bar{N}_n' = \begin{pmatrix} 1 & 0 \\ 0 & c_n \end{pmatrix} \bar{N}_n = \begin{pmatrix} c_n^{-1} \\ -z R_n \end{pmatrix} + O(z^2) \qquad \text{as } z \to 0 \quad (3.2.91c)$$

$$\bar{\mu}_n' = \begin{pmatrix} 1 & 0 \\ 0 & c_n \end{pmatrix} \bar{\mu}_n = \begin{pmatrix} z Q_{n-1} \\ c_n \end{pmatrix} + O(z^2) \qquad \text{as } z \to 0. \quad (3.2.91d)$$

These modified functions also satisfy the jump conditions (3.2.70a)–(3.2.70b) on $|z| = 1$, that is,

$$\mu_n'(z) - \bar{N}_n'(z) = z^{-2n}\rho(z)N_n'(z) \qquad (3.2.92a)$$

$$\bar{\mu}_n'(z) - N_n'(z) = z^{2n}\bar{\rho}(z)\bar{N}_n'(z). \qquad (3.2.92b)$$

Moreover, the poles of $\mu_n'(z)$ and $\bar{\mu}_n'(z)$ are the same as the poles of $\mu_n(z)$ and $\bar{\mu}_n(z)$, respectively, and the residues of these poles are determined by the relations

$$\text{Res}(\mu_n'; z_j) = z_j^{-2n} C_j N_n'(z_j), \qquad \text{Res}(\bar{\mu}_n'; \bar{z}_j) = \bar{z}_j^{2n} \bar{C}_j \bar{N}_n'(\bar{z}_j), \quad (3.2.93)$$

which follow from (3.2.73a)–(3.2.73b).

Case of no poles

Let us consider first the case when there are no discrete eigenvalues, that is, μ_n' and $\bar{\mu}_n'$ have no poles. Introducing the 2×2 matrices

$$\mathbf{m}_n(z) = \left(\mu_n'(z),\, N_n'(z)\right), \qquad \bar{\mathbf{m}}_n(z) = \left(\bar{N}_n'(z),\, \bar{\mu}_n'(z)\right) \quad (3.2.94)$$

with $\mathbf{m}_n(z)$ analytic outside the unit circle $|z| = 1$ and $\bar{\mathbf{m}}_n(z)$ analytic inside, we can write the jump conditions (3.2.92a)–(3.2.92b) as

$$\mathbf{m}_n(z) - \bar{\mathbf{m}}_n(z) = \bar{\mathbf{m}}_n(z) \mathbf{V}_n(z) \qquad |z| = 1, \quad (3.2.95)$$

where

$$\mathbf{V}_n(z) = \begin{pmatrix} -\rho(z)\bar{\rho}(z) & -z^{2n}\bar{\rho}(z) \\ z^{-2n}\rho(z) & 0 \end{pmatrix} \quad (3.2.96)$$

and

$$\mathbf{m}_n(z) \to \mathbf{I} \qquad \text{as } |z| \to \infty. \quad (3.2.97)$$

Therefore (3.2.95) can be regarded as a generalized Riemann–Hilbert boundary-value problem on $|z| = 1$ with boundary conditions given by (3.2.97).

We consider the integral operators

$$\bar{P}(f)(z) = \lim_{\substack{\zeta \to z \\ |\zeta| < 1}} \frac{1}{2\pi i} \oint_{|w|=1} \frac{f(w)}{w - \zeta} dw \quad (3.2.98a)$$

$$P(f)(z) = \lim_{\substack{\zeta \to z \\ |\zeta| > 1}} \frac{1}{2\pi i} \oint_{|w|=1} \frac{f(w)}{w - \zeta} dw \quad (3.2.98b)$$

defined for $|z| < 1$ and $|z| > 1$, respectively, for any function $f(w)$ continuous on $|w| = 1$. These are the projection operators for functions analytic inside and outside, respectively, the unit circle on the complex z-plane. They are the counterparts of the projection operators (2.2.51) for the upper and lower k-planes in the IST for the continuous NLS.

Applying $\bar{P}$ to both sides of equation (3.2.95) yields

$$\bar{\mathbf{m}}_n(z) = \mathbf{I} - \lim_{\substack{\zeta \to z \\ |\zeta|<1}} \frac{1}{2\pi i} \oint_{|w|=1} \frac{\bar{\mathbf{m}}_n(w)\mathbf{V}_n(w)}{w - \zeta} dw, \qquad (3.2.99)$$

which allows one, in principle, to find $\bar{\mathbf{m}}_n(z)$. In component form, (3.2.99) yields

$$\bar{N}'_n(z) = \begin{pmatrix} 1 \\ 0 \end{pmatrix} - \lim_{\substack{\zeta \to z \\ |\zeta|<1}} \frac{1}{2\pi i} \oint_{|w|=1} \frac{w^{-2n}\rho(w)\bar{\mu}'_n(w) - \rho(w)\bar{\rho}(w)\bar{N}'_n(w)}{w - \zeta} dw$$

$$\qquad (3.2.100)$$

$$\bar{\mu}'_n(z) = \begin{pmatrix} 0 \\ 1 \end{pmatrix} + \lim_{\substack{\zeta \to z \\ |\zeta|<1}} \frac{1}{2\pi i} \oint_{|w|=1} \frac{w^{2n}\bar{\rho}(w)\bar{N}'_n(w)}{w - \zeta} dw, \qquad (3.2.101)$$

which is a system of linear integral equations, on $|z| = 1$, for $\bar{N}'_n(z)$ and $\bar{\mu}'_n(z)$ in terms of the scattering data. The solutions of these integral equations can be analytically continued into the region $|z| < 1$. Equivalently, by applying the outside projector P to both sides of equations (3.2.92a)–(3.2.92b) we get

$$N'_n(z) = \begin{pmatrix} 0 \\ 1 \end{pmatrix} + \lim_{\substack{\zeta \to z \\ |\zeta|>1}} \frac{1}{2\pi i} \oint_{|w|=1} \frac{w^{2n}\bar{\rho}(w)\bar{N}'_n(w)}{w - \zeta} dw$$

$$\mu'_n(z) = \begin{pmatrix} 1 \\ 0 \end{pmatrix} - \lim_{\substack{\zeta \to z \\ |\zeta|>1}} \frac{1}{2\pi i} \oint_{|w|=1} \frac{w^{-2n}\rho(w)N'_n(w)}{w - \zeta} dw.$$

To recover Q_n and R_n we compute the power series expansions of the Jost functions around $z = 0$. By comparing the expansions (3.2.91c) and (3.2.91d) with the expansions about $z = 0$ of the right-hand sides of (3.2.100) and (3.2.101) or, equivalently, of (3.2.99),

$$\bar{\mathbf{m}}_n(z) = \mathbf{I} - \frac{z}{2\pi i} \oint_{|w|=1} w^{-2}\bar{\mathbf{m}}_n(w)\mathbf{V}_n(w)dw + O(z^{-2}),$$

we obtain

$$Q_n = \frac{1}{2\pi i} \oint_{|w|=1} w^{2n}\bar{\rho}(w)\bar{N}'^{(1)}_{n+1}(w)dw$$

$$R_n = \frac{1}{2\pi i} \oint_{|w|=1} \left(w^{-2(n+1)}\rho(w)\bar{\mu}'^{(2)}_n(w) - w^{-2}\rho(w)\bar{\rho}(w)\bar{N}'^{(2)}_n(w) \right) dw,$$

where $\bar{N}'_n(z)$ and $\bar{\mu}'_n(z)$ are determined by the linear system (3.2.100)–(3.2.101) and the superscript (j) for $j = 1, 2$ denotes the j-th component of the corresponding vector. Hence, the formulation of the inverse problem is complete. When $R_n = Q_n^*$, there are no discrete eigenvalues. Using the symmetry relation (3.2.85) between the reflection coefficients ρ and $\bar{\rho}$ (cf. Symmetry 3.2), we can write the potential as

$$Q_n = \frac{1}{2\pi i} \oint_{|w|=1} w^{2n} \rho^*(w) \bar{N}_{n+1}^{\prime(1)}(w) dw.$$

Note that we recover the potentials from the Jost functions that are defined in the region $|z| \leq 1$, namely, $\bar{\mu}'_n(z)$ and $\bar{N}'_n(z)$, because there are no terms in the z expansions of $\mu'_n(z)$ and $N'_n(z)$ that are simple functions of Q_n and R_n.

Case of poles

The method of solution requires an extra step if $\mu'_n(z)$ and $\bar{\mu}'_n(z)$ have poles. As before, we apply $\bar{P}$ to both sides of (3.2.92a) and P to both sides of (3.2.92b). Taking into account the analytic properties of the eigenfunctions and of $a(z), \bar{a}(z)$ we obtain

$$\bar{N}'_n(z) = \binom{1}{0} + \sum_{j=1}^{J} C_j z_j^{-2n} \left[\frac{1}{z - z_j} N'_n(z_j) + \frac{1}{z + z_j} N'_n(-z_j) \right] \quad (3.2.102a)$$

$$- \lim_{\substack{\zeta \to z \\ |\zeta| < 1}} \frac{1}{2\pi i} \oint_{|w|=1} \frac{w^{-2n} \rho(w) N'_n(w)}{w - \zeta} dw$$

$$N'_n(z) = \binom{0}{1} + \sum_{j=1}^{\bar{J}} \bar{C}_j \bar{z}_j^{2n} \left[\frac{1}{z - \bar{z}_j} \bar{N}'_n(\bar{z}_j) + \frac{1}{z + \bar{z}_j} \bar{N}'_n(-\bar{z}_j) \right] \quad (3.2.102b)$$

$$+ \lim_{\substack{\zeta \to z \\ |\zeta| > 1}} \frac{1}{2\pi i} \oint_{|w|=1} \frac{w^{2n} \bar{\rho}(w) \bar{N}'_n(w)}{w - \zeta} dw,$$

where $N'_n(z_j)$ is $N'_n(z)$ evaluated at the eigenvalue z_j, $N'_n(-z_j)$ is $N'_n(z)$ evaluated at the eigenvalue $-z_j$, and similarly for $\bar{N}'_n(\bar{z}_j)$ and $\bar{N}'_n(-\bar{z}_j)$. Here we have included the fact that the eigenvalues arise in pairs $\pm z_j$ in $|z| > 1$ and $\pm \bar{z}_\ell$ in $|z| < 1$ and that the corresponding norming constants satisfy (3.2.74)–(3.2.75). Equations (3.2.102a)–(3.2.102b) constitute a system of linear integral equations on $|z| = 1$. This system depends on the vectors $\left\{ N'_n(z_j), N'_n(-z_j) \right\}_{j=1}^{J}$ and $\left\{ \bar{N}'_n(\bar{z}_j), \bar{N}'_n(-\bar{z}_j) \right\}_{j=1}^{\bar{J}}$. We obtain expressions for these vectors by evaluating (3.2.102a) at the points $\pm \bar{z}_j$ and (3.2.102b) at the points $\pm z_j$. This results in a linear algebraic–integral system composed of (3.2.102a)–(3.2.102b) and

$$\bar{N}'_n(\bar{z}_j) = \begin{pmatrix} 1 \\ 0 \end{pmatrix} + \sum_{k=1}^{J} C_k z_k^{-2n} \left[\frac{1}{\bar{z}_j - z_k} N'_n(z_k) + \frac{1}{\bar{z}_j + z_k} N'_n(-z_k) \right]$$
$$- \frac{1}{2\pi i} \oint_{|w|=1} \frac{w^{-2n} \rho(w) N'_n(w)}{w - \bar{z}_j} dw \qquad (3.2.102c)$$

$$\bar{N}'_n(-\bar{z}_j) = \begin{pmatrix} 1 \\ 0 \end{pmatrix} - \sum_{k=1}^{J} C_k z_k^{-2n} \left[\frac{1}{\bar{z}_j + z_k} N'_n(z_k) + \frac{1}{\bar{z}_j - z_k} N'_n(-z_k) \right]$$
$$- \frac{1}{2\pi i} \oint_{|w|=1} \frac{w^{-2n} \rho(w) N'_n(w)}{w + \bar{z}_j} dw \qquad (3.2.102d)$$

$$N'_n(z_j) = \begin{pmatrix} 0 \\ 1 \end{pmatrix} + \sum_{k=1}^{\bar{J}} \bar{C}_k \bar{z}_k^{2n} \left[\frac{1}{z_j - \bar{z}_k} \bar{N}'_n(\bar{z}_k) + \frac{1}{z_j + \bar{z}_k} \bar{N}'_n(-\bar{z}_k) \right]$$
$$+ \frac{1}{2\pi i} \oint_{|w|=1} \frac{w^{2n} \bar{\rho}(w) \bar{N}'_n(w)}{w - z_j} dw \qquad (3.2.102e)$$

$$N'_n(-z_j) = \begin{pmatrix} 0 \\ 1 \end{pmatrix} - \sum_{k=1}^{\bar{J}} \bar{C}_k \bar{z}_k^{2n} \left[\frac{1}{z_j + \bar{z}_k} \bar{N}'_n(\bar{z}_k) + \frac{1}{z_j - \bar{z}_k} \bar{N}'_n(-\bar{z}_k) \right]$$
$$+ \frac{1}{2\pi i} \oint_{|w|=1} \frac{w^{2n} \bar{\rho}(w) \bar{N}'_n(w)}{w + z_j} dw, \qquad (3.2.102f)$$

where (3.2.102c)–(3.2.102d) hold for each eigenvalue $\{\bar{z}_j\}_{j=1}^{\bar{J}}$ and (3.2.102e)–(3.2.102f) hold for each eigenvalue $\{z_j\}_{j=1}^{J}$.

As in the case where there are no discrete eigenvalues, we can recover R_n from the power series expansion of $\bar{N}'_n(z)$. There is, however, no easy way to obtain the potential from $N'_n(z)$. Instead, we apply $\bar{P}$ to both sides of (3.2.92b) to obtain the representation

$$\bar{\mu}'_n(z) = \begin{pmatrix} 0 \\ 1 \end{pmatrix} + \sum_{j=1}^{\bar{J}} \bar{C}_j \bar{z}_j^{2n} \left[\frac{1}{z - \bar{z}_j} \bar{N}'_n(\bar{z}_j) + \frac{1}{z + \bar{z}_j} \bar{N}'_n(-\bar{z}_j) \right]$$
$$+ \frac{1}{2\pi i} \oint_{|w|=1} \frac{w^{2n} \bar{\rho}(w) \bar{N}'_n(w)}{w - z} dw. \qquad (3.2.103)$$

Now, by comparing the power series expansions of the right-hand sides of (3.2.102a) and (3.2.103) to the expansions (3.2.91c) and (3.2.91d), respectively, we obtain

$$R_n = 2 \sum_{j=1}^{J} C_j z_j^{-2(n+1)} N_n''^{(2)}(z_j)$$

$$+ \frac{1}{2\pi i} \oint_{|w|=1} w^{-2(n+1)} \rho(w) N_n''^{(2)}(w) dw \qquad (3.2.104a)$$

$$Q_{n-1} = -2 \sum_{j=1}^{\bar{J}} \bar{C}_j \bar{z}_j^{2(n-1)} \bar{N}_n''^{(1)}(\bar{z}_j)$$

$$+ \frac{1}{2\pi i} \oint_{|w|=1} w^{2(n-1)} \bar{\rho}(w) \bar{N}_n''^{(1)}(w) dw, \qquad (3.2.104b)$$

where we have used that for the solutions of the system (3.2.102a)–(3.2.102f) the following relations hold:

$$N_n''^{(1)}(-z) = -N_n''^{(1)}(z), \qquad N_n''^{(2)}(-z) = N_n''^{(2)}(z)$$
$$\bar{N}_n''^{(1)}(-z) = \bar{N}_n''^{(1)}(z), \qquad \bar{N}_n''^{(2)}(-z) = -\bar{N}_n''^{(2)}(z).$$

Note that from (3.2.91d), (3.2.103) it also follows that

$$c_n = 1 - 2 \sum_{j=1}^{\bar{J}} \bar{C}_j \bar{z}_j^{2n-1} \bar{N}_n''^{(2)}(\bar{z}_j) + \frac{1}{2\pi i} \oint_{|w|=1} w^{2n-1} \bar{\rho}(w) \bar{N}_n''^{(2)}(w) dw.$$

$$(3.2.105)$$

If the potentials satisfy the symmetry condition $R_n = -Q_n^*$, then the scattering data satisfy the relations (3.2.85)–(3.2.87) and (3.2.104b) becomes

$$Q_n = -2 \sum_{j=1}^{\bar{J}} (z_j^*)^{-2(n+1)} C_j^* \bar{N}_{n+1}''^{(1)}(1/z_j^*) - \frac{1}{2\pi i} \oint_{|w|=1} w^{2n} \rho^*(w) \bar{N}_{n+1}''^{(1)}(w) dw.$$

Reflectionless potentials

In the case where the scattering data comprise proper eigenvalues but $\rho(z) = \bar{\rho}(z) = 0$ on $|z| = 1$, the algebraic–integral system (3.2.102c)–(3.2.102f) reduces to the linear algebraic system

$$\bar{N}_n'(\bar{z}_j) = \begin{pmatrix} 1 \\ 0 \end{pmatrix} + \sum_{k=1}^{J} C_k z_k^{-2n} \left[\frac{1}{\bar{z}_j - z_k} N_n'(z_k) + \frac{1}{\bar{z}_j + z_k} N_n'(-z_k) \right]$$

$$\bar{N}_n'(-\bar{z}_j) = \begin{pmatrix} 1 \\ 0 \end{pmatrix} - \sum_{k=1}^{J} C_k z_k^{-2n} \left[\frac{1}{\bar{z}_j + z_k} N_n'(z_k) + \frac{1}{\bar{z}_j - z_k} N_n'(-z_k) \right]$$

$$N_n'(z_j) = \begin{pmatrix} 0 \\ 1 \end{pmatrix} + \sum_{k=1}^{\bar{J}} \bar{C}_k \bar{z}_k^{2n} \left[\frac{1}{z_j - \bar{z}_k} \bar{N}_n'(\bar{z}_k) + \frac{1}{z_j + \bar{z}_k} \bar{N}_n'(-\bar{z}_k) \right]$$

$$N_n'(-z_j) = \begin{pmatrix} 0 \\ 1 \end{pmatrix} - \sum_{k=1}^{\bar{J}} \bar{C}_k \bar{z}_k^{2n} \left[\frac{1}{z_j + \bar{z}_k} \bar{N}_n'(\bar{z}_k) + \frac{1}{z_j - \bar{z}_k} \bar{N}_n'(-\bar{z}_k) \right].$$

Moreover, the potentials are given by

$$Q_{n-1} = -2 \sum_{j=1}^{\bar{J}} \bar{C}_j \bar{z}_j^{2(n-1)} \bar{N}_n'^{(1)}(\bar{z}_j)$$

$$R_n = 2 \sum_{j=1}^{J} C_j z_j^{-2(n+1)} N_n'^{(2)}(z_j).$$

If there is one quartet of eigenvalues $\{\pm z_1, \pm \bar{z}_1\}$ with $|z_1| > 1$ and $|\bar{z}_1| < 1$, then we can solve for $\bar{N}_n'$, N_n', obtaining in particular

$$\bar{N}_n'^{(1)}(\bar{z}_1) = \left[1 + 4C_1 \bar{C}_1 \frac{\bar{z}_1^{-2(n-1)} \bar{z}_1^{2n}}{(z_1^2 - \bar{z}_1^2)^2} \right]^{-1}$$

$$N_n'^{(2)}(z_1) = \left[1 + 4C_1 \bar{C}_1 \frac{z_1^{-2n} \bar{z}_1^{2(n+1)}}{(z_1^2 - \bar{z}_1^2)^2} \right]^{-1}.$$

Then it follows that

$$Q_n = -\frac{2D_1 \bar{z}_1^{2(n+1)}}{1 + 4C_1 D_1 (z_1^2 - \bar{z}_1^2)^{-2} z_1^{-2n} \bar{z}_1^{2(n+2)}} \qquad (3.2.106a)$$

$$R_n = \frac{2C_1 z_1^{-2(n+1)}}{1 + 4C_1 D_1 (z_1^2 - \bar{z}_1^2)^{-2} z_1^{-2n} \bar{z}_1^{2(n+2)}}, \qquad (3.2.106b)$$

where we introduced the modified norming constant

$$D_1 = \bar{z}_1^{-2} \bar{C}_1. \qquad (3.2.107)$$

Note that from (3.2.105) it also follows that

$$c_{-\infty} = \lim_{n \to -\infty} c_n = z_1^2 \bar{z}_1^{-2}. \qquad (3.2.108)$$

In order to obtain a nonsingular potential, we impose the following symmetry:

Symmetry 3.3 The scattering data are such that (i) the eigenvalues satisfy the relation $\bar{z}_1 = \frac{1}{z_1^*}$, (ii) the product of the norming constants C_1 and D_1 is real, and, moreover, (iii) $C_1 D_1 > 0$.

Then, with the substitution $z_1 = e^{(\alpha_1 + i\beta_1)}$, the expressions (3.2.106a)–(3.2.106b) can be written in the form

$$Q_n = -\frac{D_1}{(C_1 D_1)^{1/2}} \sinh(2\alpha_1)\, e^{2i(n+1)\beta_1} \operatorname{sech}(2\alpha_1(n+1) - \delta_1) \quad (3.2.109a)$$

$$R_n = \frac{C_1}{(C_1 D_1)^{1/2}} \sinh(2\alpha_1)\, e^{-2i(n+1)\beta_1} \operatorname{sech}(2\alpha_1(n+1) - \delta_1), \quad (3.2.109b)$$

where

$$\delta_1 = \log(C_1 D_1)^{1/2} - \log \sinh(2\alpha_1). \tag{3.2.110}$$

Recall that if $R_n = -Q_n^*$, then Symmetry 3.2 holds. Therefore, in particular, if $R_n = -Q_n^*$ and the associated scattering data consist of a single quartet of eigenvalues (and their respective norming constants), then, as a consequence, Symmetry 3.3 holds. However, the converse is not necessarily true. Symmetry 3.3 in the scattering data is insufficient to ensure that (3.2.106a)–(3.2.106b) satisfy the symmetry $R_n = -Q_n^*$. To obtain the symmetry $R_n = -Q_n^*$ in (3.2.106a)–(3.2.106b) we must impose the condition $|C_1| = |D_1|$ in addition to Symmetry 3.3. Equivalently, to obtain the symmetry $R_n = -Q_n^*$ we must have (i) $\bar{z}_1 = \frac{1}{z_1^*}$ and (ii) $\bar{C}_1 = z_1^2 C_1^*$.

Typically, only solutions with the symmetry $R_n = -Q_n^*$ are considered. However, we emphasize that the sech profile of the potentials (3.2.109a)–(3.2.109b) results from Symmetry 3.3 in the scattering data, and we do not need to require that $R_n = -Q_n^*$. In the scalar evolution equation – that is IDNLS – Symmetry 3.3 is only slightly more general than Symmetry 3.2. Nevertheless, we have shown that the sech envelope potentials exist in a more general setting (i.e., when $R_n \neq -Q_n^*$).

We recall that, if the potentials are not constrained to satisfy the symmetry $R_n = -Q_n^*$, then, to ensure that the IST procedure is well defined, we must separately impose the condition $(1 - R_n Q_n) \neq 0$ for all n. However, we point out that this second condition is time-invariant if $\|R\|_1, \|Q\|_1 < \infty$.

Gel'fand–Levitan–Marchenko equations

Like in the continuous case, we can also provide a reconstruction for the potentials by means of Gel'fand–Levitan–Marchenko integral equations. Indeed, let us represent the eigenfunctions ψ_n and $\bar{\psi}_n$ in terms of triangular kernels,

$$\psi_n(z) = \sum_{j=n}^{+\infty} z^{-j} K(n, j) \qquad |z| > 1 \tag{3.2.111a}$$

$$\bar{\psi}_n(z) = \sum_{j=n}^{+\infty} z^j \bar{K}(n, j) \qquad |z| < 1, \qquad (3.2.111b)$$

where $K(n, j) = \left(K^{(1)}(n, j), K^{(2)}(n, j)\right)^T$ and $\bar{K}(n, j) = \left(\bar{K}^{(1)}(n, j), \bar{K}^{(2)}(n, j)\right)^T$, and write equations (3.2.59a)–(3.2.59b) in the form

$$\phi_n(z)a^{-1}(z) - \bar{\psi}_n(z) = \psi_n(z)\rho(z) \qquad (3.2.112a)$$

$$\bar{\phi}_n(z)\bar{a}^{-1}(z) - \psi_n(z) = \bar{\psi}_n(z)\bar{\rho}(z) \qquad (3.2.112b)$$

with ρ and $\bar{\rho}$ given by (3.2.69).

By applying the operator $\frac{1}{2\pi i} \oint_{|z|=1} dz\, z^{-m-1}$ for $m \geq n$ to equation (3.2.112a) and taking into account the asymptotics (3.2.49), (3.2.54), and (3.2.65a)–(3.2.65b), as well as the triangular representations (3.2.111a)–(3.2.111b), we obtain

$$\bar{K}(n, m) + \sum_{j=n}^{+\infty} K(n, j)F(m + j) = \binom{1}{0} \delta_{m,n} \qquad m \geq n, \qquad (3.2.113)$$

where

$$F(n) = \sum_{j=1}^{J} z_j^{-n-1} C_j + \frac{1}{2\pi i} \oint_{|z|=1} z^{-n-1} \rho(z) dz. \qquad (3.2.114)$$

Analogously, operating on equation (3.2.112b) with $\frac{1}{2\pi i} \oint_{|z|=1} dz\, z^{m-1}$ for $m \geq n$ yields

$$K(n, m) + \sum_{j=n}^{+\infty} \bar{K}(n, j)\bar{F}(m + j) = \binom{0}{1} \delta_{m,n} \qquad m \geq n, \qquad (3.2.115)$$

where

$$\bar{F}(n) = -\sum_{j=1}^{\bar{J}} \bar{z}_j^{n-1}\bar{C}_j + \frac{1}{2\pi i} \oint_{|z|=1} z^{n-1} \bar{\rho}(z) dz. \qquad (3.2.116)$$

Equations (3.2.113) and (3.2.115) constitute the Gel'fand–Levitan–Marchenko equations. Comparing the representations (3.2.111a)–(3.2.111b) for the eigenfunctions with the asymptotics (3.2.55) and (3.2.56), we obtain, recalling (3.2.9b), the reconstruction of the potentials in terms of the kernels of the GLM equations, that is,

$$K^{(1)}(n, n) = \bar{K}^{(2)}(n, n) = 0, \qquad \bar{K}^{(1)}(n, n) = K^{(2)}(n, n) = c_n^{-1} \qquad (3.2.117)$$

$$Q_n = -\frac{K^{(1)}(n, n + 1)}{K^{(2)}(n, n)}, \qquad R_n = -\frac{\bar{K}^{(2)}(n, n + 1)}{\bar{K}^{(1)}(n, n)}, \qquad (3.2.118)$$

where, as usual, $K^{(j)}$ and $\bar{K}^{(j)}$ for $j = 1, 2$ are the j-th component of the vectors K and $\bar{K}$, respectively.

It is more convenient to write equations (3.2.113) and (3.2.115) as forced summation equations, which is accomplished if we introduce $\kappa(n, m)$ and $\bar{\kappa}(n, m)$ such that

$$\kappa(n, n) = \begin{pmatrix} 0 \\ 1 \end{pmatrix} \qquad \bar{\kappa}(n, n) = \begin{pmatrix} 1 \\ 0 \end{pmatrix} \qquad (3.2.119a)$$

and for $m > n$

$$K(n, m) = \prod_{j=n}^{+\infty} \left(1 - R_j Q_j\right) \kappa(n, m) \qquad (3.2.119b)$$

$$\bar{K}(n, m) = \prod_{j=n}^{+\infty} \left(1 - R_j Q_j\right) \bar{\kappa}(n, m). \qquad (3.2.119c)$$

Then equations (3.2.113), (3.2.115) become

$$\bar{\kappa}(n, m) + \begin{pmatrix} 0 \\ 1 \end{pmatrix} F(m + n) + \sum_{j=n+1}^{+\infty} \kappa(n, j) F(m + j) = 0 \qquad m > n$$

$$(3.2.120a)$$

$$\kappa(n, m) + \begin{pmatrix} 1 \\ 0 \end{pmatrix} \bar{F}(m + n) + \sum_{j=n+1}^{+\infty} \bar{\kappa}(n, j) \bar{F}(m + j) = 0 \qquad m > n$$

$$(3.2.120b)$$

and the potentials are obtained from

$$Q_n = -\kappa^{(1)}(n, n + 1), \qquad R_n = -\bar{\kappa}^{(2)}(n, n + 1). \qquad (3.2.121)$$

Note that the sums in (3.2.114) (resp. (3.2.116)) are performed over all the discrete eigenvalues that are outside (resp. inside) the unit circle. Since these eigenvalues are paired and the corresponding norming constants satisfy (3.2.74)–(3.2.75), the GLM equations can be simplified as follows:

$$\bar{\kappa}(n, m) + \begin{pmatrix} 0 \\ 1 \end{pmatrix} F_R(m + n) + \sum_{\substack{j=n \\ j+m=\text{odd}}}^{+\infty} \kappa(n, j) F_R(m + j) = 0 \qquad m > n$$

$$(3.2.122a)$$

$$\kappa(n, m) + \begin{pmatrix} 1 \\ 0 \end{pmatrix} \bar{F}_R(m + n) + \sum_{\substack{j=n \\ j+m=\text{odd}}}^{+\infty} \bar{\kappa}(n, j) \bar{F}_R(m + j) = 0 \qquad m > n,$$

$$(3.2.122b)$$

where

$$F_R(n) = \begin{cases} 2\sum_{j=1}^J z_j^{-n-1} C_j + \frac{1}{\pi i}\int_{C_R} z^{-n-1}\rho(z)dz & n = \text{odd} \\ 0 & n = \text{even} \end{cases} \quad (3.2.123)$$

$$\bar{F}_R(n) = \begin{cases} -2\sum_{j=1}^{\bar{J}} \bar{z}_j^{-n-1} \bar{C}_j + \frac{1}{\pi i}\int_{C_R} z^{-n-1}\bar{\rho}(z)dz & n = \text{odd} \\ 0 & n = \text{even} \end{cases} \quad (3.2.124)$$

and C_R denotes the right half of the unit circle.

If the symmetry $R_n = \mp Q_n^*$ holds, then, from (3.2.80)–(3.2.83) and (3.2.84), it follows that

$$\bar{F}(m) = \mp F^*(m) \qquad (3.2.125)$$

$$\bar{K}(n, m) = \begin{pmatrix} K^{(2)}(n, m) \\ \mp K^{(1)}(n, m) \end{pmatrix}^*. \qquad (3.2.126)$$

In this case, equations (3.2.113), (3.2.115) solving the inverse problem reduce to

$$K^{(2)}(n, m) \mp \sum_{j'=n}^{+\infty} \sum_{j''=n}^{+\infty} K^{(2)}(n, j'')F(j' + j'')F^*(j' + m) = \delta_{n,m} \quad (3.2.127)$$

$$K^{(1)}(n, m) = \sum_{j'=n}^{+\infty} K^{(2)*}(n, j')F^*(j' + m), \qquad (3.2.128)$$

and the potentials are reconstructed by means of the first of (3.2.118).

Existence and uniqueness of solutions

The question of existence and uniqueness of solutions of linear summation equations can be examined by the use of the Fredholm alternative. Consider the homogeneous equations corresponding to (3.2.120a)–(3.2.120b),

$$h_1(m) + \sum_{j=n+1}^{+\infty} h_2(j)F(m + j) = 0 \qquad m > n \qquad (3.2.129a)$$

$$h_2(m) + \sum_{j=n+1}^{+\infty} h_1(j)\bar{F}(m + j) = 0 \qquad m > n, \qquad (3.2.129b)$$

and suppose $h(n) = (h_1(n), h_2(n))$ is a solution of (3.2.129a)–(3.2.129b) that vanishes identically for $m \leq n$. Multiply (3.2.129a)–(3.2.129b) by (h_1^*, h_2^*), sum over all integers m, and use

$$\sum_{m=n+1}^{\infty} |h_j(m)|^2 = \sum_{m=-\infty}^{\infty} |h_j(m)|^2 \qquad j = 1, 2,$$

to obtain

$$\sum_{m=-\infty}^{\infty} \left\{ |h_1(m)|^2 + |h_2(m)|^2 + \sum_{j=-\infty}^{\infty} \left[h_2(m)h_1^*(j)F(m+j) \right. \right.$$

$$\left. \left. + h_1(j)\, h_2^*(m)\bar{F}(m+j) \right] \right\} = 0. \tag{3.2.130}$$

If the symmetry $R_n = -Q_n^*$ holds, then, from (3.2.80)–(3.2.83) and (3.2.84), it follows that

$$\sum_{m=-\infty}^{\infty} \left\{ |h_1(m)|^2 + |h_2(m)|^2 + 2i\,\mathrm{Im} \sum_{j=-\infty}^{\infty} h_2(j)h_1^*(m)F(m+j) \right\} = 0.$$

The real and imaginary parts of the previous equation must both vanish, from which it follows that

$$h(m) \equiv 0.$$

Second, if $R_n = Q_n^*$, the symmetry (3.2.84) allows equation (3.2.130) to be written in the form

$$\sum_{m=-\infty}^{\infty} \left\{ |h_1(m)|^2 + |h_2(m)|^2 + 2\,\mathrm{Re} \sum_{j=-\infty}^{\infty} h_2(j)h_1^*(m)F(m+j) \right\} = 0. \tag{3.2.131}$$

As we pointed out previously, the scattering problem with this reduction is formally self-adjoint and there are no discrete eigenvalues. This implies that

$$F(m) = \frac{1}{2\pi i} \oint_{|z|=1} z^{-m-1} \frac{b(z)}{a(z)}\, dz. \tag{3.2.132}$$

A function $h(n)$ of a discrete variable n assuming integer values, and its discrete Fourier transform $\hat{h}(z)$

$$\hat{h}(z) = \sum_{n=-\infty}^{\infty} h(n)z^n, \qquad h(n) = \frac{1}{2\pi i} \oint_{|z|=1} \hat{h}(z)z^{n-1}\, dz,$$

satisfy a discrete version of Parseval's identity, namely,

$$\sum_{n=-\infty}^{\infty} |h(n)|^2 = \frac{1}{2\pi i} \oint_{|z|=1} z^{-1} |\hat{h}(z)|^2 \equiv \frac{1}{2\pi} \int_0^{2\pi} |\hat{h}(e^{i\theta})|^2\, d\theta.$$

Substituting these results into (3.2.131) yields

$$\int_0^{2\pi} \left\{ |\hat{h}_1(e^{i\theta})|^2 + |\hat{h}_2^*(e^{i\theta})|^2 + 2\,\mathrm{Re}\frac{b}{a}(e^{i\theta})\hat{h}_1(e^{-i\theta})\hat{h}_2^*(e^{i\theta}) \right\} d\theta = 0. \tag{3.2.133}$$

Since $|(b/a)| < 1$, we have

$$\left|2\mathrm{Re}\left[\frac{b}{a}(e^{i\theta})\hat{h}_1(e^{-i\theta})\hat{h}_2^*(e^{i\theta})\right]\right| < 2|\hat{h}_1(e^{-i\theta})||\hat{h}_2^*(e^{i\theta})|$$

$$\leq |\hat{h}_1(e^{-i\theta})|^2 + |\hat{h}_2(e^{i\theta})|^2;$$

hence

$$\hat{h}(z) \equiv 0$$

and

$$h(n) \equiv 0.$$

We conclude that when $R_n = \mp Q_n^*$, the integral equations (3.2.120a), (3.2.120b) admit no homogenous solutions but the trivial one. It remains to be shown that, like in the continuous case, the discrete Gel'fand–Levitan–Marchenko type summation equations behave like Fredholm equations.

3.2.4 Time evolution

The operator (3.2.5) determines the evolution of the Jost functions. From this we deduce the time evolution of the scattering data. Since we have assumed that $Q_n, R_n \to 0$ as $n \to \pm\infty$, then the time-dependence (3.2.5) is asymptotically of the form

$$\partial_\tau v_n = \begin{pmatrix} -i\omega & 0 \\ 0 & i\omega \end{pmatrix} v_n \qquad \text{as } n \to \pm\infty,$$

where

$$\omega = \tfrac{1}{2}\left(z - z^{-1}\right)^2. \tag{3.2.134}$$

This system has solutions that are linear combinations of the solutions $v_n^+ = \left(e^{-i\omega\tau}, 0\right)^T$ and $v_n^- = \left(0, e^{i\omega\tau}\right)^T$. However, such solutions are not compatible with the fixed boundary conditions of the Jost functions (3.2.12), and therefore we define the time-dependent functions

$$\mathcal{M}_n(z, \tau) = e^{-i\omega\tau} M_n(z, \tau), \qquad \bar{\mathcal{M}}_n(z, \tau) = e^{i\omega\tau}\bar{M}_n(z, \tau)$$

$$\mathcal{N}_n(z, \tau) = e^{i\omega\tau} N_n(z, \tau), \qquad \bar{\mathcal{N}}_n(z, \tau) = e^{-i\omega\tau}\bar{N}_n(z, \tau)$$

to be solutions of the time-dependence equation (3.2.5). These τ-dependent functions satisfy the relations

$$\mathcal{M}_n(z, \tau) = z^{-2n} e^{-2i\omega\tau} b(z, \tau)\mathcal{N}_n(z, \tau) + a(z, \tau)\bar{\mathcal{N}}_n(z, \tau) \tag{3.2.135a}$$

$$\bar{\mathcal{M}}_n(z, \tau) = z^{2n} e^{2i\omega\tau} \bar{b}(z, \tau)\bar{\mathcal{N}}_n(z, \tau) + \bar{a}(z, \tau)\mathcal{N}_n(z, \tau), \tag{3.2.135b}$$

which are obtained from the equations (3.2.59a)–(3.2.59b).

To find the expressions for the evolution of the scattering coefficients, we first differentiate (3.2.135a)–(3.2.135b) with respect to τ to obtain

$$\partial_\tau \mathcal{M}_n(z, \tau) = z^{-2n} e^{-2i\omega\tau} \left\{ b(z, \tau)\partial_\tau \mathcal{N}_n(z, \tau) + [b_\tau(z, \tau) \right.$$
$$\left. - 2i\omega b\,(z, \tau)] \mathcal{N}_n(z, \tau) \right\} + a_\tau(z, \tau)\bar{\mathcal{N}}_n(z, \tau)$$
$$+ a(z, \tau)\partial_\tau \bar{\mathcal{N}}_n(z, \tau) \qquad (3.2.136a)$$

$$\partial_\tau \bar{\mathcal{M}}_n(z, \tau) = z^{2n} e^{2i\omega\tau} \left\{ \bar{b}(z, \tau)\partial_\tau \bar{\mathcal{N}}_n(z, \tau) + \left[\bar{b}_\tau(z, \tau) \right. \right.$$
$$\left. + 2i\omega\,\bar{b}(z, \tau) \right] \bar{\mathcal{N}}_n(z, \tau) \right\} + \bar{a}_\tau(z, \tau)\mathcal{N}_n(z, \tau)$$
$$+ \bar{a}(z, \tau)\partial_\tau \mathcal{N}_n(z, \tau). \qquad (3.2.136b)$$

On the other hand, because the functions $\mathcal{M}_n(z, \tau)$, $\bar{\mathcal{M}}_n(z, \tau)$, $\mathcal{N}_n(z, \tau)$, and $\bar{\mathcal{N}}_n(z, \tau)$ satisfy (3.2.5), we have

$$\partial_\tau \mathcal{M}_n(z, \tau) = z^{-2n} e^{-2i\omega\tau} b(z, \tau)\partial_\tau \mathcal{N}_n(z, \tau) + a(z, \tau)\partial_\tau \bar{\mathcal{N}}_n(z, \tau) \qquad (3.2.137a)$$
$$\partial_\tau \bar{\mathcal{M}}_n(z, \tau) = z^{2n} e^{2i\omega\tau} \bar{b}(z, \tau)\partial_\tau \bar{\mathcal{N}}_n(z, \tau) + \bar{a}(z, \tau)\partial_\tau \mathcal{N}_n(z, \tau). \qquad (3.2.137b)$$

Comparing (3.2.136a)–(3.2.136b) with, respectively, (3.2.137a)–(3.2.137b) and examining the asymptotics of these expressions as $n \to +\infty$, one gets

$$b_\tau(z, \tau) = 2i\omega b(z, \tau) \qquad a_\tau(z, \tau) = 0$$
$$\bar{a}_\tau(z, \tau) = 0 \qquad \bar{b}_\tau(z, \tau) = -2i\omega\bar{b}(z, \tau),$$

and therefore

$$b(z, \tau) = b(z, 0)e^{2i\omega\tau} \qquad a(z, \tau) = a(z, 0) \qquad (3.2.138a)$$
$$\bar{a}(z, \tau) = \bar{a}(z, 0) \qquad \bar{b}(z, \tau) = \bar{b}(z, 0)e^{-2i\omega\tau}. \qquad (3.2.138b)$$

The evolution of the reflection coefficients is thus given by

$$\rho(z, \tau) = \rho(z, 0)e^{2i\omega\tau} \qquad (3.2.139a)$$
$$\bar{\rho}(z, \tau) = \bar{\rho}(z, 0)e^{-2i\omega\tau}. \qquad (3.2.139b)$$

From (3.2.138a) it is clear that the eigenvalues (i.e., the zeros of $a(z)$ and $\bar{a}(z)$) are constant as the solution evolves. Not only the number of eigenvalues, but also their locations, are fixed. Thus, the eigenvalues are time-independent discrete states of the evolution.

The norming constants, however, are not fixed. Their evolution is obtained analogously and is given by

$$C_j(\tau) = C_j(0)e^{2i\omega_j\tau}, \qquad \bar{C}_j(\tau) = \bar{C}_j(0)e^{-2i\bar{\omega}_j\tau}, \qquad (3.2.140)$$

where

$$\omega_j = \frac{1}{2}\left(z_j - z_j^{-1} \right)^2, \qquad \bar{\omega}_j = \frac{1}{2}\left(\bar{z}_j - \bar{z}_j^{-1} \right)^2. \qquad (3.2.141)$$

The expressions for the evolution of the scattering data allow one to solve the initial-value problem for the system (3.1.2a)–(3.1.2b). The procedure is the following: (i) The scattering data are calculated from the initial time (e.g., at $\tau = 0$); (ii) the scattering data at later time (say, $\tau = \tau_1$) are determined by the formulas (3.2.139a)–(3.2.139b) and (3.2.140); and (iii) the solution at $\tau = \tau_1$ is constructed from the scattering data.

3.3 Soliton solutions

The soliton solutions of (3.2.7a)–(3.2.7b) are the reflectionless potentials (i.e., corresponding to $\rho(z) = \bar{\rho}(z) = 0$ on $|z| = 1$) where the eigenvalues appear in sets of four. The scattering data of a J-soliton solution are composed of:

(i) the $4J$ eigenvalues $\pm z_j = \pm e^{\alpha_j + i\beta_j}$ and $\pm \bar{z}_j = \pm \frac{1}{z_j^*}$, where $|z_j| > 1$ and $j = 1, \dots, J$;
(ii) the associated $2J$ norming constants $C_j(\tau)$, $D_j(\tau)$, where

$$D_j(\tau) = \bar{z}_j^{-2} \bar{C}_j(\tau) \qquad (3.3.142)$$

and $C_j(0)D_j(0) \in \mathbb{R}$, $C_j(0)D_j(0) > 0$.

Note that the evolution of the norming constants given by (3.2.140) also assures that $C_j(\tau)D_j(\tau) \in \mathbb{R}$, $C_j(\tau)D_j(\tau) > 0$ for $\tau \neq 0$.

The one-soliton solution is obtained from (3.2.109a)–(3.2.109b) by taking into account the explicit time-dependence as given by (3.2.140), that is,

$$Q_n(\tau) = -\frac{D_1(0)}{(C_1(0)D_1(0))^{1/2}} e^{i(2\beta_1(n+1) - 2w\tau)}$$

$$\times \sinh(2\alpha_1)\operatorname{sech}\left(2\alpha_1(n+1) - 2v\tau - \delta\right) \qquad (3.3.143a)$$

$$R_n(\tau) = \frac{C_1(0)}{(C_1(0)D_1(0))^{1/2}} e^{-i(2\beta_1(n+1) - 2w\tau)}$$

$$\times \sinh(2\alpha_1)\operatorname{sech}\left(2\alpha_1(n+1) - 2v\tau - \delta\right), \qquad (3.3.143b)$$

where

$$v = -\sinh(2\alpha_1)\sin(2\beta_1), \qquad w = \cosh(2\alpha_1)\cos(2\beta_1) - 1 \qquad (3.3.144a)$$

$$\delta = \log\left(C_1(0)D_1(0)\right)^{1/2} - \log\sinh(2\alpha_1). \qquad (3.3.144b)$$

Each of the (3.3.143a), (3.3.143b) represents a localized traveling wave with a single peak that is modulated by a complex carrier phase. This is the one-soliton solution obtained in [11].

Recall that, in order to derive the IDNLS (3.1.2a)–(3.1.2b) and the associated pair of linear operators (3.2.4), (3.2.5) from the NLS (2.1.1) and the associated pair of linear operators (2.2.3)–(2.2.4), we let

$$Q_n = hq_n = hq(nh), \qquad R_n = hr_n = hr(nh), \qquad \tau = h^{-2}t, \quad (3.3.145a)$$

and

$$z_1 = e^{-ik_1 h}, \qquad C_1(0) = ih\tilde{C}_1(0), \qquad D_1(0) = -ih\tilde{D}_1(0). \quad (3.3.145b)$$

By substituting these expressions into (3.3.143a)–(3.3.143b) we obtain

$$q_n(t) = icAe^{-2i(\xi hn - \tilde{w}t) - i\psi} \operatorname{sech}(2\eta h(n+1) - 2\tilde{v}t - d) \quad (3.3.146a)$$
$$r_n(t) = ic^{-1}Ae^{2i(\xi hn - \tilde{w}t) + i\psi} \operatorname{sech}(2\eta h(n+1) - 2\tilde{v}t - d), \quad (3.3.146b)$$

where

$$A = \frac{\sinh(2\eta h)}{h}, \qquad \tilde{w} = \frac{1 - \cosh(2\eta h)\cos(2\xi h)}{h^2}, \qquad \tilde{v} = \frac{\sinh(2\eta h)\sin(2\xi h)}{h^2}$$

$$(3.3.147a)$$

$$d = \log(\tilde{C}_1(0)\tilde{D}_1(0))^{\frac{1}{2}} - \log \frac{\sinh(2\eta h)}{h} \quad (3.3.147b)$$

and

$$k_1 = \xi + i\eta, \qquad \eta = \frac{\alpha_1}{h} = \frac{\log|z_1|}{h}, \qquad \xi = -\frac{\beta_1}{h} = -\frac{\arg z_1}{h},$$

$$(3.3.147c)$$

$$c = \frac{|D_1(0)|^{1/2}}{|C_1(0)|^{1/2}}, \qquad \psi = \arg C_1(0) + 2\xi h. \quad (3.3.147d)$$

The expressions (3.3.146a)–(3.3.146b) give the one-soliton solution of the version of the IDNLS that explicitly contains h, that is, the system (3.2.7a)–(3.2.7b).

In the limit $h \to 0$, $nh \to x$, (3.3.146a)–(3.3.146b) become

$$q(x,t) = 2\eta c\, e^{-2i\xi x + 4i(\xi^2 - \eta^2)t - i\psi_0} \operatorname{sech}(2\eta x - 8\xi\eta t - \delta_0) \quad (3.3.148a)$$

$$r(x,t) = 2\eta c^{-1} e^{2i\xi x - 4i(\xi^2 - \eta^2)t + i\psi_0} \operatorname{sech}(2\eta x - 8\xi\eta t - \delta_0) \quad (3.3.148b)$$

$$\delta_0 = \log(\tilde{C}_1(0)\tilde{D}_1(0))^{\frac{1}{2}} - \log 2\eta, \qquad \psi_0 = \arg\tilde{C}_1(0) - \frac{\pi}{2}, \quad (3.3.149)$$

which is the one-soliton solution of the continuous problem (2.1.2a)–(2.1.2b). A similar result holds for the multisoliton solutions. Note that the IDNLS equation (3.1.1) is obtained in the reduction $r_n = \mp q_n^*$. In this case, from (3.2.87) and (3.3.142), it follows that $\frac{D_1}{C_1} = \pm \frac{C_1^*}{\tilde{C}_1}$.

We point out some differences between the discrete soliton and the soliton solution of the PDE. Both (3.3.146a)–(3.3.146b) and (3.3.148a)–(3.3.148b) are composed of traveling waves with a sech profile that is modulated by a complex carrier wave. Moreover, in both the PDE and the lattice soliton, the velocity of the traveling wave depends on the spatial frequency of the carrier. However, this has different consequences for the lattice and the PDE:

1. On the lattice there is a maximum spatial frequency (i.e., $|\xi| \leq \frac{2\pi}{h}$), which leads to a maximum speed for the soliton on the lattice (i.e., $|v| \leq \frac{\sinh(2\eta h)}{h^2}$), while there is no such upper bound on soliton velocity in the PDE.
2. In the PDE soliton, the velocity depends linearly on the spatial frequency. In contrast, on the lattice, the velocity goes like $\sin(2\xi h)$. Hence, the speed increases with $|\xi|$ for small $|\xi|$ but then decreases to zero as $|\xi|$ increases further. As a consequence, unlike the PDE, it is possible, on the lattice, to have two solitons with the same velocity even though they have different spatial carrier frequencies.

Finally, we remark that in the literature the solitons of the NLS and the IDNLS are generally considered only when $r = -q^*$ and $r_n = -q_n^*$. In this case, $c = 1$ and (3.3.148a)–(3.3.148b) give the NLS soliton (2.3.88). Here, however, we have shown that one can consider the soliton in slightly more general conditions: Even though $r_n \neq -q_n^*$, the potentials are still such that (i) the eigenvalues appear in quartets $\pm z_j$, $\pm \frac{1}{z_j^*}$ and (ii) the products of the associated norming constants is real and positive (see Symmetry 3.3). The characteristic localized traveling-wave solution is a result of a symmetry in the eigenvalues of the solution. Because these eigenvalues are time-independent, this symmetry (or lack thereof) is a conserved characteristic of solutions in the nonreduced system (3.3.146a)–(3.3.146b).

The problem of a multisoliton collision can be investigated by looking at the asymptotic states as $\tau \to \pm\infty$, proceeding in a similar way as in the continuous case. Consider a pure J-soliton solution for $R_n = -Q_n^*$ and assume, without loss of generality, $v_1 < v_2 < \ldots < v_J$. Then, for $\tau \to \pm\infty$, the potential breaks up into individual solitons of the form (3.3.143a), and as $\tau \to -\infty$ the discrete solitons are distributed along the n-axis in the order corresponding to $n_J, n_{J-1}, \ldots, n_1$; the order of the soliton sequence is reversed as $\tau \to +\infty$. To determine the result of the interaction among solitons, we trace the passage of the eigenfunctions through the asymptotic states. We denote the soliton coordinates at the instant of time τ by $n_j(\tau)$ ($|\tau|$ is assumed large enough so that one can talk about individual solitons). If $\tau \to -\infty$, then $n_J \ll n_{J-1} \ll \ldots \ll n_1$. The function $\phi_n(z_j)$ has the form $\phi_n(z_j) \sim z_j^n (1 \ , \ 0)^T$ when $n \ll n_J$. After passing through the J-th soliton, it will be of the form $\phi_n(z_j) \sim a_J(z_j) z_j^n \, (1, \, 0)^T$, where $a_J(z)$ is the transmission coefficient relative to the J-th soliton. Repeating the argument, we find

$$\phi_n(z_j) \sim z_j^n \prod_{l=j+1}^{J} a_l(z_j) \begin{pmatrix} 1 \\ 0 \end{pmatrix} \qquad n_{j+1} \ll n \ll n_j.$$

Upon passing through the j-th soliton, since the corresponding state is a bound state, we get

$$\phi_n(z_j) \sim z_j^{-n} S_j \prod_{l=j+1}^{J} a_l(z_j) \begin{pmatrix} 0 \\ 1 \end{pmatrix} \qquad n_j \ll n \ll n_{j-1}. \qquad (3.3.150)$$

On the other hand, starting from $n \gg n_1$ and proceeding in a similar way, we find for the eigenfunction ψ_n the asymptotic behavior:

$$\psi_n(z_j) \sim z_j^{-n} \prod_{l=1}^{j-1} \left(z_l^{-2} \bar{z}_l^2 \right) a_l(z_j) \begin{pmatrix} 0 \\ 1 \end{pmatrix} \qquad n_j \ll n \ll n_{j-1}, \qquad (3.3.151)$$

where we have used (3.2.62a) and (3.2.63a) to get the transmission coefficient for the right eigenfunctions, as well as the result (3.2.108). Comparing (3.3.150) and (3.3.151) and recalling (3.2.72), we get

$$S_j \prod_{l=j+1}^{J} a_l(z_j) = b_j \prod_{l=1}^{j-1} a_l(z_j). \qquad (3.3.152)$$

It is convenient to write S_j as

$$S_j(\tau) = e^{i(z_j - z_j^{-1})^2 \tau + 2\delta_j + i\psi_j}$$

so that (3.2.72) yields

$$b_j(\tau) \sim e^{i(z_j - z_j^{-1})^2 \tau + 2\delta_j^- + i\psi_j^-} \prod_{l=j+1}^{J} a_l(z_j) \prod_{m=1}^{j-1} \left(z_m^{-2} \bar{z}_m^2 \right) a_m(z_j)^{-1} \qquad \tau \to -\infty,$$

where δ_j^-, ψ_j^- denote the asymptotics of the real functions δ_j, ψ_j as $\tau \to -\infty$.

Proceeding in a similar fashion as $\tau \to +\infty$ and taking into account that the order of the soliton sequence is reversed, we get

$$b_j(\tau) \sim e^{i(z_j - z_j^{-1})^2 \tau + 2\delta_j^+ + i\psi_j^+} \prod_{l=1}^{j-1} a_l(z_j) \prod_{m=j+1}^{J} \left(z_m^{-2} \bar{z}_m^2 \right) a_m(z_j)^{-1} \qquad \tau \to +\infty.$$

Therefore we conclude that

$$e^{2(\delta_j^+ - \delta_j^-) + i(\psi_j^+ - \psi_j^-)} = \prod_{l=1}^{j-1} a_l(z_j)^2 \prod_{m=j+1}^{J} \left(z_m^{-2} \bar{z}_m^2 \right) a_m(z_j)^{-2}$$

or, explicitly, using the formula (3.2.90) for the pure one-soliton transmission coefficient,

$$
e^{2(\delta_j^+ - \delta_j^-) + i(\psi_j^+ - \psi_j^-)} = \prod_{l=1}^{j-1} \left(\frac{z_j^2 - z_l^2}{z_j^2 - (z_l^*)^{-2}} \right)^2 \prod_{m=j+1}^{J} (z_m^{-2} \bar{z}_m^2) \left(\frac{z_j^2 - (z_m^*)^{-2}}{z_j^2 - z_m^2} \right)^2.
$$

(3.3.153)

According to (3.3.144b), this last formula provides the phase shift of the j-th soliton on the transition between the asymptotic states $\tau \to \pm\infty$ due to the interaction with the other solitons.

For instance, in the two-soliton case we have

$$
\delta_1^+ - \delta_1^- = -(\delta_2^+ - \delta_2^-) = \log \left| \frac{(z_1^2 - \bar{z}_2^2)(z_2^2 - \bar{z}_1^2)}{(z_1^2 - z_2^2)(\bar{z}_1^2 - \bar{z}_2^2)} \right|
$$

(3.3.154a)

$$
\psi_1^+ - \psi_1^- = \arg \left[z_2^2 \bar{z}_2^{-2} \frac{(z_1^2 - \bar{z}_2^2)(\bar{z}_1^2 - \bar{z}_2^2)}{(z_1^2 - z_2^2)(\bar{z}_1^2 - \bar{z}_2^2)} \right].
$$

(3.3.154b)

Note that letting $z_1 = e^{-ihk_1}$, $z_2 = e^{-ihk_2}$, in the continuous limit (i.e., for $h \to 0$) these expressions give back (2.3.97a)–(2.3.97b).

3.4 Conserved quantities and Hamiltonian structure

We showed that the scattering coefficient $a(z)$ is time-independent. Since $a(z)$ is analytic for $|z| > 1$, it admits a Laurent expansion whose coefficients are constants of the motion as well. From the representation (3.2.66a) for $a(z)$, it follows that the quantities

$$
\sum_{n=-\infty}^{+\infty} Q_n M_n^{(2),-2j+1}
$$

(3.4.155)

are conserved for any integer $j \geq 0$, and the coefficients $M_n^{(2),-j}$ of the asymptotic expansion of $M_n^{(2)}(z)$ at large z can be calculated iteratively from (3.2.47)–(3.2.48). For instance, the first two constants of the motion are given by

$$
\Gamma_1 = \sum_{n=-\infty}^{+\infty} Q_n R_{n-1}, \qquad \Gamma_2 = \sum_{n=-\infty}^{+\infty} \left\{ Q_n R_{n-2} - \frac{1}{2} (Q_n R_{n-1})^2 \right\}.
$$

(3.4.156)

Note that we subtracted Γ_1 from the expression for Γ_2 obtained from (3.4.155) for $j = 2$.

The scattering coefficient $\bar{a}(z)$ is also a constant of the motion, and proceeding exactly as before one can obtain a second set of conserved quantities given by

$$\sum_{n=-\infty}^{+\infty} R_n \bar{M}_n^{(1),2j-1} \qquad (3.4.157)$$

for any $j \geq 1$. This yields

$$\bar{\Gamma}_1 = \sum_{n=-\infty}^{+\infty} Q_{n-1} R_n, \qquad \bar{\Gamma}_2 = \sum_{n=-\infty}^{+\infty} \left\{ Q_{n-2} R_n - \frac{1}{2} (Q_{n-1} R_n)^2 \right\}. \qquad (3.4.158)$$

By taking into account the τ-dependence of the scattering coefficients (3.2.138a), we show that the determinant of the scattering matrix is a constant of the motion, that is,

$$\det \begin{pmatrix} a(z,\tau) & \bar{b}(z,\tau) \\ b(z,\tau) & \bar{a}(z,\tau) \end{pmatrix} = \det \begin{pmatrix} a(z,0) & \bar{b}(z,0) \\ b(z,0) & \bar{a}(z,0), \end{pmatrix}.$$

Then, from (3.2.60), it follows that

$$c_{-\infty}(\tau) = c_{-\infty}(0) = \prod_{n=-\infty}^{+\infty} (1 - R_n Q_n). \qquad (3.4.159)$$

Note that when the potentials satisfy the symmetry condition $R_n = \mp Q_n^*$, the first constants of the motion in (3.4.156) and (3.4.158) become

$$\Gamma_1 = \mp \sum_{n=-\infty}^{+\infty} Q_n Q_{n-1}^*, \qquad \bar{\Gamma}_1 = \Gamma_1^*. \qquad (3.4.160)$$

Moreover, from (3.4.159) follows that

$$c_{-\infty} = \prod_{n=-\infty}^{+\infty} \left(1 \pm |Q_n|^2 \right) \qquad (3.4.161)$$

is also a conserved quantity.

The system of equations (3.1.2a)–(3.1.2b) is a Hamiltonian system [85], [114] with

coordinates (q) : $Q_n(\tau)$ (3.4.162a)

momenta (p) : $R_n(\tau)$ (3.4.162b)

Hamiltonian (H) : $\displaystyle -\sum_{n=-\infty}^{+\infty} R_n (Q_{n+1} + Q_{n-1}) - 2 \sum_{n=-\infty}^{+\infty} \log (1 - R_n Q_n)$

$$(3.4.162c)$$

with the noncanonical (i.e., nonconstant) brackets

$$\{Q_m(\tau), R_n(\tau)\} = i\,(1 - R_n(\tau)Q_n(\tau))\,\delta_{n,m} \tag{3.4.163a}$$

$$\{Q_n(\tau), Q_m(\tau)\} = \{Q_n(\tau), R_m(\tau)\} = 0. \tag{3.4.163b}$$

Unlike the continuous case (cf. (2.4.114a)), since the bracket (3.4.163a) is not constant, the Jacobi identity (2.4.109) does not follow trivially from its inherent skew-symmetry but has to be checked separately. Note that the Hamiltonian is given by the conserved quantities Γ_1, $\bar{\Gamma}_1$ and $c_{-\infty}$ in (3.4.156), (3.4.158), and (3.4.159).

In the case when the initial data satisfy the additional constraints $R_n = \mp Q_n^*$, we identify coordinates Q_n and momenta Q_n^*. The corresponding Hamiltonian and brackets are given by

$$H = \pm \sum_{n=-\infty}^{+\infty} Q_n^*\,(Q_{n+1} + Q_{n-1}) \pm 2 \sum_{n=-\infty}^{+\infty} \log\left(1 \pm |Q_n|^2\right) \tag{3.4.164}$$

$$\{Q_m, Q_n^*\} = i\left(1 \pm |Q_n|^2\right)\delta_{n,m} \tag{3.4.165}$$

$$\{Q_n, Q_m\} = \{Q_n^*, Q_m^*\} = 0. \tag{3.4.166}$$

Chapter 4

Matrix nonlinear Schrödinger
equation (MNLS)

4.1 Overview

The inverse scattering transform for the two-component VNLS system

$$iq_t^{(1)} = q_{xx}^{(1)} + 2(|q^{(1)}|^2 + |q^{(2)}|^2)q^{(1)}$$

$$iq_t^{(2)} = q_{xx}^{(2)} + 2(|q^{(1)}|^2 + |q^{(2)}|^2)q^{(2)}$$

and the dynamics of the soliton interactions were first developed in [120]. The IST of the somewhat more general matrix system

$$i\mathbf{Q}_t = \mathbf{Q}_{xx} - 2\mathbf{Q}\mathbf{R}\mathbf{Q} \tag{4.1.1a}$$

$$-i\mathbf{R}_t = \mathbf{R}_{xx} - 2\mathbf{R}\mathbf{Q}\mathbf{R}, \tag{4.1.1b}$$

where $\mathbf{Q}$ is an $N \times M$ matrix and $\mathbf{R}$ is an $M \times N$ matrix, is a straightforward generalization of the IST for the two-component VNLS system and provides a parallel with the IST for the integrable discrete matrix system (cf. Chapter 5). Hence, in this chapter we develop the IST on the infinite line for the system (4.1.1a)–(4.1.1b), including the symmetry induced by the reduction $\mathbf{R} = \mp\mathbf{Q}^H$ as a special case.

In its basic outline, the IST for the system (4.1.1a)–(4.1.1b) follows the same steps as the IST for the scalar NLS (cf. Chapter 2). However, since scalars are replaced with matrices, some extra care is needed in the calculations. Nevertheless, the individual steps of the IST for the system (4.1.1a)–(4.1.1b) can be mapped to the steps of the IST for the NLS. In particular, the treatment of the direct problem follows [6], [21]. The inverse problem is formulated as a Riemann–Hilbert boundary-value problem, following [6]. The Gel'fand–Levitan–Marchenko integral equations are derived from the RH problem. We also show that under the reduction $\mathbf{R} = \mp\mathbf{Q}^H$ the GLM equations have no homogeneous solutions.

While the IST is similar, the effect of the vector generalization that takes one from the NLS to the VNLS manifests itself in the dynamics of the soliton interactions. In addition to the phase parameters of the scalar NLS solitons (the location of the peak and the overall complex phase), the solitons of the VNLS have a *vector* polarization that has no counterpart in the scalar NLS. In fact, a single soliton of the VNLS can be described as a scalar NLS-type soliton multiplied by a polarization vector. Although a single VNLS soliton is, in effect, governed by the scalar NLS, the vector nature of the solitons (i.e., the associated polarization vector) affects the interaction of the vector solitons with one another.

When vector solitons interact, they shift one another's polarization vectors. This polarization shift distinguishes vector–soliton interactions from scalar–soliton interactions. Like the scalar-solitons interactions governed by the NLS, vector–soliton interactions governed by the VNLS induce shifts in the location of the peak and the overall complex phase of the interacting solitons while leaving the amplitude and velocity of the individual solitons invariant. However, the polarization shift (that occurs in addition to the other phase shifts) in vector-soliton interactions is an element that distinguishes vector solitons from scalar solitons.

Finally, in this chapter we describe how the VNLS can be understood as a Hamiltonian system with infinitely many conserved quantities. These conserved quantities, including the Hamiltonian of the VNLS, are derived by making use of the IST machinery that is developed here for the VNLS.

4.2 The inverse scattering transform for MNLS

4.2.1 Operator pair

The Lax pair for the matrix nonlinear Schrödinger system (4.1.1a)–(4.1.1b) is naturally obtained from the matrix generalization of the linear system (2.2.3)–(2.2.4), that is,

$$\mathbf{v}_x = \begin{pmatrix} -ik\mathbf{I}_N & \mathbf{Q} \\ \mathbf{R} & ik\mathbf{I}_M \end{pmatrix} \mathbf{v} \qquad (4.2.2)$$

and

$$\mathbf{v}_t = \begin{pmatrix} 2ik^2\mathbf{I}_N + i\mathbf{QR} & -2k\mathbf{Q} - i\mathbf{Q}_x \\ -2k\mathbf{r} + i\mathbf{R}_x & -2ik^2\mathbf{I}_M - i\mathbf{RQ} \end{pmatrix} \mathbf{v}, \qquad (4.2.3)$$

where $\mathbf{v}$ is an M-component vector; $\mathbf{Q} = \mathbf{Q}(x, t)$ is an $N \times M$ matrix; $\mathbf{R} = \mathbf{R}(x, t)$ is an $M \times N$ matrix; and $\mathbf{I}_N$, $\mathbf{I}_M$ are the $N \times N$ and $M \times M$ identity matrices, respectively. Indeed, the compatibility condition (i.e., the equality of the mixed derivatives $\mathbf{v}_{xt} = \mathbf{v}_{tx}$) is equivalent to the statement that $\mathbf{Q}$ and $\mathbf{R}$ satisfy the evolution equations (4.1.1a)–(4.1.1b) if k, the scattering parameter, is independent of x and t. Under the reduction $\mathbf{R} = \mp \mathbf{Q}^H$, the system (4.1.1a)–(4.1.1b) corresponds to the single (matrix) PDE

$$ i\mathbf{Q}_t = \mathbf{Q}_{xx} \pm 2\mathbf{Q}\mathbf{Q}^H\mathbf{Q}, \tag{4.2.4} $$

and if $\mathbf{Q} \equiv \mathbf{q}$ is either a row vector ($N = 1$) or a column vector ($M = 1$), it reduces to VNLS (1.3.4), that is,

$$ i\mathbf{q}_t = \mathbf{q}_{xx} \pm 2\|\mathbf{q}\|^2 \mathbf{q}. \tag{4.2.5} $$

Moreover, the system (4.1.1a)–(4.1.1b) reduces to (2.1.2a)–(2.1.2b) for $N = M = 1$; hence the contents of Chapter 2 regarding the scalar NLS equation can be obtained from the results of the present chapter with this prescription.

As before, we refer to the equation with the x derivative, equation (4.2.2), as the scattering problem and the equation with the t derivative, equation (4.2.3), as the time-dependence.

4.2.2 Direct scattering problem

Jost functions and integral equations

We refer to solutions of the scattering problem (4.2.2) as eigenfunctions with respect to the parameter k. When the potentials $\mathbf{Q}, \mathbf{R} \to 0$ rapidly as $x \to \pm\infty$, the eigenfunctions are asymptotic to the solutions of

$$ \mathbf{v}_x = \begin{pmatrix} -ik\mathbf{I}_N & \mathbf{0} \\ \mathbf{0} & ik\mathbf{I}_M \end{pmatrix} \mathbf{v} $$

as $|x| \to \infty$. The solutions of this differential equation have the bases

$$ \phi(x,k) \sim \begin{pmatrix} \mathbf{I}_N \\ \mathbf{0} \end{pmatrix} e^{-ikx}, \qquad \bar{\phi}(x,k) \sim \begin{pmatrix} \mathbf{0} \\ \mathbf{I}_M \end{pmatrix} e^{ikx} \qquad \text{as } x \to -\infty \tag{4.2.6a} $$

$$ \psi(x,k) \sim \begin{pmatrix} \mathbf{0} \\ \mathbf{I}_M \end{pmatrix} e^{ikx}, \qquad \bar{\psi}(x,k) \sim \begin{pmatrix} \mathbf{I}_N \\ \mathbf{0} \end{pmatrix} e^{-ikx} \qquad \text{as } x \to +\infty, \tag{4.2.6b} $$

where $\mathbf{I}_N, \mathbf{I}_M$ are the $N \times N$ and $M \times M$ identity matrices, respectively, and $\phi, \bar{\phi}, \psi, \bar{\psi}$ are matrix-valued functions of the following dimensions:

$$ \phi(x,k): \quad (N+M) \times N, \qquad \bar{\phi}(x,k): \quad (N+M) \times M $$
$$ \psi(x,k): \quad (N+M) \times M, \qquad \bar{\psi}(x,k): \quad (N+M) \times N. $$

In the following, we will refer to ϕ and $\bar{\phi}$ as "left" eigenfunctions, since they are given by fixing their asymptotics at $x \to -\infty$, that is, from the "left," as opposed to the "right" eigenfunctions ψ and $\bar{\psi}$, which are fixed by their behavior at $x \to +\infty$, that is, from the "right."

As done earlier and in the following analysis, it is convenient to consider functions with constant boundary conditions. Hence, we define the Jost functions as follows:

$$\mathbf{M}(x, k) = e^{ikx}\phi(x, k), \qquad \bar{\mathbf{M}}(x, k) = e^{-ikx}\bar{\phi}(x, k), \qquad (4.2.7a)$$

$$\mathbf{N}(x, k) = e^{-ikx}\psi(x, k), \qquad \bar{\mathbf{N}}(x, k) = e^{ikx}\bar{\psi}(x, k). \qquad (4.2.7b)$$

If the scattering problem (4.2.2) is rewritten as

$$\mathbf{v}_x = \left(ik\mathbf{J} + \tilde{\mathbf{Q}}\right)\mathbf{v}, \qquad (4.2.8)$$

where

$$\mathbf{J} = \begin{pmatrix} -\mathbf{I}_N & \mathbf{0} \\ \mathbf{0} & \mathbf{I}_M \end{pmatrix}, \qquad \tilde{\mathbf{Q}} = \begin{pmatrix} \mathbf{0} & \mathbf{Q} \\ \mathbf{R} & \mathbf{0} \end{pmatrix}, \qquad (4.2.9)$$

then the Jost functions $\mathbf{M}(x, k)$ and $\bar{\mathbf{N}}(x, k)$ are solutions of the differential equation

$$\chi_x(x, k) = ik(\mathbf{J} + \mathbf{I}_{N+M})\chi(x, k) + \left(\tilde{\mathbf{Q}}\chi\right)(x, k) \qquad (4.2.10a)$$

while $\mathbf{N}(x, k)$ and $\bar{\mathbf{M}}(x, k)$ satisfy

$$\tilde{\chi}_x(x, k) = ik(\mathbf{J} - \mathbf{I}_{N+M})\tilde{\chi}(x, k) + \left(\tilde{\mathbf{Q}}\tilde{\chi}\right)(x, k) \qquad (4.2.10b)$$

with the constant boundary conditions

$$\mathbf{M}(x, k) \to \begin{pmatrix} \mathbf{I}_N \\ \mathbf{0} \end{pmatrix}, \qquad \bar{\mathbf{M}}(x, k) \to \begin{pmatrix} \mathbf{0} \\ \mathbf{I}_M \end{pmatrix} \qquad \text{as } x \to -\infty \qquad (4.2.11a)$$

$$\mathbf{N}(x, k) \to \begin{pmatrix} \mathbf{0} \\ \mathbf{I}_M \end{pmatrix}, \qquad \bar{\mathbf{N}}(x, k) \to \begin{pmatrix} \mathbf{I}_N \\ \mathbf{0} \end{pmatrix} \qquad \text{as } x \to +\infty. \qquad (4.2.11b)$$

It is convenient to introduce the following notation: An $(N + M) \times J$ matrix $\mathbf{A}$ will be denoted as

$$\mathbf{A} = \begin{pmatrix} \mathbf{A}^{(\text{up})} \\ \mathbf{A}^{(\text{dn})} \end{pmatrix},$$

where the superscripts "up" and "dn" indicate, respectively, the top N rows and the bottom M rows of matrix $\mathbf{A}$ (i.e., $\mathbf{A}^{(\text{up})}$ is the $N \times J$ upper block and $\mathbf{A}^{(\text{dn})}$ is the lower $M \times J$ block of matrix $\mathbf{A}$).

Solutions of the differential equations (4.2.10a)–(4.2.10b) can be represented by means of the following integral equations:

$$\chi(x, k) = \mathbf{w} + \int_{-\infty}^{+\infty} \mathbf{G}(x - x', k) \left(\tilde{\mathbf{Q}}\chi\right)(x', k) dx'$$

$$\tilde{\chi}(x, k) = \tilde{\mathbf{w}} + \int_{-\infty}^{+\infty} \tilde{\mathbf{G}}(x - x', k) \left(\tilde{\mathbf{Q}}\tilde{\chi}\right)(x', k) dx'$$

or, in component form,

$$\chi^{m_j}(x, k) = (w)^{m_j} + \int_{-\infty}^{+\infty} \sum_{\ell=1}^{N+M} (G)^{m\ell}(x - x', k) \left(\tilde{\mathbf{Q}}\chi\right)_{\ell j} (x', k) dx'$$

$$m = 1, \ldots, N + M, \quad j = 1, \ldots, N$$

$$\tilde{\chi}^{m_j}(x, k) = (\tilde{w})^{m_j} + \int_{-\infty}^{+\infty} \sum_{\ell=1}^{N+M} (\tilde{G})^{m\ell}(x - x', k) \left(\tilde{\mathbf{Q}}\tilde{\chi}\right)_{\ell j} (x', k) dx'$$

$$m = 1, \ldots, N + M, \quad j = 1, \ldots, M,$$

where

$$\mathbf{w} = \begin{pmatrix} \mathbf{w}^{(\text{up})} \\ \mathbf{0} \end{pmatrix}, \qquad \tilde{\mathbf{w}} = \begin{pmatrix} \mathbf{0} \\ \tilde{\mathbf{w}}^{(\text{dn})} \end{pmatrix}$$

and the Green's functions $\mathbf{G}(x, k)$ and $\tilde{\mathbf{G}}(x, k)$ are $(N + M) \times (N + M)$ matrices satisfying the differential equations

$$\mathcal{L}_0 \mathbf{G}(x, k) = \delta(x) \mathbf{I}_{N+M}, \qquad \tilde{\mathcal{L}}_0 \tilde{\mathbf{G}}(x, k) = \delta(x) \mathbf{I}_{N+M},$$

where

$$\mathcal{L}_0 = \mathbf{I}_{N+M} \partial_x - ik(\mathbf{J} + \mathbf{I}_{N+M}), \qquad \tilde{\mathcal{L}}_0 = \mathbf{I}_{N+M} \partial_x - ik(\mathbf{J} - \mathbf{I}_{N+M}).$$

The Green's functions are not unique, and, as we show in the following, the choice of the Green's function and the choice of the inhomogeneous terms $\mathbf{w}$ and $\tilde{\mathbf{w}}$ together determine the Jost function and its analytic properties. In analogy with the scalar case (cf. Chapter 2), we use the Fourier transform method and find

$$\mathbf{G}_{\pm}(x, k) = \frac{1}{2\pi i} \int_{C_{\pm}} \begin{pmatrix} p^{-1}\mathbf{I}_N & \mathbf{0} \\ \mathbf{0} & (p - 2k)^{-1}\mathbf{I}_M \end{pmatrix} e^{ipx} dp$$

$$\tilde{\mathbf{G}}_{\pm}(x, k) = \frac{1}{2\pi i} \int_{C_{\pm}} \begin{pmatrix} (p + 2k)^{-1}\mathbf{I}_N & \mathbf{0} \\ \mathbf{0} & p^{-1}\mathbf{I}_M \end{pmatrix} e^{ipx} dp,$$

where $C_{\pm}$ are the contours from $-\infty$ to $+\infty$ that, respectively, pass below and above both the singularities at $p = 0$ and $p = 2k$ (see Figure 2.1).

Therefore we have

$$\mathbf{G}_\pm(x, k) = \pm\theta(\pm x)\begin{pmatrix} \mathbf{I}_N & \mathbf{0} \\ \mathbf{0} & e^{2ikx}\mathbf{I}_M \end{pmatrix},$$

$$\tilde{\mathbf{G}}_\pm(x, k) = \mp\theta(\mp x)\begin{pmatrix} e^{-2ikx}\mathbf{I}_N & \mathbf{0} \\ \mathbf{0} & \mathbf{I}_M \end{pmatrix}. \qquad (4.2.12)$$

The "+" functions are analytic in the upper half-plane of k and the "−" functions are analytic in the lower half-plane. Taking into account the boundary conditions (4.2.11a)–(4.2.11b), we get the following integral equations for the Jost solutions:

$$\mathbf{M}(x, k) = \begin{pmatrix} \mathbf{I}_N \\ \mathbf{0} \end{pmatrix} + \int_{-\infty}^{+\infty} \mathbf{G}_+(x - x', k)\,(\tilde{\mathbf{Q}}\mathbf{M})\,(x', k)dx' \qquad (4.2.13a)$$

$$\mathbf{N}(x, k) = \begin{pmatrix} \mathbf{0} \\ \mathbf{I}_M \end{pmatrix} + \int_{-\infty}^{+\infty} \tilde{\mathbf{G}}_+(x - x', k)\,(\tilde{\mathbf{Q}}\mathbf{N})\,(x', k)dx' \qquad (4.2.13b)$$

$$\bar{\mathbf{M}}(x, k) = \begin{pmatrix} \mathbf{0} \\ \mathbf{I}_M \end{pmatrix} + \int_{-\infty}^{+\infty} \tilde{\mathbf{G}}_-(x - x', k)\,(\tilde{\mathbf{Q}}\bar{\mathbf{M}})\,(x', k)dx' \qquad (4.2.13c)$$

$$\bar{\mathbf{N}}(x, k) = \begin{pmatrix} \mathbf{I}_N \\ \mathbf{0} \end{pmatrix} + \int_{-\infty}^{+\infty} \mathbf{G}_-(x - x', k)\,(\tilde{\mathbf{Q}}\bar{\mathbf{N}})\,(x', k)dx'. \qquad (4.2.13d)$$

Equations (4.2.13a)–(4.2.13d) are Volterra integral equations. The results of Lemma 2.1 can be generalized to the matrix case to show that if $\mathbf{Q}, \mathbf{R} \in L^1(\mathbb{R})$ with respect to any matrix norm, that is,

$$\|\mathbf{Q}\|_1 = \int_{-\infty}^{+\infty} \|\mathbf{Q}_a(x)\|\,dx < +\infty$$

$$\|\mathbf{R}\|_1 = \int_{-\infty}^{+\infty} \|\mathbf{R}_a(x)\|\,dx < +\infty,$$

where $\|\cdot\|_a$ is any matrix norm, then the Neumann series of the integral equations for $\mathbf{M}$ and $\mathbf{N}$ converge absolutely and uniformly (in x and k) in the upper k-plane, while the Neumann series of the integral equations for $\bar{\mathbf{M}}$ and $\bar{\mathbf{N}}$ converge absolutely and uniformly (in x and k) in the lower k-plane. This fact immediately implies that the Jost functions $\mathbf{M}(x, k)$ and $\mathbf{N}(x, k)$ are analytic functions of k for Im $k > 0$ and continuous for Im $k \geq 0$ and $\bar{\mathbf{M}}(x, k)$, $\bar{\mathbf{N}}(x, k)$ are analytic functions of k for Im $k < 0$ and continuous for Im $k \leq 0$.

Lemma 4.1 *If $\mathbf{Q}, \mathbf{R} \in L^1(\mathbb{R})$, then $\mathbf{M}(x, k), \mathbf{N}(x, k)$ defined by (4.2.13a)–(4.2.13b) are analytic functions of k for Im $k > 0$ and continuous for Im $k \geq 0$, while $\bar{\mathbf{M}}(x, k)$, $\bar{\mathbf{N}}(x, k)$ defined by (4.2.13c)–(4.2.13d) are analytic functions of k for Im $k < 0$ and continuous for Im $k \leq 0$. Moreover, these solutions are unique in the space of continuous functions.*

Proof We prove the result for $\mathbf{M}(x, k)$. The Neumann series

$$\mathbf{M}(x, k) = \sum_{j=0}^{\infty} \mathbf{C}_j(x, k), \qquad (4.2.14)$$

where

$$\mathbf{C}_0(x, k) = \begin{pmatrix} \mathbf{I}_N \\ \mathbf{0} \end{pmatrix}$$

$$\mathbf{C}_{j+1}(x, k) = \int_{-\infty}^{+\infty} \mathbf{G}_+(x - x', k) \left(\tilde{\mathbf{Q}} \mathbf{C}_j \right)(x', k) dx' \qquad j \geq 0, \quad (4.2.15)$$

is, formally, a solution of the integral equation (4.2.13a). In upper/lower component form

$$\mathbf{C}_{j+1}^{(\text{up})}(x, k) = \int_{-\infty}^{x} \mathbf{Q}(x') \mathbf{C}_j^{(\text{dn})}(x', k) dx' \qquad (4.2.16\text{a})$$

$$\mathbf{C}_{j+1}^{(\text{dn})}(x, k) = \int_{-\infty}^{x} e^{2ik(x-x')} \mathbf{R}(x') \mathbf{C}_j^{(\text{up})}(x', k) dx'. \qquad (4.2.16\text{b})$$

Because $\mathbf{C}_0^{(\text{dn})} = 0$ we have for any $j \geq 0$

$$\mathbf{C}_{2j+1}^{(\text{up})} = 0, \qquad \mathbf{C}_{2j}^{(\text{dn})} = 0.$$

Using the identities

$$\frac{1}{j!} \int_{-\infty}^{x} |f(\xi)| \left[\int_{-\infty}^{\xi} |f(\xi')| d\xi' \right]^j d\xi$$

$$= \frac{1}{(j+1)!} \int_{-\infty}^{x} \frac{d}{d\xi} \left[\int_{-\infty}^{\xi} |f(\xi')| d\xi' \right]^{j+1} d\xi$$

$$= \frac{1}{(j+1)!} \left[\int_{-\infty}^{x} |f(\xi)| d\xi \right]^{j+1},$$

where $f \in L^1(\mathbb{R})$, one can show from (4.2.16a)–(4.2.16b) by induction on j that for Im $k \geq 0$

$$\left\| \mathbf{C}_{2j+1}^{(\text{dn})}(x, k) \right\|_a \leq \frac{\left(\int_{-\infty}^{x} \|\mathbf{Q}_a(x')\| dx' \right)^j \left(\int_{-\infty}^{x} \|\mathbf{R}_a(x')\| dx' \right)^{j+1}}{j!}$$

$$\left\| \mathbf{C}_{2j}^{(\text{up})}(x, k) \right\|_a \leq \frac{\left(\int_{-\infty}^{x} \|\mathbf{Q}_a(x')\| dx' \right)^j \left(\int_{-\infty}^{x} \|\mathbf{R}_a(x')\| dx' \right)^j}{j!}.$$

Therefore, if $\mathbf{Q}, \mathbf{R} \in L^1(\mathbb{R})$, the series (4.2.14) is bounded by a uniformly convergent power series, which proves that the Neumann series (4.2.14) is itself uniformly convergent for Im $k \geq 0$. Therefore $\mathbf{M}(x, k)$ is analytic for Im $k > 0$, continuous for Im $k \geq 0$, and unique in the space of continuous functions (cf. Chapter 2, Lemma 2.1).

The proofs for the other eigenfunctions are similar.

From the integral equations (4.2.13a)–(4.2.13d) we can compute the asymptotic expansion for large k of the Jost functions. Integration by parts yields, in the respective half-planes,

$$\mathbf{M}(x, k) = \begin{pmatrix} \mathbf{I}_N - \frac{1}{2ik} \int_{-\infty}^{x} \mathbf{Q}(x')\mathbf{R}(x')dx' \\ -\frac{1}{2ik}\mathbf{R}(x) \end{pmatrix} + O(k^{-2}) \qquad (4.2.17a)$$

$$\bar{\mathbf{N}}(x, k) = \begin{pmatrix} \mathbf{I}_N + \frac{1}{2ik} \int_{x}^{+\infty} \mathbf{Q}(x')\mathbf{R}(x')dx' \\ -\frac{1}{2ik}\mathbf{R}(x) \end{pmatrix} + O(k^{-2}) \qquad (4.2.17b)$$

$$\mathbf{N}(x, k) = \begin{pmatrix} \frac{1}{2ik}\mathbf{Q}(x) \\ \mathbf{I}_M - \frac{1}{2ik} \int_{x}^{+\infty} \mathbf{R}(x')\mathbf{Q}(x')dx' \end{pmatrix} + O(k^{-2}) \qquad (4.2.17c)$$

$$\bar{\mathbf{M}}(x, k) = \begin{pmatrix} \frac{1}{2ik}\mathbf{Q}(x) \\ \mathbf{I}_M + \frac{1}{2ik} \int_{-\infty}^{x} \mathbf{R}(x')\mathbf{Q}(x')dx' \end{pmatrix} + O(k^{-2}). \qquad (4.2.17d)$$

Scattering data

The two matrix eigenfunctions with fixed boundary conditions as $x \to -\infty$ are linearly independent, as are the two matrix eigenfunctions with fixed boundary conditions as $x \to +\infty$. For any system of differential equations

$$\frac{dv}{dx} = \mathbf{A}v,$$

where $\mathbf{A} = \left(A_{ij}\right)_{i,j=1,\dots,n}$ and $v(x) = \left(v^{(1)}(x), \dots, v^{(n)}(x)\right)^T$, the Wronskian of the set of solutions $v_1, \dots, v_n$ defined as

$$W(v_1, \dots, v_n) = \det \begin{pmatrix} v_1^{(1)} & \cdots & v_1^{(n)} \\ \cdots & \cdots & \cdots \\ v_n^{(1)} & \cdots & v_n^{(n)} \end{pmatrix},$$

satisfies the differential equation

$$\frac{d}{dx} W(v_1, \dots, v_n) = \text{tr}\mathbf{A} \, W(v_1, \dots, v_n).$$

Therefore, if $\mathbf{u}(x, k)$ and $\mathbf{v}(x, k)$ are any two solutions of (4.2.2), the Wronskian of $\mathbf{u}$ and $\mathbf{v}$ is given by

$$W(\mathbf{u}(x, k), \mathbf{v}(x, k)) = \det\left(\mathbf{u}(x, k), \mathbf{v}(x, k)\right), \qquad (4.2.18)$$

and it satisfies

$$\frac{d}{dx} W(\mathbf{u}, \mathbf{v}) = ik \, (M - N) \, W(\mathbf{u}, \mathbf{v}). \qquad (4.2.19)$$

Taking into account the asymptotics (4.2.6a)–(4.2.6b), (4.2.19) yields

$$W\left(\phi(x, k), \bar{\phi}(x, k)\right) = e^{ik(M-N)x} \qquad (4.2.20a)$$

$$W\left(\psi(x, k), \bar{\psi}(x, k)\right) = -e^{ik(M-N)x}, \qquad (4.2.20b)$$

which proves that the functions ϕ, $\bar{\phi}$ are linearly independent, as are ψ and $\bar{\psi}$. Therefore can write $\phi(x, k)$ and $\bar{\phi}(x, k)$ as linear combinations of $\psi(x, k)$ and $\bar{\psi}(x, k)$, or vice versa. The coefficients of these linear combinations depend on k. Hence, we can write

$$\phi(x, k) = \psi(x, k)\mathbf{b}(k) + \bar{\psi}(x, k)\mathbf{a}(k) \qquad (4.2.21a)$$

$$\bar{\phi}(x, k) = \psi(x, k)\bar{\mathbf{a}}(k) + \bar{\psi}(x, k)\bar{\mathbf{b}}(k), \qquad (4.2.21b)$$

where $\mathbf{a}(k)$ and $\bar{\mathbf{a}}(k)$ are square matrices, $N \times N$ and $M \times M$, respectively, while $\mathbf{b}(k)$ is an $M \times N$ matrix and $\bar{\mathbf{b}}(k)$ is an $N \times M$ matrix. The relations (4.2.21a)–(4.2.21b) hold for any k such that all four eigenfunctions exist. In particular, they hold for Im $k = 0$. Let us introduce the scattering matrix

$$\mathbf{S}(\xi) = \begin{pmatrix} \mathbf{a}(\xi) & \bar{\mathbf{b}}(\xi) \\ \mathbf{b}(\xi) & \bar{\mathbf{a}}(\xi) \end{pmatrix} \qquad \xi \in \mathbb{R}$$

in terms of which (4.2.21a)–(4.2.21b) can be written as

$$\left(\phi(x, \xi), \bar{\phi}(x, \xi)\right) = \left(\bar{\psi}(x, \xi), \psi(x, \xi)\right) \mathbf{S}(\xi).$$

Making use of equations (4.2.20a)–(4.2.20b) and (4.2.21a)–(4.2.21b), we conclude that the scattering matrix is unimodular, that is,

$$\det \mathbf{S}(\xi) \equiv \det \begin{pmatrix} \mathbf{a}(\xi) & \bar{\mathbf{b}}(\xi) \\ \mathbf{b}(\xi) & \bar{\mathbf{a}}(\xi) \end{pmatrix} = 1 \qquad \xi \in \mathbb{R}. \qquad (4.2.22)$$

It also convenient to introduce the "right" scattering coefficients expressing ψ and $\bar{\psi}$ as linear combinations of ϕ and $\bar{\phi}$, that is,

$$\psi(x, k) = \phi(x, k)\mathbf{d}(k) + \bar{\phi}(x, k)\mathbf{c}(k) \qquad (4.2.23a)$$

$$\bar{\psi}(x, k) = \phi(x, k)\bar{\mathbf{c}}(k) + \bar{\phi}(x, k)\bar{\mathbf{d}}(k). \qquad (4.2.23b)$$

The scattering coefficients can be related to the Wronskian of the Jost solutions in the following way:

$$W(\phi(x, k), \psi(x, k))$$

$$= W\left(\left(\bar{\psi}(x, k), \psi(x, k)\right) \begin{pmatrix} \mathbf{a}(k) \\ \mathbf{b}(k) \end{pmatrix}, \left(\bar{\psi}(x, k), \psi(x, k)\right) \begin{pmatrix} \mathbf{0} \\ \mathbf{I}_M \end{pmatrix} \right)$$

$$= W\left((\bar{\psi}(x,k), \psi(x,k)) \begin{pmatrix} \mathbf{a}(k) & \mathbf{0} \\ \mathbf{b}(k) & \mathbf{I}_M \end{pmatrix} \right)$$

$$= W(\bar{\psi}(x,k), \psi(x,k)) \det \mathbf{a}(k) \tag{4.2.24a}$$

and similarly

$$W(\bar{\psi}(x,k), \bar{\phi}(x,k)) = W(\bar{\psi}(x,k), \psi(x,k)) \det \bar{\mathbf{a}}(k). \tag{4.2.24b}$$

From the other side, using the "right" data we get the relations

$$W(\phi(x,k), \psi(x,k)) = \det \mathbf{c}(k) W(\phi(x,k), \bar{\phi}(x,k)) \tag{4.2.24c}$$

$$W(\bar{\psi}(x,k), \bar{\phi}(x,k)) = \det \bar{\mathbf{c}}(k) W(\phi(x,k), \bar{\phi}(x,k)), \tag{4.2.24d}$$

and the comparison between (4.2.24a)–(4.2.24b) and (4.2.24c)–(4.2.24d) yields

$$\det \mathbf{a}(k) = \det \mathbf{c}(k) \tag{4.2.25a}$$

$$\det \bar{\mathbf{a}}(k) = \det \bar{\mathbf{c}}(k). \tag{4.2.25b}$$

One can also derive the following integral relationships for the scattering coefficients:

$$\mathbf{a}(k) = \mathbf{I}_N + \int_{-\infty}^{+\infty} \mathbf{Q}(x)\mathbf{M}^{(\mathrm{dn})}(x,k)dx, \tag{4.2.26a}$$

$$\mathbf{b}(k) = \int_{-\infty}^{+\infty} e^{-2ikx}\mathbf{R}(x)\mathbf{M}^{(\mathrm{up})}(x,k)dx, \tag{4.2.26b}$$

$$\bar{\mathbf{a}}(k) = \mathbf{I}_M + \int_{-\infty}^{+\infty} \mathbf{R}(x)\bar{\mathbf{M}}^{(\mathrm{up})}(x,k)dx, \tag{4.2.26c}$$

$$\bar{\mathbf{b}}(k) = \int_{-\infty}^{+\infty} e^{2ikx}\mathbf{Q}(x)\bar{\mathbf{M}}^{(\mathrm{dn})}(x,k)dx, \tag{4.2.26d}$$

and

$$\bar{\mathbf{c}}(k) = \mathbf{I}_N - \int_{-\infty}^{+\infty} \mathbf{Q}(x)\bar{\mathbf{N}}^{(\mathrm{dn})}(x,k)dx, \tag{4.2.27a}$$

$$\bar{\mathbf{d}}(k) = -\int_{-\infty}^{+\infty} e^{-2ikx}\mathbf{R}(x)\bar{\mathbf{N}}^{(\mathrm{up})}(x,k)dx, \tag{4.2.27b}$$

$$\mathbf{c}(k) = \mathbf{I}_M - \int_{-\infty}^{+\infty} \mathbf{R}(x)\mathbf{N}^{(\mathrm{up})}(x,k)dx, \tag{4.2.27c}$$

$$\mathbf{d}(k) = -\int_{-\infty}^{+\infty} e^{2ikx}\mathbf{Q}(x)\mathbf{N}^{(\mathrm{dn})}(x,k)dx. \tag{4.2.27d}$$

Indeed, introducing

$$\mathbf{\Delta}(x,k) = \mathbf{M}(x,k) - \bar{\mathbf{N}}(x,k)\mathbf{a}(k)$$

and using the integral equations (4.2.13a), (4.2.13d) for $\mathbf{M}$ and $\bar{\mathbf{N}}$, as well as the relation

$$\mathbf{G}_+(x, k) - \mathbf{G}_-(x, k) = \begin{pmatrix} \mathbf{I}_N & \mathbf{0} \\ \mathbf{0} & e^{2ikx}\mathbf{I}_M \end{pmatrix},$$

we can write

$$\Delta(x, k) - \int_{-\infty}^{+\infty} \mathbf{G}_-(x - x', k) \left(\tilde{\mathbf{Q}}\Delta\right)(x', k)dx'$$

$$= \begin{pmatrix} \mathbf{I}_N - \mathbf{a}(k) \\ \mathbf{0} \end{pmatrix} - \int_{-\infty}^{+\infty} \begin{pmatrix} \mathbf{Q}(x')\mathbf{M}^{(\mathrm{dn})}(x', k) \\ e^{2ik(x-x')}\mathbf{R}(x')\mathbf{M}^{(\mathrm{up})}(x', k) \end{pmatrix} dx'. \qquad (4.2.28)$$

From the other side, the scattering equation (4.2.21a) yields

$$\Delta(x, k) = e^{2ikx}\mathbf{N}(x, k)\mathbf{b}(k),$$

and then

$$\Delta(x, k) - \int_{-\infty}^{+\infty} \mathbf{G}_-(x - x', k) \left(\tilde{\mathbf{Q}}\Delta\right)(x', k) dx' = \begin{pmatrix} \mathbf{0} \\ \mathbf{b}(k) \end{pmatrix} e^{2ikx}, \qquad (4.2.29)$$

where we used the integral equation (4.2.13b) for $\mathbf{N}(x, k)$ as well as the identity

$$\mathbf{G}_-(x, k) = e^{2ikx}\tilde{\mathbf{G}}_+(x, k).$$

Comparing (4.2.28) and (4.2.29), we get the integral representations (4.2.26a) and (4.2.26b). Equations (4.2.26c) and (4.2.26d) are derived analogously, by considering $\bar{\Delta}(x, k) = \bar{\mathbf{M}}(x, k) - \mathbf{N}(x, k)\bar{\mathbf{a}}(k)$. Similarly, one obtains the integral representations for the "right" data (4.2.27a)–(4.2.27d).

Since $\mathbf{M}(x, k)$ and $\bar{\mathbf{M}}(x, k)$ are analytic, respectively, for $\mathrm{Im}\, k > 0$ and $\mathrm{Im}\, k < 0$, and continuous up to $\mathrm{Im}\, k = 0$, the integral representations (4.2.26a) and (4.2.26c) imply that $\mathbf{a}(k)$ is analytic in the upper k-plane and continuous for $\mathrm{Im}\, k = 0$ and $\bar{\mathbf{a}}(k)$ is analytic in the lower k-plane and continuous for $\mathrm{Im}\, k = 0$. On the other hand, $\mathbf{b}(k)$ and $\bar{\mathbf{b}}(k)$ cannot in general be continued off the real axis.

From the integral representations (4.2.26a) and (4.2.26c) and the asymptotics (4.2.17a), (4.2.17d) it also follows that

$$\mathbf{a}(k) = \mathbf{I}_N - \frac{1}{2ik} \int_{-\infty}^{+\infty} \mathbf{Q}(x)\mathbf{R}(x)dx + O(k^{-2}) \qquad (4.2.30a)$$

$$\bar{\mathbf{a}}(k) = \mathbf{I}_M + \frac{1}{2ik} \int_{-\infty}^{+\infty} \mathbf{R}(x)\mathbf{Q}(x)dx + O(k^{-2}). \qquad (4.2.30b)$$

With these asymptotics, the analytic properties of $\mathbf{a}$ and $\bar{\mathbf{a}}$ that we have already established and the additional assumptions that $\det \mathbf{a}(\xi) \neq 0$, $\det \bar{\mathbf{a}}(\xi) \neq 0$ for all $\xi \in \mathbb{R}$, we conclude that there must be finitely many zeros of $\det \mathbf{a}(k)$ in the upper k-plane and finitely many zeros of $\det \bar{\mathbf{a}}(k)$ in the lower k-plane.

Note that equations (4.2.21a) and (4.2.21b) can be written as

$$\mu(x, k) = \bar{\mathbf{N}}(x, k) + e^{2ikx}\mathbf{N}(x, k)\rho(k) \qquad (4.2.31a)$$

$$\bar{\mu}(x, k) = \mathbf{N}(x, k) + e^{-2ikx}\bar{\mathbf{N}}(x, k)\bar{\rho}(k), \qquad (4.2.31b)$$

where

$$\mu(x, k) = \mathbf{M}(x, k)\mathbf{a}^{-1}(k), \qquad \bar{\mu}(x, k) = \bar{\mathbf{M}}(x, k)\bar{\mathbf{a}}^{-1}(k) \qquad (4.2.32)$$

and the *reflection coefficients* are defined by

$$\rho(k) = \mathbf{b}(k)\mathbf{a}^{-1}(k), \qquad \bar{\rho}(k) = \bar{\mathbf{b}}(k)\bar{\mathbf{a}}^{-1}(k). \qquad (4.2.33)$$

Moreover, from (4.2.21a)–(4.2.21b) and (4.2.23a)–(4.2.23b) it follows that for any $\xi \in \mathbb{R}$

$$\begin{pmatrix} \mathbf{a}(\xi) & \bar{\mathbf{b}}(\xi) \\ \mathbf{b}(\xi) & \bar{\mathbf{a}}(\xi) \end{pmatrix}^{-1} = \begin{pmatrix} \bar{\mathbf{c}}(\xi) & \mathbf{d}(\xi) \\ \bar{\mathbf{d}}(\xi) & \mathbf{c}(\xi) \end{pmatrix},$$

and therefore, under the assumptions $\det \mathbf{a}(\xi) \neq 0$, $\det \bar{\mathbf{a}}(\xi) \neq 0$ for all $\xi \in \mathbb{R}$, the following relations between "left" and "right" scattering data hold:

$$(\mathbf{I}_N - \bar{\rho}(\xi)\rho(\xi))\,\mathbf{a}(\xi)\bar{\mathbf{c}}(\xi) = \mathbf{I}_N \qquad (4.2.34a)$$

$$(\mathbf{I}_M - \rho(\xi)\bar{\rho}(\xi))\,\bar{\mathbf{a}}(\xi)\mathbf{c}(\xi) = \mathbf{I}_M. \qquad (4.2.34b)$$

Proper eigenvalues and norming constants

As before, we define a proper eigenvalue to be a (complex) value of k for which the scattering problem (4.2.2) admits a bounded solution that decays as $x \to \pm\infty$.

If $\operatorname{Im} k > 0$, from the asymptotics (4.2.6a)–(4.2.6b) it follows that $\phi(x, k)$ decays as $x \to -\infty$ while $\psi(x, k)$ decays as $x \to +\infty$. Therefore, for $k_j = \xi_j + i\eta_j$ in the upper k-plane to be an eigenvalue, it must be that one of the solutions is in the span of the other one, that is,

$$W(\phi(x, k_j), \psi(x, k_j)) = 0.$$

Similarly, $\bar{k}_j$ in the lower k-plane is an eigenvalue if, and only if,

$$W(\bar{\phi}(x, \bar{k}_j), \bar{\psi}(x, \bar{k}_j)) = 0.$$

From the other side, equations (4.2.24a)–(4.2.24b) show that the eigenvalues in the upper k-plane are the points $k = k_j$ such that $\det \mathbf{a}(k_j) = 0$ and the eigenvalues in the lower k-plane are the zeros $\bar{k}_j$ of $\det \bar{\mathbf{a}}(k)$. There are no proper eigenvalues on the real k-axis because none of the basis eigenfunctions vanishes as $x \to \pm\infty$ if $\text{Im } k = 0$, and we also assume that $\det \mathbf{a}(\xi) \neq 0$, $\det \bar{\mathbf{a}}(\xi) \neq 0$ for any $\xi \in \mathbb{R}$.

From equations (4.2.32) it follows that $\mu(x, k)$ is a meromorphic function of k with poles at the zeros of $\det \mathbf{a}(k)$ and $\bar{\mu}(x, k)$ is a meromorphic function with poles at the zeros of $\det \bar{\mathbf{a}}(k)$. Let us assume that $\det \mathbf{a}(k)$ has J simple zeros at the points $\left\{k_j = \xi_j + i\eta_j : \eta_j > 0\right\}_{j=1}^{J}$ and $\det \bar{a}(k)$ has $\bar{J}$ simple zeros at $\left\{\bar{k}_\ell = \bar{\xi}_\ell + i\bar{\eta}_\ell : \bar{\eta}_\ell < 0\right\}_{\ell=1}^{\bar{J}}$. Let $\boldsymbol{\alpha}(k)$ be the cofactor matrix of $\mathbf{a}(k)$ and let us introduce the notation $a(k) = \det \mathbf{a}(k)$. Then, from equation (4.2.31a), it follows that

$$\text{Res}\left(\mu; k_j\right) = \lim_{k \to k_j} \left(k - k_j\right) \mu(x, k)$$

$$= e^{2ik_j x} \mathbf{N}\left(x, k_j\right) \frac{1}{a'(k_j)} \mathbf{b}(k_j) \boldsymbol{\alpha}(k_j)$$

$$= e^{2ik_j x} \mathbf{N}\left(x, k_j\right) \mathbf{C}_j, \qquad (4.2.35)$$

where $'$ denotes the derivative with respect to the parameter k and for any $j = 1, \ldots, J$

$$\mathbf{C}_j = \frac{1}{a'(k_j)} \mathbf{b}(k_j) \boldsymbol{\alpha}(k_j) \qquad (4.2.36)$$

is an $M \times N$ matrix that is referred to as the norming constant associated with the discrete eigenvalue k_j. By a similar procedure, we can find the expression for the residues of the poles of $\bar{\mu}(x, k)$,

$$\text{Res}\left(\bar{\mu}; \bar{k}_\ell\right) = e^{-2i\bar{k}_\ell x} \bar{\mathbf{N}}\left(x, \bar{k}_\ell\right) \bar{\mathbf{C}}_\ell, \qquad (4.2.37)$$

where for any $\ell = 1, \ldots, \bar{J}$

$$\bar{\mathbf{C}}_\ell = \frac{1}{\bar{a}'(\bar{k}_\ell)} \bar{\mathbf{b}}(\bar{k}_\ell) \bar{\boldsymbol{\alpha}}(\bar{k}_\ell) \qquad (4.2.38)$$

is an $N \times M$ matrix (as before, $\bar{a}(k) = \det \bar{\mathbf{a}}(k)$ and $\bar{\boldsymbol{\alpha}}(k)$ is the cofactor matrix of $\bar{\mathbf{a}}(k)$).

To summarize, in the general case the eigenvalues $\left\{k_j : \text{Im } k_j > 0\right\}_{j=1}^{J} \cup \left\{\bar{k}_\ell : \text{Im } \bar{k}_\ell < 0\right\}_{\ell=1}^{\bar{J}}$ and the associated norming constants $\left\{\mathbf{C}_j\right\}_{j=1}^{J} \cup \left\{\bar{\mathbf{C}}_\ell\right\}_{\ell=1}^{\bar{J}}$, together with the reflection coefficients $\{\rho(\xi), \bar{\rho}(\xi) : \xi \in \mathbb{R}\}$ given by (4.2.33), constitute the scattering data.

Symmetry reductions

The M-component VNLS equation is a special case of the system (4.1.1a)–(4.1.1b) where $\mathbf{Q} \equiv \mathbf{q}$ is an M-component row vector (i.e., $N = 1$) and

$\mathbf{R} \equiv \mathbf{r} = \mp \mathbf{q}^H$. More generally, we can consider the symmetry

$$\mathbf{R} = \mp \mathbf{Q}^H \tag{4.2.39}$$

for any matrix $\mathbf{Q}$.

Note that, as in the scalar case, when $\mathbf{R} = \mathbf{Q}^H$ the scattering operator is Hermitian, and therefore in this reduction the scattering problem (4.2.2) with potentials decaying rapidly enough as $x \to \pm\infty$ does not admit discrete eigenvalues off the real k-axis.

The symmetries (4.2.39) in the potentials induce symmetries in the scattering data. To determine such symmetries, we consider the matrix-valued functions defined in the upper k-plane:

$$\mathbf{f}_\pm(x, k) = \left[\sigma_\pm \bar{\phi}(x, k^*) \right]^H \phi(x, k),$$

where

$$\sigma_\pm = \begin{pmatrix} \mathbf{I}_N & \mathbf{0} \\ \mathbf{0} & \pm \mathbf{I}_M \end{pmatrix}. \tag{4.2.40}$$

From the scattering problem (4.2.2) it follows that under the symmetry (4.2.39)

$$\frac{\partial}{\partial x} \mathbf{f}_\pm(x, k) = \mathbf{0},$$

and equating the asymptotic behavior of $\mathbf{f}_\pm$ as $x \to +\infty$ and $x \to -\infty$ following from (4.2.6a)–(4.2.6b) and (4.2.21a)–(4.2.21b), we find that

$$\bar{\mathbf{a}}^H(k^*)\mathbf{b}(k) \pm \bar{\mathbf{b}}^H(k^*)\mathbf{a}(k) = \mathbf{0}.$$

Consequently

$$\bar{\rho}^H(k^*) = \mp \rho(k) \tag{4.2.41}$$

for any $k \in \mathbb{R}$. An analogous argument yields that $\mathbf{g}_\pm(x, k) = \left[\sigma_\pm \bar{\phi}(x, k^*) \right]^H \psi(x, k)$ and $\mathbf{h}_\pm(x, k) = \left[\sigma_\pm \bar{\psi}(x, k^*) \right]^H \phi(x, k)$, defined in the upper k-plane, are independent of x. Therefore, comparing the asymptotics of $\mathbf{g}_\pm$ and $\mathbf{h}_\pm$ as $|x| \to \infty$ following from (4.2.6a)–(4.2.6b) and (4.2.23a)–(4.2.23b), we get

$$\bar{\mathbf{a}}^H(k^*) = \mathbf{c}(k) \tag{4.2.42a}$$

$$\bar{\mathbf{c}}^H(k^*) = \mathbf{a}(k) \tag{4.2.42b}$$

and, in particular,

$$\det \mathbf{c}(k) = \left(\det \bar{\mathbf{a}}(k^*) \right)^*. \tag{4.2.43}$$

Taking into account (4.2.25a)–(4.2.25b), under the reduction (4.2.39), we conclude that

$$\det \bar{\mathbf{a}}(k^*) = (\det \mathbf{a}(k))^* , \qquad (4.2.44)$$

which implies that k_j in the upper k-plane is an eigenvalue if, and only if,

$$\bar{k}_j = k_j^* \qquad (4.2.45)$$

is an eigenvalue in the lower k-plane. Consequently, $J = \bar{J}$, that is, the number of eigenvalues in the upper k-plane is equal to the number of eigenvalues in the lower k-plane. As far as the associated norming constants are concerned, from (4.2.36), (4.2.38) and (4.2.41), (4.2.45) it follows that

$$\bar{\mathbf{C}}_j = \mp \mathbf{C}_j^H \qquad (4.2.46)$$

for $j = 1, \dots, J$.

The comparison of the asymptotic behavior of the functions $\hat{\mathbf{f}}_\pm(x, k) = \left[\boldsymbol{\sigma}_\pm \bar{\phi}(x, k^*)\right]^H \bar{\phi}(x, k)$ and $\hat{\mathbf{g}}_\pm(x, k) = [\boldsymbol{\sigma}_\pm \phi(x, k^*)]^H \phi(x, k)$ as $x \to \pm\infty$ yields the following characterization equations for the scattering data:

$$\mathbf{a}^H\left(k^*\right)\mathbf{a}(k) = \mathbf{I}_N \mp \mathbf{b}^H\left(k^*\right)\mathbf{b}(k) \qquad \operatorname{Im} k = 0 \qquad (4.2.47a)$$

$$\bar{\mathbf{a}}^H\left(k^*\right)\bar{\mathbf{a}}(k) = \mathbf{I}_M \mp \bar{\mathbf{b}}^H\left(k^*\right)\bar{\mathbf{b}}(k) \qquad \operatorname{Im} k = 0, \qquad (4.2.47b)$$

which can be considered as defining a matrix Riemann-Hilbert problem with zeros (matrix factorization problem) for $\mathbf{a}(k)$ and $\bar{\mathbf{a}}(k)$ with jump given across the real axis and boundary conditions given by (4.2.30a)–(4.2.30b).

Trace formula

We assume that $a(k) = \det \mathbf{a}(k)$ and $\bar{a}(k) = \det \bar{\mathbf{a}}(k)$ have the simple zeros $\{k_j : \operatorname{Im} k_j > 0\}_{j=1}^J$ and $\{\bar{k}_j : \operatorname{Im} \bar{k}_j < 0\}_{j=1}^{\bar{J}}$, respectively, and define

$$\alpha(k) = \prod_{m=1}^{J} \frac{k - k_m^*}{k - k_m} a(k), \qquad \bar{\alpha}(k) = \prod_{m=1}^{\bar{J}} \frac{k - \bar{k}_m^*}{k - \bar{k}_m} \bar{a}(k). \qquad (4.2.48)$$

$\alpha(k)$ is analytic in the upper k-plane, where it has no zeros, while $\bar{\alpha}(k)$ is analytic in the lower k-plane, where it has no zeros; moreover, due to (4.2.30a)–(4.2.30b), $\alpha(k), \bar{\alpha}(k) \to 1$ as $|k| \to \infty$ in the proper half-plane. Therefore we have

$$\log a(k) = \sum_{m=1}^{J} \log\left(\frac{k - k_m}{k - k_m^*}\right) + \frac{1}{2\pi i} \int_{-\infty}^{+\infty} \frac{\log\left(\alpha(\xi)\bar{\alpha}(\xi)\right)}{\xi - k} d\xi, \qquad \operatorname{Im} k > 0$$

$$(4.2.49a)$$

$$\log \bar{a}(k) = \sum_{m=1}^{J} \log\left(\frac{k - \bar{k}_m}{k - \bar{k}_m^*}\right) - \frac{1}{2\pi i} \int_{-\infty}^{+\infty} \frac{\log\left(\alpha(\xi)\bar{\alpha}(\xi)\right)}{\xi - k} d\xi \qquad \operatorname{Im} k < 0.$$

$$(4.2.49b)$$

If the symmetry (4.2.39) holds, from (4.2.44) and (4.2.46) it follows that $\alpha(\xi)\bar\alpha(\xi) = a(\xi)\bar a(\xi) = |a(\xi)|^2$, and, using (4.2.34a), $|a(\xi)|^2 = \det(\mathbf{I}_N - \bar\rho(\xi)\rho(\xi))^{-1}$. Hence, equation (4.2.49b) allows one to recover $a(k)$, $\bar a(k)$ for any k in the proper half-plane from knowledge of $\{k_j, \ \mathrm{Im}\ k_j > 0\}_{j=1}^{J}$, $\{\bar k_j, \ \mathrm{Im}\ \bar k_j > 0\}_{j=1}^{J}$, and $\alpha(\xi)\bar\alpha(\xi) = \det\left(\mathbf{I}_N \pm \rho^H(\xi)\rho(\xi)\right)^{-1}$ for $\xi \in \mathbb{R}$.

The problem of reconstructing the matrices $\mathbf{a}(k)$ and $\bar{\mathbf{a}}(k)$ is more complicated. We do note, however, that from (4.2.26a) and (4.2.26c) and the inverse scattering below, the (matrix) scattering coefficients $\mathbf{a}(k)$ and $\bar{\mathbf{a}}(k)$ can be reconstructed, in principle, from the scattering data $\{k_j, \ \mathbf{C}_j\}_{j=1}^{J}$, $\{\bar k_j, \ \bar{\mathbf{C}}_j\}_{j=1}^{J}$ and $\{\rho(\xi), \bar\rho(\xi) : \xi \in \mathbb{R}\}$. In the inverse problem we use this method to obtain $\mathbf{a}(k)$ and $\bar{\mathbf{a}}(k)$ for the special case of a pure one-soliton potential.

4.2.3 Inverse scattering problem

The inverse problem is the construction of a map from the scattering data back to the potentials. We start with (i) the reflection coefficients $\rho(\xi)$ and $\bar\rho(\xi)$ for $\xi \in \mathbb{R}$, (ii) the discrete eigenvalues $\{k_j, \ \mathrm{Im}\ k_j > 0\}_{j=1}^{J}$ and $\{\bar k_j, \ \mathrm{Im}\ \bar k_j < 0\}_{j=1}^{J}$, and (iii) the corresponding norming constants $\{\mathbf{C}_j\}_{j=1}^{J}$ and$\{\bar{\mathbf{C}}_j\}_{j=1}^{J}$. First we use these data to recover the Jost functions. Then, we recover the potentials in terms of these Jost functions.

In the previous section, we showed that the functions $\mathbf{N}(x, k)$ and $\bar{\mathbf{N}}(x, k)$ exist and are analytic in the regions $\mathrm{Im}\ k > 0$ and $\mathrm{Im}\ k < 0$, respectively, if $\mathbf{Q}, \mathbf{R} \in L^1(\mathbb{R})$. Similarly, under the same conditions on the potentials, the functions $\mu(x, k)$ and $\bar\mu(x, k)$ defined by (4.2.32) are meromorphic in the regions $\mathrm{Im}\ k > 0$ and $\mathrm{Im}\ k < 0$, respectively. Therefore, in the inverse problem we assume the unknown functions are sectionally meromorphic. With this assumption, equations (4.2.31a)–(4.2.31b) can be considered to be the jump conditions of a Riemann–Hilbert problem. As before, to recover the sectionally meromorphic functions from the scattering data, we convert the Riemann–Hilbert problem to a system of linear integral equations by the use of projection operators.

Case of no poles

We begin by solving the Riemann–Hilbert problem in the case where μ and $\bar\mu$ have no poles. Introducing the $(N + M) \times (N + M)$ matrices

$$\mathbf{m}_+(x, k) = (\mu(x, k), \mathbf{N}(x, k)), \qquad \mathbf{m}_-(x, k) = \left(\bar{\mathbf{N}}(x, k), \bar\mu(x, k)\right),$$
(4.2.50)

we can write the "jump" conditions (4.2.31a)–(4.2.31b) as

$$\mathbf{m}_+(x, k) - \mathbf{m}_-(x, k) = \mathbf{m}_-(x, k)\mathbf{V}(x, k),$$
(4.2.51)

where

$$\mathbf{V}(x,k) = \begin{pmatrix} -\rho(k)\bar{\rho}(k) & -\bar{\rho}(k)e^{-2ikx} \\ \rho(k)e^{2ikx} & 0 \end{pmatrix} \tag{4.2.52}$$

and

$$\mathbf{m}_{\pm}(x,k) \to \mathbf{I}_{N+M}$$

as $|k| \to \infty$ in the proper half-plane. Applying the projector P^- defined in (2.2.51) to equation (4.2.51) yields

$$\mathbf{m}_-(x,k) = \mathbf{I}_{N+M} + \frac{1}{2\pi i} \int_{-\infty}^{+\infty} \frac{\mathbf{m}_-(x,\xi)\mathbf{V}(x,\xi)}{\xi - (k - i0)} d\xi, \tag{4.2.53}$$

which allows one, in principle, to find $\mathbf{m}_-(x,k)$. Note that, as $|k| \to \infty$,

$$\mathbf{m}_-(x,k) = \mathbf{I} - \frac{1}{2\pi i k} \int_{-\infty}^{+\infty} \mathbf{m}_-(x,\xi)\mathbf{V}(x,\xi)d\xi + O(k^{-2}). \tag{4.2.54}$$

Taking into account the asymptotics (4.2.17a)–(4.2.17d) and the definitions (4.2.50), (4.2.52), from equation (4.2.54) we obtain the reconstruction of the potentials in terms of the scattering data, that is,

$$\mathbf{R}(x) = \frac{1}{\pi} \int_{-\infty}^{+\infty} e^{2ikx} \mathbf{N}^{(\mathrm{dn})}(x,k)\rho(k)dk \tag{4.2.55a}$$

$$\mathbf{Q}(x) = \frac{1}{\pi} \int_{-\infty}^{+\infty} e^{-2ikx} \bar{\mathbf{N}}^{(\mathrm{up})}(x,k)\bar{\rho}(k)dk. \tag{4.2.55b}$$

Case of poles

Suppose now that the potentials are such that $a(k) = \det \mathbf{a}(k)$ and $\bar{a}(k) = \det \bar{\mathbf{a}}(k)$ have a finite number of simple zeros in the regions Im $k > 0$ and Im $k < 0$, respectively, which we denote as $\{k_j, \text{Im } k_j > 0\}_{j=1}^J$ and $\{\bar{k}_j, \text{Im } \bar{k}_j < 0\}_{j=1}^J$. We further assume that $a(\xi) \neq 0, \bar{a}(\xi) \neq 0$ for any $\xi \in \mathbb{R}$. As before, we apply P^- to both sides of (4.2.31a) and P^+ to both sides of (4.2.31b). Taking into account the analytic properties of the Jost functions and the asymptotics (4.2.17a)–(4.2.17d), as well as (4.2.35), we obtain

$$\bar{\mathbf{N}}(x,k) = \begin{pmatrix} \mathbf{I}_N \\ \mathbf{0} \end{pmatrix} + \sum_{j=1}^J \frac{e^{2ik_jx}}{(k-k_j)} \mathbf{N}_j(x)\mathbf{C}_j$$

$$+ \frac{1}{2\pi i} \int_{-\infty}^{+\infty} \frac{e^{2i\xi x}}{\xi - (k - i0)} \mathbf{N}(x,\xi)\rho(\xi)d\xi \tag{4.2.56a}$$

$$\mathbf{N}(x, k) = \begin{pmatrix} \mathbf{0} \\ \mathbf{I}_M \end{pmatrix} + \sum_{j=1}^{\bar{J}} \frac{e^{-2i\bar{k}_j x}}{(k - \bar{k}_j)} \bar{\mathbf{N}}_j(x)\bar{\mathbf{C}}_j$$

$$-\frac{1}{2\pi i} \int_{-\infty}^{+\infty} \frac{e^{-2i\xi x}}{\xi - (k + i0)} \bar{\mathbf{N}}(x, \xi)\bar{\rho}(\xi)d\xi, \qquad (4.2.56b)$$

where, as before, $\mathbf{N}_j(x) = \mathbf{N}(x, k_j)$ and $\bar{\mathbf{N}}_j(x) = \bar{\mathbf{N}}(x, \bar{k}_j)$. To close the system, we evaluate equation (4.2.56a) at $k = \bar{k}_\ell$ for any $\ell = 1, \ldots, \bar{J}$ and (4.2.56b) at $k = k_j$ for any $j = 1, \ldots, J$, thus getting

$$\bar{\mathbf{N}}_\ell(x) = \begin{pmatrix} \mathbf{I}_N \\ \mathbf{0} \end{pmatrix} + \sum_{j=1}^{J} \frac{e^{2ik_j x}}{(\bar{k}_\ell - k_j)} \mathbf{N}_j(x)\mathbf{C}_j + \frac{1}{2\pi i} \int_{-\infty}^{+\infty} \frac{e^{2i\xi x}}{\xi - \bar{k}_\ell} \mathbf{N}(x, \xi)\rho(\xi)d\xi$$

$$(4.2.56c)$$

$$\mathbf{N}_j(x) = \begin{pmatrix} \mathbf{0} \\ \mathbf{I}_M \end{pmatrix} + \sum_{m=1}^{\bar{J}} \frac{e^{-2i\bar{k}_m x}}{(k_j - \bar{k}_m)} \bar{\mathbf{N}}_m(x)\bar{\mathbf{C}}_m$$

$$-\frac{1}{2\pi i} \int_{-\infty}^{+\infty} \frac{e^{-2i\xi x}}{\xi - k_j} \bar{\mathbf{N}}(x, \xi)\bar{\rho}(\xi)d\xi. \qquad (4.2.56d)$$

Equations (4.2.56a)–(4.2.56c) constitute a linear algebraic–integral system of equations that, in principle, solve the inverse problem for the eigenfunctions $\mathbf{N}(x, k)$ and $\bar{\mathbf{N}}(x, k)$.

By comparing the asymptotic expansions at large k of the right-hand sides of (4.2.56a) and (4.2.56b) to the expansions (4.2.17b) and (4.2.17c), respectively, we obtain

$$\mathbf{R}(x) = -2i \sum_{j=1}^{J} e^{2ik_j x} \mathbf{N}_j^{(dn)}(x)\mathbf{C}_j + \frac{1}{\pi} \int_{-\infty}^{+\infty} e^{2i\xi x} \mathbf{N}^{(dn)}(x, \xi)\rho(\xi)d\xi$$

$$(4.2.57a)$$

$$\mathbf{Q}(x) = 2i \sum_{j=1}^{\bar{J}} e^{-2i\bar{k}_j x} \bar{\mathbf{N}}_j^{(up)}(x)\bar{\mathbf{C}}_j + \frac{1}{\pi} \int_{-\infty}^{+\infty} e^{-2i\xi x} \bar{\mathbf{N}}^{(up)}(x, \xi)\bar{\rho}(\xi)d\xi,$$

$$(4.2.57b)$$

which reconstruct the potentials and thus complete the formulation of the inverse problem.

If the potentials decay rapidly enough at infinity, so that $\rho(k)$ can be analytically continued above all poles $\{k_j, \ \text{Im} \ k_j > 0\}_{j=1}^{J}$ and $\bar{\rho}(k)$ can be analytically continued below all poles $\{\bar{k}_j, \ \text{Im} \ \bar{k}_j < 0\}_{j=1}^{\bar{J}}$, then the system of equations

(4.2.56a)–(4.2.56d) can be simplified as follows:

$$\bar{\mathbf{N}}(x,k) = \begin{pmatrix} \mathbf{I}_N \\ \mathbf{0} \end{pmatrix} + \frac{1}{2\pi i} \int_{C_0} \frac{e^{2i\xi x}}{\xi - k} \mathbf{N}(x,\xi)\rho(\xi)d\xi \qquad (4.2.58a)$$

$$\mathbf{N}(x,k) = \begin{pmatrix} \mathbf{0} \\ \mathbf{I}_M \end{pmatrix} - \frac{1}{2\pi i} \int_{\bar{C}_0} \frac{e^{-2i\xi x}}{\xi - k} \bar{\mathbf{N}}(x,\xi)\bar{\rho}(\xi)d\xi, \qquad (4.2.58b)$$

where C_0 is a contour from $-\infty$ to $+\infty$ that passes above all zeros of $a(k)$ and $\bar{C}_0$ is a contour from $-\infty$ to $+\infty$ that passes below all zeros of $\bar{a}(k)$. Under the same hypothesis, equations (4.2.57a)–(4.2.57b) can be written as

$$\mathbf{R}(x) = \frac{1}{\pi} \int_{C_0} e^{2i\xi x} \mathbf{N}^{(\mathrm{dn})}(x,\xi)\rho(\xi)d\xi \qquad (4.2.59a)$$

$$\mathbf{Q}(x) = \frac{1}{\pi} \int_{\bar{C}_0} e^{-2i\xi x} \bar{\mathbf{N}}^{(\mathrm{up})}(x,\xi)\bar{\rho}(\xi)d\xi. \qquad (4.2.59b)$$

Gel'fand–Levitan–Marchenko equations

We can also provide a reconstruction for the potentials by developing the Gel'fand–Levitan–Marchenko integral equations, instead of using the projection operators (cf. [21]). In analogy with the scalar case, we represent the eigenfunctions in terms of triangular kernels,

$$\mathbf{N}(x,k) = \begin{pmatrix} \mathbf{0} \\ \mathbf{I}_M \end{pmatrix} + \int_x^{+\infty} \mathbf{K}(x,s)e^{-ik(x-s)}ds \qquad s > x, \quad \mathrm{Im}\, k > 0$$

$$(4.2.60a)$$

$$\bar{\mathbf{N}}(x,k) = \begin{pmatrix} \mathbf{I}_N \\ \mathbf{0} \end{pmatrix} + \int_x^{+\infty} \bar{\mathbf{K}}(x,s)e^{ik(x-s)}ds \qquad s > x, \quad \mathrm{Im}\, k < 0,$$

$$(4.2.60b)$$

where $\mathbf{K}$ and $\bar{\mathbf{K}}$ are $(N+M) \times M$ and $(N+M) \times N$ matrices, respectively.

Applying the operator $\frac{1}{2\pi} \int_{-\infty}^{+\infty} dk\, e^{-ik(x-y)}$ for $y > x$ to equation (4.2.58a), we get

$$\bar{\mathbf{K}}(x,y) + \begin{pmatrix} \mathbf{0} \\ \mathbf{I}_M \end{pmatrix} \mathbf{F}(x+y) + \int_x^{+\infty} \mathbf{K}(x,s)\mathbf{F}(s+y)ds = 0, \qquad (4.2.61a)$$

where

$$\mathbf{F}(x) = \frac{1}{2\pi} \int_{C_0} \rho(\xi)e^{i\xi x}d\xi = \frac{1}{2\pi} \int_{-\infty}^{+\infty} \rho(\xi)e^{i\xi x}d\xi - i\sum_{j=1}^{J} \mathbf{C}_j e^{ik_j x}.$$

$$(4.2.61b)$$

Analogously, operating on equation (4.2.58b) with $\frac{1}{2\pi}\int_{-\infty}^{+\infty}dk\,e^{ik(x-y)}$ for $y > x$ gives

$$\mathbf{K}(x, y) + \begin{pmatrix} \mathbf{I}_N \\ \mathbf{0} \end{pmatrix} \bar{\mathbf{F}}(x+y) + \int_x^{+\infty} \bar{\mathbf{K}}(x, s)\bar{\mathbf{F}}(s+y)ds = 0, \qquad (4.2.61c)$$

where

$$\bar{\mathbf{F}}(x) = \frac{1}{2\pi}\int_{\bar{C}_0}\bar{\rho}(\xi)e^{-i\xi x}d\xi = \frac{1}{2\pi}\int_{-\infty}^{+\infty}\bar{\rho}(\xi)e^{-i\xi x}d\xi + i\sum_{j=1}^{\bar{J}}\bar{\mathbf{C}}_j e^{-i\bar{k}_j x}.$$

$$(4.2.61d)$$

Equations (4.2.61a) and (4.2.61c) constitute the Gel'fand–Levitan–Marchenko integral equations.

Inserting the representations (4.2.60a)–(4.2.60b) for the eigenfunctions into equations (4.2.59a)–(4.2.59b), we obtain the reconstruction of the potentials in terms of the kernels of the GLM equations, that is,

$$\mathbf{Q}(x) = -2\mathbf{K}^{(\mathrm{dn})}(x, x), \qquad \mathbf{R}(x) = -2\bar{\mathbf{K}}^{(\mathrm{up})}(x, x). \qquad (4.2.62)$$

If the symmetry (4.2.39) between the potentials holds, then, taking into account (4.2.41)–(4.2.46), we also have the symmetry

$$\bar{\mathbf{F}}(x) = \mp\mathbf{F}^H(x). \qquad (4.2.63)$$

Existence and uniqueness of solution

The question of existence and uniqueness of solutions of linear integral equations is usually examined by the use of the Fredholm alternative. For example, under suitable assumptions on the decay of the potentials, the restriction $\mathbf{R} = \mp\mathbf{Q}^H$ is sufficient to guarantee that the solutions of (4.2.61a) and (4.2.61c) exist and are unique. To show this, consider the corresponding homogeneous equations ($y > x$)

$$\mathbf{h}_1(y) + \int_x^{+\infty} \mathbf{h}_2(s)\mathbf{F}(s+y)ds = 0 \qquad (4.2.64a)$$

$$\mathbf{h}_2(y) + \int_x^{+\infty} \mathbf{h}_1(s)\bar{\mathbf{F}}(s+y)ds = 0, \qquad (4.2.64b)$$

where $\mathbf{h}_1$ is an $(N+M)\times N$ matrix; $\mathbf{h}_2$ is $(N+M)\times M$; and $\mathbf{F}$, $\bar{\mathbf{F}}$ are $M\times N$ and $N\times M$, respectively, given by (4.2.61b) and (4.2.61d). Suppose

$\mathbf{h}(y) = (\mathbf{h}_1(y), \mathbf{h}_2(y))$ is a solution of (4.2.64a)–(4.2.64b) that vanishes identically for $y < x$ and write for each matrix element

$$h_1^{(i,j)}(y) + \sum_{k=1}^{M} \int_x^{+\infty} h_2^{(i,k)}(s) F^{(k,j)}(s+y) ds = 0$$

$$i = 1, \ldots, (N+M), \quad j = 1, \ldots, N \tag{4.2.65a}$$

$$h_2^{(i,j)}(y) + \sum_{k=1}^{N} \int_x^{+\infty} h_1^{(i,k)}(s) \bar{F}^{(k,j)}(s+y) ds = 0$$

$$i = 1, \ldots, (N+M), \quad j = 1, \ldots, M. \tag{4.2.65b}$$

We multiply (4.2.65a) by $(h_1^{(i,j)}(y))^*$ and (4.2.65b) by $(h_2^{(i,j)}(y))^*$, sum over all i, j, and integrate with respect to y to obtain

$$\int_{-\infty}^{\infty} \left\{ \sum_{i=1}^{N+M} \sum_{j=1}^{N} |h_1^{(i,j)}(y)|^2 + \sum_{i=1}^{N+M} \sum_{j=1}^{M} |h_2^{(i,j)}(y)|^2 \right.$$

$$+ \int_{-\infty}^{\infty} \left[\sum_{i=1}^{N+M} \sum_{j=1}^{N} \sum_{k=1}^{M} \left(h_1^{(i,j)}(y) \right)^* h_2^{(i,k)}(s) F^{(k,j)}(s+y) \right.$$

$$\left. + \sum_{i=1}^{N+M} \sum_{j=1}^{M} \sum_{k=1}^{N} \left(h_2^{(i,j)}(y) \right)^* h_1^{(i,k)}(s) \bar{F}^{(k,j)}(s+y) \right] ds \right\} dy = 0. \tag{4.2.66}$$

If $\mathbf{R} = -\mathbf{Q}^H$, then the symmetry condition (4.2.63) allows (4.2.66) to be written as

$$0 = \int_{-\infty}^{\infty} \left\{ \sum_{i=1}^{N+M} \sum_{j=1}^{N} |h_1^{(i,j)}(y)|^2 + \sum_{i=1}^{N+M} \sum_{j=1}^{M} |h_2^{(i,j)}(y)|^2 \right.$$

$$\left. + 2i \mathrm{Im} \int_{-\infty}^{\infty} \sum_{i=1}^{N+M} \sum_{j=1}^{N} \sum_{k=1}^{M} \left(h_1^{(i,j)}(y) \right)^* h_2^{(i,k)}(s) F^{(k,j)}(s+y) ds \right\} dy.$$

$$\tag{4.2.67}$$

The real and imaginary parts of (4.2.67) must both vanish, from which it follows that

$$\mathbf{h}_1(y) \equiv \mathbf{0}, \qquad \mathbf{h}_2(y) \equiv \mathbf{0}.$$

Moreover, if

$$\mathbf{R}(x) = \mathbf{Q}^H(x), \tag{4.2.68}$$

using (4.2.63), equation (4.2.66) then becomes

$$
0 = \int_{-\infty}^{\infty} \left\{ \sum_{i=1}^{N+M} \sum_{j=1}^{N} |h_1^{(i,j)}(y)|^2 + \sum_{i=1}^{N+M} \sum_{j=1}^{M} |h_2^{(i,j)}(y)|^2 \right.
$$

$$
\left. + 2\mathrm{Re} \int_{-\infty}^{\infty} \sum_{i=1}^{N+M} \sum_{j=1}^{N} \sum_{k=1}^{M} \left(h_1^{(i,j)}(y) \right)^* h_2^{(i,k)}(s) F^{(k,j)}(s+y) ds \right\} dy.
$$

(4.2.69)

Moreover, in this case the scattering problem is formally self-adjoint, and there are no discrete eigenvalues; therefore (4.2.61b) reduces to

$$
\mathbf{F}(x) = \frac{1}{2\pi} \int_{-\infty}^{+\infty} \rho(\xi) e^{i\xi x} d\xi.
$$

(4.2.70)

Writing (4.2.69) in the Fourier transform space and using Parseval's identity (2.2.75) yields

$$
0 = \int_{-\infty}^{\infty} \left\{ \sum_{i=1}^{N+M} \sum_{j=1}^{N} |\hat{h}_1^{(i,j)}(\xi)|^2 + \sum_{i=1}^{N+M} \sum_{j=1}^{M} |\hat{h}_2^{(i,j)}(\xi)|^2 \right.
$$

$$
\left. + 2\mathrm{Re} \int_{-\infty}^{\infty} \sum_{i=1}^{N+M} \sum_{j=1}^{N} \sum_{k=1}^{M} \left(\hat{h}_1^{(i,j)}(\xi) \right)^* \hat{h}_2^{(i,k)}(-\xi) \rho^{(k,j)}(\xi) \right\} d\xi. \quad (4.2.71)
$$

Under the reduction (4.2.68), from (4.2.34a) and the symmetry relations (4.2.41) and (4.2.42b), it follows that

$$
|\det \mathbf{a}(\xi)|^2 \det \left(\mathbf{I}_N - \rho^H(\xi) \rho(\xi) \right) = 1.
$$

For the M-component VNLS equation, $N = 1$ and ρ is an M-component column vector; therefore the latter equation yields $1 - ||\rho||^2 = |a|^{-2}$, and, consequently, $|\rho^{(j,1)}(\xi)| \equiv |\rho^{(j)}(\xi)| < 1$ for any $j = 1, \ldots, M$. Then, for any i, k,

$$
\left| 2\mathrm{Re} \left[\hat{h}_1^{(i,1)}(-\xi) \left(\hat{h}_2^{(i,k)}(\xi) \right)^* \rho^{(k,1)}(\xi) \right] \right| < 2|\hat{h}_1^{(i,1)}(-\xi)||\hat{h}_2^{(i,k)}(\xi)|
$$

$$
\leq |\hat{h}_1^{(i,1)}(-\xi)|^2 + |\hat{h}_2^{(i,k)}(\xi)|^2;
$$

and hence from (4.2.71) it follows that $\hat{\mathbf{h}}_1(\xi) \equiv \mathbf{0}$, $\hat{\mathbf{h}}_2(\xi) \equiv \mathbf{0}$ and therefore

$$
\mathbf{h}_1(y) \equiv \mathbf{0}, \qquad \mathbf{h}_2(y) \equiv \mathbf{0}.
$$

We conclude that when $\mathbf{R} = \mp \mathbf{Q}^H$, for $N = 1$ in the defocusing (+) case the integral equations (4.2.61a) and (4.2.61c) admit no homogenous solutions but the trivial one. The complete study of the inverse scattering problem is outside the scope of this work. However, we can say that when the kernel $\mathbf{F}(x + y)$ is compact, as it would be when the class of potentials is suitably restricted, then the Gel'fand–Levitan–Marchenko equations are Fredholm. In this case, the Fredholm alternative applies and the latter equation implies that the solution of (4.2.61a) and (4.2.61c) exists and is unique.

4.2.4 Time evolution

The operator (4.2.3) that fixes the evolution of the Jost functions can be written as

$$\partial_t \mathbf{v} = \begin{pmatrix} \mathbf{A} & \mathbf{B} \\ \mathbf{C} & \mathbf{D} \end{pmatrix} \mathbf{v}, \tag{4.2.72}$$

where $\mathbf{B}, \mathbf{C} \to 0$ as $x \to \pm\infty$ (since we have assumed that $\mathbf{Q}, \mathbf{R} \to 0$ as $x \to \pm\infty$). Then the time-dependence must asymptotically satisfy

$$\partial_t \mathbf{v} = \begin{pmatrix} A_\infty \mathbf{I}_N & 0 \\ 0 & -A_\infty \mathbf{I}_M \end{pmatrix} \mathbf{v} \quad \text{as } x \to \pm\infty, \tag{4.2.73}$$

where we introduced $A_\infty = 2ik^2$ such that

$$\lim_{|x|\to\infty} \mathbf{A}(x,k) = A_\infty \mathbf{I}_N = 2ik^2 \mathbf{I}_N, \qquad \lim_{|x|\to\infty} \mathbf{D}(x,k) = -A_\infty \mathbf{I}_M = -2ik^2 \mathbf{I}_M.$$

The system (4.2.73) has solutions that are linear combinations of

$$\mathbf{v}^+ = \begin{pmatrix} e^{A_\infty t} \mathbf{I}_N \\ 0 \end{pmatrix}, \qquad \mathbf{v}^- = \begin{pmatrix} 0 \\ e^{-A_\infty t} \mathbf{I}_M \end{pmatrix}.$$

However, such solutions are not compatible with the fixed boundary conditions of the Jost functions (4.2.6a)–(4.2.6b). We define time-dependent functions

$$\mathbf{\Phi}(x,t) = e^{A_\infty t} \phi(x,t), \qquad \bar{\mathbf{\Phi}}(x,t) = e^{-A_\infty t} \bar{\phi}(x,t) \tag{4.2.74a}$$

$$\mathbf{\Psi}(x,t) = e^{-A_\infty t} \psi(x,t), \qquad \bar{\mathbf{\Psi}}(x,t) = e^{A_\infty t} \bar{\psi}(x,t) \tag{4.2.74b}$$

to be solutions of the time-differential equation (4.2.72). Then the evolution equations for ϕ and $\bar{\phi}$ become

$$\partial_t \phi = \begin{pmatrix} \mathbf{A} - A_\infty \mathbf{I}_N & \mathbf{B} \\ \mathbf{C} & \mathbf{D} - A_\infty \mathbf{I}_M \end{pmatrix} \phi,$$

$$\partial_t \bar{\phi} = \begin{pmatrix} \mathbf{A} + A_\infty \mathbf{I}_N & \mathbf{B} \\ \mathbf{C} & \mathbf{D} + A_\infty \mathbf{I}_M \end{pmatrix} \bar{\phi}, \tag{4.2.75}$$

so that, taking into account equations (4.2.21a)–(4.2.21b) and evaluating (4.2.75) as $x \to +\infty$, one obtains

$$\partial_t \mathbf{a}(k) = 0, \qquad \partial_t \bar{\mathbf{a}}(k) = 0$$

$$\partial_t \mathbf{b}(k) = -2A_\infty \mathbf{b}(k), \qquad \partial_t \bar{\mathbf{b}}(k) = 2A_\infty \bar{\mathbf{b}}(k)$$

or, explicitly,

$$\mathbf{a}(k, t) = \mathbf{a}(k, 0), \qquad \bar{\mathbf{a}}(k, t) = \bar{\mathbf{a}}(k, 0) \tag{4.2.76a}$$

$$\mathbf{b}(k, t) = e^{-4ik^2 t}\mathbf{b}(k, 0), \qquad \bar{\mathbf{b}}(k, t) = e^{4ik^2 t}\bar{\mathbf{b}}(k, 0). \tag{4.2.76b}$$

From (4.2.76a) it is clear that the eigenvalues (i.e., the zeros of $a(k) \equiv \det \mathbf{a}(k)$ and $\bar{a}(k) \equiv \bar{\mathbf{a}}(k)$) are constant as the solutions evolve. Not only the number of eigenvalues, but also their locations, are fixed. Thus, the eigenvalues are time-independent discrete states of the evolution. On the other hand, the evolution of the reflection coefficients (4.2.33) is given, like in the scalar case, by

$$\rho(k, t) = \rho(k, 0)e^{-4ik^2 t}, \qquad \bar{\rho}(k, t) = \bar{\rho}(k, 0)e^{4ik^2 t}, \tag{4.2.77}$$

and this also gives the evolution of the norming constants,

$$\mathbf{C}_j(t) = \mathbf{C}_j(0)e^{-4ik_j^2 t}, \qquad \bar{\mathbf{C}}_j(t) = \bar{\mathbf{C}}_j(0)e^{4i\bar{k}_j^2 t}. \tag{4.2.78}$$

4.3 Soliton solutions

4.3.1 One-soliton solution

In the case where the scattering data comprise proper eigenvalues but $\rho(\xi) = \bar{\rho}(\xi) = 0$ for $\xi \in \mathbb{R}$, the algebraic–integral system (4.2.56a)–(4.2.56d) reduces to the linear algebraic system

$$\bar{\mathbf{N}}_l(x) = \begin{pmatrix} \mathbf{I}_N \\ 0 \end{pmatrix} + \sum_{j=1}^{J} \frac{e^{2ik_j x}}{(\bar{k}_l - k_j)} \mathbf{N}_j(x)\mathbf{C}_j \qquad l = 1, \dots, \bar{J} \tag{4.3.79a}$$

$$\mathbf{N}_j(x) = \begin{pmatrix} 0 \\ \mathbf{I}_M \end{pmatrix} + \sum_{m=1}^{\bar{J}} \frac{e^{-2i\bar{k}_m x}}{(k_j - \bar{k}_m)} \bar{\mathbf{N}}_m(x)\bar{\mathbf{C}}_m \qquad j = 1, \dots, J. \tag{4.3.79b}$$

The one-soliton solution is obtained for $J = \bar{J} = 1$. In the relevant physical case, when the potentials satisfy the symmetry $\mathbf{R} = -\mathbf{Q}^H$, taking into account (4.2.45) and (4.2.46), we get

$$\bar{\mathbf{N}}_1(x) = \left[\mathbf{I}_N + \frac{e^{-4\eta x}}{4\eta^2}\mathbf{C}_1^H \mathbf{C}_1\right]^{-1} \begin{pmatrix} \mathbf{I}_N \\ -\frac{e^{2ik_1 x}}{2i\eta}\mathbf{C}_1 \end{pmatrix} \tag{4.3.80}$$

$$\mathbf{N}_1(x) = \begin{pmatrix} 0 \\ \mathbf{I}_M \end{pmatrix} - \frac{e^{-2ik_1^* x}}{2i\eta} \left[\mathbf{I}_N + \frac{e^{-4\eta x}}{4\eta^2}\mathbf{C}_1^H \mathbf{C}_1\right]^{-1} \begin{pmatrix} \mathbf{C}_1^H \\ -\frac{e^{2ik_1 x}}{2i\eta}\mathbf{C}_1 \mathbf{C}_1^H \end{pmatrix}, \tag{4.3.81}$$

where $k_1 = \xi + i\eta$, $\eta > 0$. The vector case is obtained for $N = 1$ in which case the matrix $\mathbf{C}_1$ reduces to an M-component column vector, while $\bar{\mathbf{C}}_1 \equiv -\mathbf{C}_1^H$.

From (4.2.57b) with $\mathbf{Q} = \mathbf{q}$, it then follows that

$$\mathbf{q}(x) = -2i\eta e^{-2i\xi x}\text{sech}(2\eta x - 2\delta)\frac{\mathbf{C}_1^H}{\|\mathbf{C}_1\|}, \qquad (4.3.82)$$

where

$$e^{2\delta} = \frac{\|\mathbf{C}_1\|}{2\eta}, \qquad \|\mathbf{C}_1\|^2 = \mathbf{C}_1^H\mathbf{C}_1. \qquad (4.3.83)$$

Introducing the explicit time-dependence of $\mathbf{C}_1$ as given by (4.2.78), one finally gets the one-soliton solution of the VNLS equation,

$$\mathbf{q}(x, t) = \mathbf{p}\, 2\eta\, e^{-2i\xi x + 4i(\xi^2 - \eta^2)t - i\pi/2}\text{sech}(2\eta x - 8\xi\eta t - 2\delta_0), \qquad (4.3.84)$$

where

$$e^{2\delta_0} = \frac{\|\mathbf{C}_1(0)\|}{2\eta}, \qquad \mathbf{p} = \frac{\mathbf{C}_1^H(0)}{\|\mathbf{C}_1(0)\|}. \qquad (4.3.85)$$

It is evident that the one-soliton solution of the VNLS is of the form

$$\mathbf{q}(x, t) = \mathbf{p}\, q(x, t), \qquad (4.3.86)$$

where q is the one-soliton solution of the scalar NLS (cf. (2.3.88)). Thus an individual soliton of the VNLS is fundamentally governed by the scalar NLS. However, the vector soliton (4.3.84) is also characterized by a polarization, that is, the vector $\mathbf{p}$, and the vector nature of the solution affects the dynamics when solitons with different polarizations interact. These interactions are described in Section 4.3.4.

4.3.2 Transmission coefficients for the pure one-soliton potential

To obtain the transmission coefficients relative to the pure one-soliton solution corresponding to the pairs $\{k_1, \mathbf{C}_1\}$ and $\{\bar{k}_1, \bar{\mathbf{C}}_1\}$ of a discrete eigenvalue/norming constant, we insert (4.2.57a)–(4.2.57b) for $J = \bar{J} = 1$ and $\rho(\xi) = \bar{\rho}(\xi) = 0$ for any real ξ into (4.2.27a) and (4.2.27c) and use (4.2.56a)–(4.2.56b) to get

$$\bar{\mathbf{c}}_1(k) = \mathbf{I}_N - \frac{2i}{k - k_1}\int_{-\infty}^{+\infty} e^{2i(k_1 - \bar{k}_1)x}\bar{\mathbf{N}}_1^{(\text{up})}(x)\bar{\mathbf{C}}_1\mathbf{N}_1^{(\text{dn})}(x)\mathbf{C}_1 dx \qquad (4.3.87)$$

$$\mathbf{c}_1(k) = \mathbf{I}_M + \frac{2i}{k - \bar{k}_1}\int_{-\infty}^{+\infty} e^{2i(k_1 - \bar{k}_1)x}\mathbf{N}_1^{(\text{dn})}(x)\mathbf{C}_1\bar{\mathbf{N}}_1^{(\text{up})}(x)\bar{\mathbf{C}}_1 dx. \qquad (4.3.88)$$

For the sake of simplicity we now restrict ourselves to the case of the M-component VNLS equation, that is, we take $N = 1$, and $\mathbf{r} = -\mathbf{q}^H$. Then $\mathbf{C}_1$ is an M-component column vector, $\bar{\mathbf{C}}_1 = -\mathbf{C}_1^H$, and therefore $\bar{\mathbf{c}}_1(k)$ reduces to a scalar, while $\mathbf{c}_1(k)$ is an $M \times M$ matrix. Taking into account the symmetry (4.2.45) and substituting (4.3.80)–(4.3.81), we find that

$$\mathbf{c}_1(k) = \mathbf{I}_M - \frac{k_1 - k_1^*}{k - k_1^*} \frac{1}{\|\mathbf{C}_1\|^2} \mathbf{C}_1 \mathbf{C}_1^H \qquad (4.3.89a)$$

$$\bar{c}_1(k) = \bar{c}_1(k) = \frac{k - k_1^*}{k - k_1}. \qquad (4.3.89b)$$

From the symmetries (4.2.42a)–(4.2.42b) it then follows that

$$\mathbf{a}_1(k) = a(k) = \frac{k - k_1}{k - k_1^*} \qquad (4.3.89c)$$

$$\bar{\mathbf{a}}_1(k) = \mathbf{I}_M + \frac{k_1 - k_1^*}{k - k_1} \frac{1}{\|\mathbf{C}_1\|^2} \mathbf{C}_1 \mathbf{C}_1^H. \qquad (4.3.89d)$$

Note that, in terms of the unit polarization vector (4.3.85), equations (4.3.89a) and (4.3.89d) can be written as

$$\mathbf{c}_1(k) = \mathbf{I}_M - \frac{k_1 - k_1^*}{k - k_1^*} \mathbf{p}_1^H \mathbf{p}_1 \qquad (4.3.90a)$$

$$\bar{\mathbf{a}}_1(k) = \mathbf{I}_M + \frac{k_1 - k_1^*}{k - k_1} \mathbf{p}_1^H \mathbf{p}_1. \qquad (4.3.90b)$$

4.3.3 Vector symmetry

The vector equation (4.2.5) reduces to the NLS (2.1.1) under the reduction

$$\mathbf{q}(x, t) = \mathbf{p} \, q(x, t), \qquad (4.3.91)$$

where $\mathbf{p} = \left(p^{(1)}, \ldots, p^{(M)} \right)$ is a constant M-component vector such that

$$\|\mathbf{p}\|^2 = |p^{(1)}|^2 + \cdots + |p^{(M)}|^2 = 1 \qquad (4.3.92)$$

and q is a scalar function of x and t. As a consequence, any solution of the scalar NLS generates a family of solutions of the VNLS parametrized by the choice of $\mathbf{p}$, to which we refer as reduction solutions. The vector $\mathbf{p}$ is usually called the polarization of the reduction solution.

The fact that the polarization of a reduction solution is arbitrary (up to the restriction $\|\mathbf{p}\|^2 = 1$) is a manifestation of a symmetry of the VNLS. The dependent variable transformation

$$\tilde{\mathbf{q}} = \mathbf{q}\mathbf{U}, \qquad (4.3.93)$$

where $\mathbf{U}$ is any unitary matrix, leaves the VNLS invariant. Thus, if $\mathbf{q}$ is a reduction solution of the form (4.3.91), then $\tilde{\mathbf{q}}$ is also a reduction solution with $\tilde{\mathbf{p}} = \mathbf{p}\mathbf{U}$ and $\tilde{q} = q$. The transformation (4.3.93) can be interpreted as a change in the basis of the vector space of the dependent variable. For any reduction solution, the polarization depends entirely on the choice of the basis, and by changing the basis via transformations of the form (4.3.93) one can arbitrarily change the polarization.

We refer to the squared absolute value of a component of the polarization as the intensity of that component. By definition, the sum of the intensities, that is, the total intensity, is 1. However, the distribution of intensity among the components of the vector of any given reduction solution depends on the choice of the polarization basis.

4.3.4 Vector–soliton interactions

We consider the solution of the VNLS ($\mathbf{r} = -\mathbf{q}^H$ and $N = 1$) corresponding to the scattering data

$$\left\{ k_j = \xi_j + i\eta_j, \ \eta_j > 0, \ \mathbf{C}_j \right\}_{j=1}^J \cup \left\{ \rho(\xi) = 0 \text{ for } \xi \in \mathbb{R} \right\}.$$

To get the pure J-soliton solution, one can in principle solve the linear system (4.3.79a)–(4.3.79b). The problem of a J-soliton collision can be investigated by looking at the asymptotic states as $t \to \pm\infty$ (cf. [120]). For a generic multiple-soliton solution of the VNLS, with the individual solitons traveling at different velocities, in the backward ($t \to -\infty$) and forward ($t \to +\infty$) long-time limits such a solution asymptotically breaks up into individual solitons,

$$\mathbf{q}^\pm \sim \sum_{j=1}^J \mathbf{p}_j^\pm \, q_j^\pm \qquad t \to \pm\infty, \tag{4.3.94}$$

where $\mathbf{p}_j^\pm$ is a complex unit vector and $q_j^\pm$ is a one-soliton solution of the NLS such that q_j^- and q_j^+ are characterized by the same amplitude and velocity. Let us fix the values of the soliton parameters for $t \to -\infty$, that is, for each eigenvalue $k_j = \xi_j + i\eta_j$ we assign a vector $\mathbf{S}_j^-$ (see (4.3.84)–(4.3.85)) that completely determines $\mathbf{q}_j^-$. For $t \to +\infty$ we denote the corresponding vectors by $\mathbf{S}_j^+$. For definiteness, let us assume that $\xi_1 < \xi_2 < \cdots < \xi_J$. Then, as $t \to -\infty$, the solitons are distributed along the x-axis in the order corresponding to $\xi_J, \xi_{J-1}, \ldots, \xi_1$; the order of the soliton sequence is reversed as $t \to +\infty$. To determine the result of the interaction among solitons, that is, to calculate $\mathbf{S}_j^+$ given $\mathbf{S}_j^-$, we trace the passage of the eigenfunctions through the asymptotic states [120]. We denote the "soliton coordinates" (i.e., the center of the solitons) at the instant of time t by $x_j(t)$ ($|t|$ is assumed large enough so that one can talk about individual solitons). If $t \to -\infty$, then $x_J \ll x_{J-1} \ll \cdots \ll x_1$. The function $\phi(x, k_j)$ has the form

$$\phi(x, k_j) \sim e^{-ik_j x} \begin{pmatrix} \mathbf{I}_N \\ \mathbf{0} \end{pmatrix} \qquad x \ll x_J.$$

After passing through the J-th soliton it will become

$$\phi(x, k_j) \sim e^{-ik_j x} \begin{pmatrix} \mathbf{I}_N \\ \mathbf{0} \end{pmatrix} \mathbf{a}_J(k_j),$$

where $\mathbf{a}_J(k)$ is the transmission coefficient relative to the J-th soliton. Repeating the argument (see Appendix B for details), we find

$$\phi(x, k_j) \sim e^{-ik_j x} \begin{pmatrix} \mathbf{I}_N \\ \mathbf{0} \end{pmatrix} \prod_{\substack{l=j+1 \\ \text{right}}}^{J} \mathbf{a}_l(k_j) \qquad x_{j+1} \ll x \ll x_j,$$

where the notation "right" indicates that the product is performed such that the matrix with index ℓ occurs to the right of the matrix with index $\ell - 1$, that is,

$$\prod_{\substack{l=j+1 \\ \text{right}}}^{J} \mathbf{a}_l(k_j) = \mathbf{a}_{j+1}(k_j)\mathbf{a}_{j+2}(k_j)\dots \mathbf{a}_J(k_j).$$

Upon passing through the j-th soliton, since the corresponding state is a bound state, we get

$$\phi(x, k_j) \sim e^{ik_j x} \begin{pmatrix} \mathbf{0} \\ \mathbf{I}_M \end{pmatrix} \mathbf{S}_j^- \prod_{\substack{l=j+1 \\ \text{right}}}^{J} \mathbf{a}_l(k_j) \qquad x_j \ll x \ll x_{j-1}. \tag{4.3.95}$$

On the other hand, starting from $x \gg x_1$ and proceeding in a similar way, we find for the eigenfunction ψ the following asymptotic behavior:

$$\psi(x, k_j) \sim e^{ik_j x} \begin{pmatrix} \mathbf{0} \\ \mathbf{I}_M \end{pmatrix} \prod_{\substack{l=1 \\ \text{left}}}^{j-1} \mathbf{c}_l(k_j, \mathbf{S}_l^-) \qquad x_j \ll x \ll x_{j-1}, \tag{4.3.96}$$

where the additional argument in $\mathbf{c}_l$ is to take into account the dependence on the norming constant or, equivalently, on the polarization vector (cf. (4.3.89a), (4.3.90a)). The notation "left" indicates that the product is performed such that the matrix with index ℓ occurs to the left of the matrix with index $\ell - 1$, that is,

$$\prod_{\substack{l=1 \\ \text{left}}}^{j-1} \mathbf{c}_l(k_j, \mathbf{S}_l^-) = \mathbf{c}_{j-1}(k_j, \mathbf{S}_{j-1}^-)\mathbf{c}_{j-2}(k_j, \mathbf{S}_{j-2}^-)\dots \mathbf{c}_1(k_j, \mathbf{S}_1^-).$$

Note that we could use (4.2.23a) and the symmetry (4.2.42a) relating $\mathbf{c}(k)$ to $\bar{\mathbf{a}}^H(k^*)$ to express (4.3.96) in terms of "left" scattering data.

Since we restrict ourselves to the case $N = 1$, the $\mathbf{a}_l(k_j)$ are scalars and therefore we write them simply as $a_l(k_j)$; moreover, equation (4.2.35) yields

$$\phi(x, k_j) = \psi(x, k_j)\mathbf{C}_j\, a'(k_j).$$

Comparing (4.3.95) and (4.3.96), we get

$$\prod_{l=j+1}^{J} a_l(k_j)\, \mathbf{S}_j^- = \prod_{\substack{l=1 \\ \text{left}}}^{j-1} \mathbf{c}_l(k_j, \mathbf{S}_l^-)\, \mathbf{C}_j\, a'(k_j), \tag{4.3.97}$$

and therefore

$$\mathbf{C}_j(t)\, a'(k_j) \sim \prod_{l=j+1}^{J} a_l(k_j) \prod_{\substack{l=1 \\ \text{right}}}^{j-1} \left(\mathbf{c}_l(k_j, \mathbf{S}_l^-)\right)^{-1} \mathbf{S}_j^- \qquad t \to -\infty. \tag{4.3.98a}$$

Proceeding in a similar fashion as $t \to +\infty$ and taking into account that the order of the soliton sequence is reversed, we get

$$\mathbf{C}_j(t)\, a'(k_j) \sim \prod_{l=1}^{j-1} a_l(k_j) \prod_{\substack{l=j+1 \\ \text{left}}}^{J} \left(\mathbf{c}_l(k_j, \mathbf{S}_l^+)\right)^{-1} \mathbf{S}_j^+ \qquad t \to +\infty. \tag{4.3.98b}$$

Taking into account (4.2.78), from these two representations (4.3.98a)–(4.3.98b) for $\mathbf{C}_j(t)$ we obtain

$$\mathbf{S}_J^+ = \prod_{l=1}^{J-1} \frac{1}{a_l(k_J)} \prod_{\substack{l=1 \\ \text{right}}}^{J-1} \left(\mathbf{c}_l(k_J, \mathbf{S}_l^-)\right)^{-1} \mathbf{S}_J^- \tag{4.3.99a}$$

$$\mathbf{S}_j^+ = \prod_{l=1}^{j-1} \frac{1}{a_l(k_j)} \prod_{l=j+1}^{J} a_l(k_j) \prod_{\substack{l=j+1 \\ \text{right}}}^{J} \mathbf{c}_l(k_j, \mathbf{S}_l^+) \prod_{\substack{l=1 \\ \text{right}}}^{j+1} \left(\mathbf{c}_l(k_j, \mathbf{S}_l^-)\right)^{-1} \mathbf{S}_j^-$$

$$j = 1, \ldots, J - 1. \tag{4.3.99b}$$

The relations (4.3.99a) and (4.3.99b) solve the problem of a J–soliton collision. Indeed, given $\mathbf{S}_l^-$ for $l = 1, \ldots, J$, (4.3.99a) allows one to find $\mathbf{S}_J^+$, and since the expression (4.3.99b) for $\mathbf{S}_j^+$ with $j < J$ depends, through the $\mathbf{c}$'s, on $\mathbf{S}_l^+$ for $l > j$, one can find iteratively $\mathbf{S}_j^+$ for any j.

Two-soliton interaction

Let us consider the interaction of two solitons in more detail. In this case, from (4.3.99a)–(4.3.99b) it follows that

$$\mathbf{S}_2^+ = \frac{1}{a_1(k_2)} \left(\mathbf{c}_1(k_2, \mathbf{S}_1^-)\right)^{-1} \mathbf{S}_2^- \tag{4.3.100a}$$

$$\mathbf{S}_1^+ = a_2(k_1)\, \mathbf{c}_2(k_1, \mathbf{S}_2^+)\, \mathbf{S}_1^-. \tag{4.3.100b}$$

Note that the formulas (4.3.100a)–(4.3.100b) are not symmetric with respect to the exchange of the subscripts 1 and 2. However, such a notation expresses the invariance of the system under the substitution $t \to -t$, $\mathbf{q} \to \mathbf{q}^*$ (here a "fast" soliton becomes a "slow" soliton, and vice versa). Taking into account the explicit expressions for a_j and $\mathbf{c}_j$ in the pure soliton case, given by (4.3.89c) and (4.3.98a), one can solve (4.3.100a) for $\mathbf{S}_2^+$ and then substitute it into the right-hand side of (4.3.100b) to get $\mathbf{S}_1^+$. Indeed, let us introduce, according to (4.3.85) and (4.3.94),

$$\mathbf{p}_j^\pm = \frac{\left(\mathbf{S}_j^\pm\right)^H}{\left\|\mathbf{S}_j^\pm\right\|} \qquad j = 1, 2. \tag{4.3.101}$$

Then, using (4.3.89c), (4.3.90a), we obtain

$$a_j(k) = \frac{k - k_j}{k - k_j^*} \tag{4.3.102a}$$

$$\mathbf{c}_j(k, \mathbf{p}_j) = \mathbf{I}_M - \frac{k_j - k_j^*}{k - k_j^*}\mathbf{p}_j^H \mathbf{p}_j \tag{4.3.102b}$$

$$\mathbf{c}_j^{-1}(k, \mathbf{p}_j) = \mathbf{I}_M + \frac{k_j - k_j^*}{k - k_j}\mathbf{p}_j^H \mathbf{p}_j. \tag{4.3.102c}$$

It is convenient to define

$$\chi^2 = \frac{\left\|\mathbf{S}_2^+\right\|^2}{\left\|\mathbf{S}_2^-\right\|^2} \equiv \frac{1}{\left\|\mathbf{S}_2^-\right\|^2}\left(\mathbf{S}_2^+\right)^H \mathbf{S}_2^+, \tag{4.3.103}$$

which, according to (4.3.100a), is given by

$$\chi^2 = |a_1(k_2)|^{-2}\, \mathbf{p}_2^-\left(\mathbf{c}_1^{-1}(k_1, \mathbf{p}_1^-)\right)^H \mathbf{c}_1^{-1}(k_1, \mathbf{p}_1^-)\mathbf{p}_2^{-H}. \tag{4.3.104}$$

Inserting (4.3.102a) and (4.3.102c), we obtain

$$\chi^2 = \left|\frac{k_1 - k_2^*}{k_1 - k_2}\right|^2\left[1 + \frac{\left(k_1^* - k_1\right)\left(k_2 - k_2^*\right)}{|k_1 - k_2|^2}\left|\mathbf{p}_1^{-*}\cdot\mathbf{p}_2^-\right|^2\right], \tag{4.3.105}$$

and one can check that also

$$\chi^2 = \frac{\left\|\mathbf{S}_1^-\right\|^2}{\left\|\mathbf{S}_1^+\right\|^2}.$$

From (4.3.100a)–(4.3.100b) it then follows that

$$\mathbf{p}_2^+ = \frac{1}{\chi}\frac{1}{a_1^*(k_2)}\mathbf{p}_2^-\left(\mathbf{c}_1^{-1}(k_2, \mathbf{p}_1^-)\right)^H \tag{4.3.106a}$$

$$\mathbf{p}_1^+ = \chi\, a_2^*(k_1)\mathbf{p}_1^-\left(\mathbf{c}_2(k_1, \mathbf{p}_2^+)\right)^H; \tag{4.3.106b}$$

and, according to the explicit expressions (4.3.102a)–(4.3.102c) for the trans-mission coefficients, the unit polarization vectors after the interaction, $\mathbf{p}_j^+$, $j = 1, 2$, can be expressed in terms of the parameters characterizing the collid-ing solitons as follows:

$$\mathbf{p}_2^+ = \frac{1}{\chi} \frac{k_1 - k_2^*}{k_1^* - k_2^*} \left[\mathbf{p}_2^- + \frac{k_1^* - k_1}{k_2^* - k_1^*} \left(\mathbf{p}_1^{-*} \cdot \mathbf{p}_2^- \right) \mathbf{p}_1^- \right] \qquad (4.3.107a)$$

$$\mathbf{p}_1^+ = \frac{1}{\chi} \frac{k_1 - k_2^*}{k_1 - k_2} \left[\mathbf{p}_1^- + \frac{k_2^* - k_2}{k_2 - k_1} \left(\mathbf{p}_2^{-*} \cdot \mathbf{p}_1^- \right) \mathbf{p}_2^- \right], \qquad (4.3.107b)$$

with χ given by (4.3.105).

Note that, as in the scalar case, the centers of the solitons shift during the collision process. However, in contrast to the scalar soliton interaction, these shifts depend both on k_1, k_2 and on the polarization vectors $\mathbf{p}_1^-$, $\mathbf{p}_2^-$. Indeed, if we denote by $\delta_1^\pm$, $\delta_2^\pm$ the centers of the two solitons before/after the interaction, from (4.3.100a)–(4.3.100b) it follows that

$$e^{2\delta_1^\pm} = \frac{\|\mathbf{S}_1^\pm\|}{2\eta_1} \qquad e^{2\delta_2^\pm} = \frac{\|\mathbf{S}_2^\pm\|}{2\eta_2};$$

hence

$$e^{2(\delta_2^+ - \delta_2^-)} = \frac{\|\mathbf{S}_2^+\|}{\|\mathbf{S}_2^-\|} \equiv \chi \qquad e^{2(\delta_1^+ - \delta_1^-)} = \frac{\|\mathbf{S}_1^+\|}{\|\mathbf{S}_1^-\|} \equiv \frac{1}{\chi}.$$

As a consequence, the relative separation $x_{1,2}^\pm$ between the soliton centers also varies during the collision. Precisely,

$$x_{1,2}^- = \frac{\delta_2^-}{\eta_2} - \frac{\delta_1^-}{\eta_1} \equiv \frac{2\delta_2^-}{i \left(k_2 - k_2^* \right)} - \frac{2\delta_1^-}{i \left(k_1 - k_1^* \right)}$$

$$x_{1,2}^+ = \frac{\delta_2^+}{\eta_2} - \frac{\delta_1^+}{\eta_1} \equiv \frac{2\delta_2^+}{i \left(k_2 - k_2^* \right)} - \frac{2\delta_1^+}{i \left(k_1 - k_1^* \right)}$$

and

$$\Delta x = x_{1,2}^+ - x_{1,2}^- = \frac{2 \left(\delta_2^+ - \delta_2^- \right)}{i \left(k_2 - k_2^* \right)} - \frac{2 \left(\delta_1^+ - \delta_1^- \right)}{i \left(k_1 - k_1^* \right)} = \frac{\eta_1 + \eta_2}{2\eta_1\eta_2} \log \chi,$$

with χ given by (4.3.105) and $k_j = \xi_j + i\eta_j$ for $j = 1, 2$.

In general, when the solitons interact the intensity distributions of the in-dividual solitons change. Only in the case when $\mathbf{p}_2^- = e^{i\theta}\mathbf{p}_1^-$ and in the case when $\mathbf{p}_1^{-*} \cdot \mathbf{p}_2^- = 0$, that is, when the soliton polarizations are either parallel or orthogonal, is the amplitude of each individual component of the solitons conserved.

Indeed, without loss of generality, one can parametrize the polarization vectors of the solitons as

$$\mathbf{p}_1^- = (1,\ 0)\,, \qquad \mathbf{p}_2^- = \left(\rho e^{i\varphi},\ \sqrt{1-\rho^2}\right) \qquad (4.3.108)$$

with $0 \le \rho \le 1$ and $0 \le \varphi \le 2\pi$. Then, from (4.3.107a)–(4.3.107b), the polarization vectors after the interaction are such that, for instance,

$$\left|p_1^{+\,(1)}\right|^2 = \frac{1}{\chi^2}\left|\frac{k_1 - k_2^*}{k_1 - k_2}\right|^2 \left|1 + \frac{k_2^* - k_2}{k_2 - k_1}\rho^2\right|^2 \qquad (4.3.109a)$$

$$\left|p_2^{+\,(1)}\right|^2 = \frac{1}{\chi^2}\left|\frac{k_1 - k_2^*}{k_1 - k_2}\right|^4 \rho^2 \qquad (4.3.109b)$$

$$\chi^2 = \left|\frac{k_1 - k_2^*}{k_1 - k_2}\right|^2 \left[1 + \frac{\left(k_1^* - k_1\right)\left(k_2 - k_2^*\right)}{|k_1 - k_2|^2}\rho^2\right], \qquad (4.3.109c)$$

where, as usual, $p_1^{+\,(j)}$ for $j = 1, 2$ denotes the j-th component of the corresponding polarization vector.

In the polarization basis (4.3.108), we always have $|p_1^{-\,(1)}|^2 = 1$; thus the quantity $|p_1^{+\,(1)}|^2$ measures the shift in the intensity of soliton 1 that results from the interaction of the solitons. Specifically, $0 \le |p_1^{+\,(1)}|^2 \le 1$, where $|p_1^{+\,(1)}|^2 = 1$ indicates no change in the intensity distribution of soliton 1 and $|p_1^{+\,(1)}|^2 = 0$ indicates that all the intensity of soliton 1 is shifted to an orthogonal polarization by the soliton interaction. In Figure 4.1 we plot $|p_j^{+\,(1)}|^2$ for $j = 1, 2$ as functions of ρ^2 for a specific choice of the eigenvalues.

Multisoliton interactions

Take $J = 3$ solitons and assume

$$v_1 < v_2 < v_3, \qquad (4.3.110)$$

so that the solitons are distributed along the x-axis according to $3 - 2 - 1$ as $t \to -\infty$ and $1 - 2 - 3$ as $t \to \infty$. The formulas for a three-soliton interaction, under the assumption (4.3.110) for the velocities, are given by (4.3.99a)–(4.3.99b):

$$S_3^+ = (a_1(k_3)a_2(k_3))^{-1}\left(\mathbf{c}_1(k_3, S_1^-)\right)^{-1}\left(\mathbf{c}_2(k_3, S_2^-)\right)^{-1} S_3^- \qquad (4.3.111a)$$

$$S_2^+ = \frac{a_3(k_2)}{a_1(k_2)}\mathbf{c}_3(k_2, S_3^+)\left(\mathbf{c}_1(k_2, S_1^-)\right)^{-1} S_2^- \qquad (4.3.111b)$$

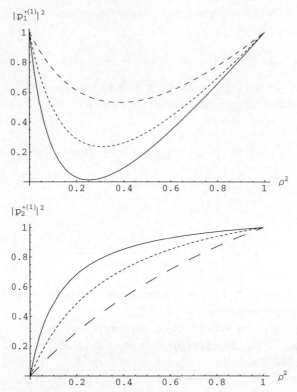

Figure 4.1: Intensity shift induced by two-soliton interaction of the two-component VNLS. The soliton parameters are $k_j = -b_j + ia_j$ with $a_1 = 1$, $a_2 = 2$, $b_1 = \Delta b$, $b_2 = -\Delta b$ and $\Delta b = 1$ (dashed), $\Delta b = 0.5$ (dotted), $\Delta b = 0.1$ (solid). The initial polarizations are given by (4.3.108).

$$\mathbf{S}_1^+ = a_2(k_1)a_3(k_1)\, \mathbf{c}_2(k_1, \mathbf{S}_2^+)\, \mathbf{c}_3(k_1, \mathbf{S}_3^+)\, \mathbf{S}_1^-. \qquad (4.3.111c)$$

According to equations (4.3.100a)–(4.3.100b) for a pairwise interaction, (4.3.111a)–(4.3.111c) can be looked at in terms of compositions of pairwise interactions, namely: (4.3.111a) corresponds to 3 interacting first with 2 and then with 1 (eventually 2 will interact with 1, but this is not relevant as far as $\mathbf{S}_3^+$ is concerned); (4.3.111b) corresponds to 2 interacting first with 1 and then with 3 (but since in $\mathbf{c}_3$ we have $\mathbf{S}_3^+$, this means, as it should be from a physical point of view, that before $2 \leftrightarrow 3$, the interaction $3 \leftrightarrow 1$ has to take place); (4.3.111b) corresponds to 3 interacting with 2 (because again in $\mathbf{c}_3$ we have $\mathbf{S}_3^+$), then 1 interacting with 3, and finally 1 interacting with 2.

To clarify these statements, let us denote by $\mathbf{S}_{\{j,\ell\}}$ the polarization of soliton j after the interaction with soliton ℓ. Note that, according to (4.3.100a)–(4.3.100b), $\mathbf{S}_{\{j,\ell\}} \neq \mathbf{S}_{\{\ell,j\}}$. Adopting this notation, and dropping the superscript $-$ for the polarization of the incoming solitons, that is, taking $\mathbf{S}_j^- \to \mathbf{S}_j$, the expression (4.3.111a) for the polarization of soliton 3 corresponds to

$$\mathbf{S}_3^+ = (\mathbf{c}_1(k_3, \mathbf{S}_1))^{-1} \mathbf{S}_{\{3,2\}} (a_1(k_3))^{-1} \equiv \mathbf{S}_{\{\{3,2\},1\}}, \qquad (4.3.112)$$

that is, to the composition of pairwise interactions $3 \leftrightarrow 2$, $3 \leftrightarrow 1$ (eventually, $2 \leftrightarrow 1$ will follow). On the other hand, depending on the initial positions and relative velocities of the solitons, one could have first 2 interacting with 1, giving $\mathbf{S}_{\{2,1\}}$ for soliton 2 and $\mathbf{S}_{\{1,2\}}$ for soliton 1; then 3 interacting with 1, giving $\mathbf{S}_{\{3,\{1,2\}\}}$ and $\mathbf{S}_{\{\{1,2\},3\}}$; and, finally, 3 interacting with 2. In this case, the final polarization of soliton 3 is denoted

$$\mathbf{S}_3^+ = \mathbf{S}_{\{\{3,\{1,2\}\},\{2,1\}\}}.$$

However, one can show that the final result is independent of the order of interaction, that is,

$$\mathbf{S}_{\{\{3,2\},1\}} = \mathbf{S}_{\{\{3,\{1,2\}\},\{2,1\}\}}. \qquad (4.3.113)$$

Because

$$\mathbf{S}_{\{3,\{1,2\}\}} = \big(\mathbf{c}_1(k_3, \mathbf{S}_{\{1,2\}})\big)^{-1} \mathbf{S}_3 (a_1(k_3))^{-1} \qquad (4.3.114)$$

and

$$\mathbf{S}_{\{\{3,\{1,2\}\},\{2,1\}\}} = \big(\mathbf{c}_2(k_3, \mathbf{S}_{\{2,1\}})\big)^{-1} \mathbf{S}_{\{3,\{1,2\}\}} (a_2(k_3))^{-1}, \qquad (4.3.115)$$

the condition (4.3.113) corresponds to

$$\mathbf{c}_2(k_3, \mathbf{S}_1)\mathbf{c}_1(k_3, \mathbf{S}_2) = \mathbf{c}_1(k_3, \mathbf{S}_{\{1,2\}})\mathbf{c}_2(k_3, \mathbf{S}_{\{2,1\}}) \qquad (4.3.116)$$

or, equivalently,

$$\mathbf{c}_2(k_3, \mathbf{p}_1)\mathbf{c}_1(k_3, \mathbf{p}_2) = \mathbf{c}_1(k_3, \mathbf{p}_{\{1,2\}})\mathbf{c}_2(k_3, \mathbf{p}_{\{2,1\}}). \qquad (4.3.117)$$

Taking, without loss of generality,

$$\mathbf{S}_1 = \begin{pmatrix} 1 \\ 0 \end{pmatrix}, \quad \mathbf{S}_2 = \begin{pmatrix} \rho e^{-i\varphi} \\ \sqrt{1-\rho^2} \end{pmatrix}, \qquad (4.3.118)$$

that is,

$$\mathbf{p}_1 = (1, 0), \quad \mathbf{p}_2 = \left(\rho e^{i\varphi}, \sqrt{1-\rho^2}\right),$$

we obtain from (4.3.100a)–(4.3.100b)

$$\mathbf{p}^H_{\{2,1\}} = \frac{\mathbf{S}_{\{2,1\}}}{||\mathbf{S}_{\{2,1\}}||} = \frac{1}{\chi}\frac{k_1^* - k_2}{k_1 - k_2}\left(\begin{array}{c} \frac{k_1^* - k_2}{k_1 - k_2}\rho e^{-i\varphi} \\ \sqrt{1 - \rho^2} \end{array}\right) \qquad (4.3.119a)$$

$$\mathbf{p}^H_{\{1,2\}} = \frac{\mathbf{S}_{\{1,2\}}}{||\mathbf{S}_{\{1,2\}}||} = \frac{1}{\chi}\frac{k_1^* - k_2}{k_1^* - k_2^*}\left(\begin{array}{c} 1 + \frac{k_2 - k_2^*}{k_2 - k_1^*}\rho^2 \\ \frac{k_2 - k_2^*}{k_2 - k_1^*}\rho\sqrt{1 - \rho^2}e^{i\varphi} \end{array}\right) \qquad (4.3.119b)$$

$$\chi^2 = \frac{(k_1 - k_2^*)(k_1^* - k_2)}{(k_1 - k_2)(k_1^* - k_2^*)}\tilde{\chi}^2 \qquad \tilde{\chi}^2 = 1 + \frac{(k_1^* - k_1)(k_2 - k_2^*)}{(k_1 - k_2)(k_1^* - k_2^*)}\rho^2.$$
$$(4.3.119c)$$

Then, using (4.3.102b) for $\mathbf{c}_j$ and (4.3.118) for $\mathbf{S}_1$ and $\mathbf{S}_2$, the left-hand side of (4.3.117) is given by

$$\mathbf{c}_2(k_3, \mathbf{p}_1)\mathbf{c}_1(k_3, \mathbf{p}_2) = \mathbf{I} - \frac{k_1 - k_1^*}{k_3 - k_1^*}\mathbf{p}_1^H\mathbf{p}_1 - \frac{k_2 - k_2^*}{k_3 - k_2^*}\mathbf{p}_2^H\mathbf{p}_2$$

$$+ \frac{(k_1 - k_1^*)(k_2 - k_2^*)}{(k_3 - k_1^*)(k_3 - k_2^*)}(\mathbf{p}_2^* \cdot \mathbf{p}_1)\mathbf{p}_1^H\mathbf{p}_2$$

$$= \mathbf{I} - \left(\begin{array}{cc} \frac{k_1 - k_1^*}{k_3 - k_1^*} + \frac{k_3 - k_1}{k_3 - k_1^*}\frac{k_2 - k_2^*}{k_3 - k_2^*}\rho^2 & \frac{k_2 - k_2^*}{k_3 - k_2^*}\rho\sqrt{1 - \rho^2}e^{-i\varphi} \\ \frac{k_2 - k_2^*}{k_3 - k_2^*}\frac{k_3 - k_1}{k_3 - k_1^*}\rho\sqrt{1 - \rho^2}e^{i\varphi} & \frac{k_2 - k_2^*}{k_3 - k_2^*}(1 - \rho^2) \end{array}\right).$$
$$(4.3.120)$$

On the other hand,

$$\mathbf{c}_1\left(k_3, \mathbf{p}_{\{1,2\}}\right) = \mathbf{I} - \frac{k_1 - k_1^*}{k_3 - k_1^*}\mathbf{p}^H_{\{1,2\}}\mathbf{p}_{\{1,2\}},$$

$$\mathbf{c}_2(k_3, \mathbf{p}_{\{2,1\}}) = \mathbf{I} - \frac{k_2 - k_2^*}{k_3 - k_2^*}\mathbf{p}^H_{\{2,1\}}\mathbf{p}_{\{2,1\}};$$

therefore

$$\mathbf{c}_1\left(k_3, \mathbf{p}_{\{1,2\}}\right)\mathbf{c}_2(k_3, \mathbf{p}_{\{2,1\}}) = \mathbf{I} - \frac{k_1 - k_1^*}{k_3 - k_1^*}\mathbf{p}^H_{\{1,2\}}\mathbf{p}_{\{1,2\}} - \frac{k_2 - k_2^*}{k_3 - k_2^*}\mathbf{p}^H_{\{2,1\}}\mathbf{p}_{\{2,1\}}$$

$$+ \frac{(k_1 - k_1^*)(k_2 - k_2^*)}{(k_3 - k_1^*)(k_3 - k_2^*)}(\mathbf{p}_{\{2,1\}}^* \cdot \mathbf{p}_{\{1,2\}})\mathbf{p}^H_{\{1,2\}}\mathbf{p}_{\{2,1\}}.$$
$$(4.3.121)$$

By using (4.3.119a)–(4.3.119b) for $\mathbf{p}_{\{1,2\}}$ and $\mathbf{p}_{\{2,1\}}$, one can show that indeed (4.3.121) coincides with (4.3.120). Hence, (4.3.117) implies that, as far as soliton 3 is concerned, the polarization shift is obtained via pairwise interactions and the result is independent of the order in which such interactions occur. In the same way one can check that the same result also holds for solitons 2 and 1.

The shift of the center of solition 3 due to the interaction with the other solitons is given by

$$\chi_3 = e^{2(\delta_3^+ - \delta_3^-)} = \frac{\|\mathbf{S}_3^+\|}{\|\mathbf{S}_3^-\|}.$$

Once can verify that this shift of the center of soliton 3 is the result of two consecutive pairwise interactions. Indeed, using the expression (4.3.106a)–(4.3.106b) for the pairwise collisions, one can verify that

$$\chi_3 = \chi_{\{3,2\}} \chi_{\{\{3,2\},1\}},$$

where we are using the notation $\{i, j\}$ to denote the interaction between the i-th and j-th solitons and $\{\{i, j\}, l\}$ the interaction between the i-th solition as it emerges from the collision with the j-th and the l-th soliton. Therefore $\chi_{\{i,j\}}$ is given by (4.3.105a) with $1 \to i$, $2 \to j$ and $i < j$. The same also can be shown for the shifts of centers of the other solitons.

Using these results, it follows by induction on equations (4.3.99a)–(4.3.99b), that a J-soliton collision is equivalent to the composition of $J(J - 1)/2$ pairwise interactions taking place in an arbitrary order (compatible with the choice for the soliton velocities). This is in agreement with the fact that the two-soliton polarization shift given by (4.3.107a)–(4.3.107b), is independent of the initial soliton "centers"$\delta_{0,j}$ in (4.3.84)–(4.3.85), which depend on the magnitude of $\|\mathbf{C}_j(0)\|$. Indeed, the fact that phase/polarization shifts in the three-soliton interaction are independent of the order of the pairwise soliton interactions implies that the phase/polarization shifts in the J-soliton case are similarly independent of the order of the pairwise soliton interactions. To see this from another perspective, we first consider the four-soliton case and then generalize.

We denote the slowest soliton as 1, the next slowest soliton as 2, et cetera. In the case of four solitons, the fastest soliton is denoted as 4, and, in the limit $t \to -\infty$, the solitons are arranged in the order 4321 from left to right. As time evolves, faster solitons overtake the slower solitons and shift their order through several intermediate steps (transpositions). In the limit $t \to +\infty$, the solitons are arranged in the order 1234. Prior to achieving this ultimate arrangement, the solitons can be in one of three possible arrangements, namely, 1243, 2134, or 1324. By pairwise comparison of these three arrangements, we show that the phase/polarization shifts of the solitons in limit $t \to +\infty$ (i.e., the solitons in the arrangement 1234) are independent of the penultimate order.

In both the arrangements 2134 and 1324, the fastest soliton is the rightmost soliton. Thus, in the evolution to the limit as $t \to +\infty$, this fastest soliton does not interact with the other three. We therefore can consider the arrangements 2134 and 1324 to be penultimate intermediate steps in the evolution of the solitons from the arrangement 3214 to the arrangement 1234 in the long-time limit. The only way for the solitons to evolve to the arrangement 3214 from the arrangement 4321 (the order in the limit as $t \to -\infty$) is the following: $4321 \to 3421 \to 3241 \to 3214$. Hence, the phase/polarization shifts of the solitons in the arrangement 3214 are unique. In the evolution of the solitons from the order 3214 to 1234, soliton 4 (the fastest soliton) is always the rightmost. In effect, the evolution from the arrangement 3214 to the long-time limit 1234 is a three-soliton interaction. In particular, this three-soliton interaction includes the evolution of both the arrangement 2134 and the arrangement 1324 to the long-time limit 1234. As was shown previously, for any three-soliton interaction, the shifts in the limit $t \to \infty$ are independent of the order of the transpositions. Therefore, the phase/polarization shifts in the long-time limit are the same whether the solitons evolve through the penultimate arrangement 2134 or 1324.

The arrangements 1243 and 2134 are both obtained from the arrangement 2143 by a single transposition: In the evolution $2143 \to 1243$, the two leftmost solitons interact; In the evolution $2143 \to 2134$, the two rightmost solitons interact. Similarly, both of these arrangements evolve to the arrangement 1234 as a result of a single transposition. Thus, the arrangements 1243 and 2134 are both intermediate steps in the evolution of the solitons from the arrangement 2143 to the arrangement 1234. The key fact is that the evolution from 2143 to 1234 is the result of two isolated transpositions: (i) the interchange of the two leftmost solitons and (ii) the interchange of the two rightmost solitons. Due to this isolation, the temporal order of these two interchanges must be irrelevant to the phase/polarization shifts of the solitons in the long-time limit. To see that the shifts of the solitons are unique when they are in the order 2143, we repeat the preceding argument. The evolution of the solitons from the arrangement 4321 (in the limit as $t \to -\infty$) to the order 2143 proceeds as follows: $4321 \to 4231 \to 4213 \to 2413 \to 2143$ or $4321 \to 4231 \to 2431 \to 2413 \to 2143$. Again, the solitons evolve from the arrangement 4231 to the arrangement 2413 by a pair of isolated two-soliton interactions. Hence, in the arrangement 2413 and in the uniquely following arrangement 2143, the phase/polarization shifts of the individual solitons are uniquely determined. This implies that the evolution from the penultimate arrangements 1243 and 2134 to the long-time limit 1234 result in the same ultimate phase shifts. Having shown that evolution from the penultimate arrangements 2134 and 1324 results in the same phase/polarization shifts in the limit as $t \to \infty$, we conclude that all three penultimate arrangements 1243, 2134, and 1324 result in the same phase/polarization shifts in the evolution of the solitons to the arrangement 1234 in the limit as $t \to +\infty$.

To generalize to the J-soliton case, we observe that the previous examples in fact include all possible cases. Consider any two arrangements of J solitons, denoted as B_1 and B_2, respectively, such that, for each of these arrangements, a single transposition of solitons (i.e., a faster soliton overtakes a slower soliton) results in a common arrangement, denoted as C. There are two possibilities: The respective transpositions by which arrangements B_1 and B_2 evolve to arrangement C involve either (i) two distinct pairs of solitons or (ii) two pairs that include a common soliton. Case (i) corresponds to the second case above. In this case, the soliton interactions consist of two distinct transpositions that occur in an isolated manner and are therefore insensitive to the temporal order in which they occur. Case (ii) is essentially a three-soliton interaction in which B_1 and B_2 are the two possible penultimate arrangements in the evolution to the arrangement C. As was shown previously, the phase/polarization shifts in this three-soliton interaction are independent of the order of interaction. In particular, the phase/polarization shifts in arrangement C are the same whether the penultimate arrangement is B_1 or B_2.

In general, for J solitons there can be up to $J - 1$ arrangements of solitons that evolve, each via the transposition of a single pair of solitons, to a common arrangement, C. However, when compared pairwise, these reduce to the two cases given above. Hence, each pair of arrangements can be seen as the penultimate step in the evolution from a common preceding arrangement, denoted as A, to the arrangement C. Therefore, in particular, the phases/polarizations of the soliton in arrangement C are independent of whether the immediately preceding arrangement was B_1 or B_2. Because this holds for all pairwise comparisons, we conclude that the phases/polarizations of the solitons in arrangement C are independent of the immediately preceding arrangement, regardless of the number of such possible arrangements. To see that phase/polarization shifts of the soliton induced by the evolution of the system from the arrangement in the limit as $t \to -\infty$ to the arrangement A are independent of the order of interactions, one uses the immediately preceding argument recursively, through a finite number of transpositions, to the arrangement in which the solitons are ordered from the fastest to the slowest (which is achieved as $t \to -\infty$). We conclude that, in the J-soliton case, the phase shifts between the limits as $t \to \pm\infty$ and the shifts in polarization are independent of the order in which the solitons interact.

4.4 Conserved quantities and Hamiltonian structure

We showed that the scattering coefficient $\mathbf{a}(k)$ is time-independent. Since $\mathbf{a}(k) - \mathbf{I}_N$ is analytic in the upper k-plane and $\mathbf{a}(k) - \mathbf{I}_N \to \mathbf{0}$ as $|k| \to \infty$, it admits a Laurent expansion whose coefficients are constants of the motion, as well. From the integral representation (4.2.26a) for $\mathbf{a}(k)$, it follows that the quantities

$$\Gamma_j = \int_{-\infty}^{+\infty} \mathbf{Q}(x)\mathbf{M}^{(\mathrm{dn}),-j}(x)dx \qquad (4.4.122)$$

are conserved for any integer $j \geq 0$ and the coefficients $\mathbf{M}^{(\mathrm{dn}),-j}(x)$ of the asymptotic expansion of $\mathbf{M}^{(\mathrm{dn})}(x,k)$ can be calculated iteratively from (4.2.13a). For instance, $\mathbf{M}^{(\mathrm{dn}),-1}(x)$ is given by (4.2.17a); inserting the asymptotics (4.2.17a) into (4.2.13a), one finds

$$\mathbf{M}^{(\mathrm{dn}),-2}(x) = \mathbf{R}(x)\int_{-\infty}^{x} \mathbf{Q}(\xi)\mathbf{R}(\xi)d\xi - \mathbf{R}_x(x)$$

and so on. Therefore the first constants of the motion are given by

$$\Gamma_1 = -\int_{-\infty}^{+\infty} \mathbf{Q}(x)\mathbf{R}(x)dx \qquad (4.4.123a)$$

$$\Gamma_2 = \int_{-\infty}^{+\infty} d\xi_1 \mathbf{Q}(\xi_1)\mathbf{R}(\xi_1)\left[\int_{-\infty}^{\xi_1} \mathbf{Q}(\xi_2)\mathbf{R}(\xi_2)d\xi_2\right]$$
$$- \int_{-\infty}^{+\infty} \mathbf{Q}(x)\mathbf{R}_x(x)dx, \qquad (4.4.123b)$$

$$\Gamma_3 = \int_{-\infty}^{+\infty} \mathbf{Q}(\xi_1)\left\{-\mathbf{R}(\xi_1)\int_{-\infty}^{\xi_1} \mathbf{Q}(\xi_2)\mathbf{R}(\xi_2)\left[\int_{-\infty}^{\xi_2} \mathbf{Q}(\xi_3)\mathbf{R}(\xi_3)d\xi_3\right]d\xi_2\right.$$
$$\left. + \frac{d}{d\xi_1}\left[\mathbf{R}(\xi_1)\int_{-\infty}^{\xi_1} \mathbf{Q}(\xi_2)\mathbf{R}(\xi_2)d\xi_2\right] - \frac{d^2}{d\xi_1^2}\mathbf{R}(\xi_1)\right\}d\xi_1, \qquad (4.4.123c)$$

and so on.

The scattering coefficient $\bar{\mathbf{a}}(k)$ is also a constant of the motion, and, proceeding exactly as before, one can obtain a second set of conserved quantities given by

$$\bar{\Gamma}_j = \int_{-\infty}^{+\infty} \mathbf{r}(x)\bar{\mathbf{M}}^{(\mathrm{up}),-j}(x)dx$$

for any $j \geq 1$, yielding

$$\bar{\Gamma}_1 = \int_{-\infty}^{+\infty} \mathbf{R}(x)\mathbf{Q}(x)dx, \qquad (4.4.124a)$$

$$\bar{\Gamma}_2 = \int_{-\infty}^{+\infty} d\xi_1 \mathbf{R}(\xi_1)\mathbf{Q}(\xi_1) \left[\int_{-\infty}^{\xi_1} \mathbf{R}(\xi_2)\mathbf{Q}(\xi_2)d\xi_2 \right]$$
$$- \int_{-\infty}^{+\infty} \mathbf{R}(x)\mathbf{Q}_x(x)dx, \tag{4.4.124b}$$

and so on.

Under the reductions $\mathbf{Q} = \mathbf{q}$, where $\mathbf{q}$, is an $M \times 1$ row vector and $\mathbf{R} = \mathbf{r} = \mp\mathbf{q}^H$, $\mathbf{q}(x)\mathbf{r}(x)$ is a scalar and the constants of motion (4.4.123) reduce to

$$\Gamma_1 = \pm \int \|\mathbf{q}(x)\|^2 \, dx \tag{4.4.125a}$$

$$\Gamma_2 = \pm \int \mathbf{q}(x)\mathbf{q}_x^H(x)dx \tag{4.4.125b}$$

$$\Gamma_3 = \int \left(\mp \|\mathbf{q}_x(x)\|^2 + \|\mathbf{q}(x)\|^4 \right) dx. \tag{4.4.125c}$$

The dynamical system (4.2.4) is Hamiltonian, with:

coordinates (q) :	$q^{(j)}(x, t)$	(4.4.126a)
momenta (p) :	$q^{(j)}(x, t)^*$	(4.4.126b)
Hamiltonian (H) :	$i\int_{-\infty}^{+\infty} \left(\mp \|\mathbf{q}_x(x)\|^2 + \|\mathbf{q}(x)\|^4 \right) dx$	(4.4.126c)

and the canonical brackets

$$\left\{ q^{(j)}(x, t), q^{(k)}(y, t)^* \right\} = i\delta(x - y)\delta_{j,k} \tag{4.4.127a}$$
$$\left\{ q^{(j)}(x, t), q^{(k)}(y, t) \right\} = 0. \tag{4.4.127b}$$

Note that the Hamiltonian is given by the conserved quantity (4.4.125c).

Chapter 5

Integrable discrete matrix NLS equation (IDMNLS)

5.1 Overview

The integrable discrete matrix NLS system

$$i\frac{d}{d\tau}\mathbf{Q}_n = \mathbf{Q}_{n+1} - 2\mathbf{Q}_n + \mathbf{A}\mathbf{Q}_n + \mathbf{Q}_n\mathbf{B} + \mathbf{Q}_{n-1} - \mathbf{Q}_{n+1}\mathbf{R}_n\mathbf{Q}_n - \mathbf{Q}_n\mathbf{R}_n\mathbf{Q}_{n-1}$$

(5.1.1a)

$$-i\frac{d}{d\tau}\mathbf{R}_n = \mathbf{R}_{n+1} - 2\mathbf{R}_n + \mathbf{B}\mathbf{R}_n + \mathbf{R}_n\mathbf{A} + \mathbf{R}_{n-1} - \mathbf{R}_{n+1}\mathbf{Q}_n\mathbf{R}_n - \mathbf{R}_n\mathbf{Q}_n\mathbf{R}_{n-1},$$

(5.1.1b)

where $\mathbf{Q}_n$ and $\mathbf{R}_n$ are, respectively, $N \times M$ and $M \times N$ matrices, results from the compatibility condition of the following linear equations [18], [167], [170]:

$$\mathbf{v}_{n+1} = \begin{pmatrix} z\mathbf{I}_N & \mathbf{Q}_n \\ \mathbf{R}_n & z^{-1}\mathbf{I}_M \end{pmatrix} \mathbf{v}_n$$

(5.1.2a)

$$\frac{d}{d\tau}\mathbf{v}_n = \begin{pmatrix} i\mathbf{Q}_n\mathbf{R}_{n-1} - \frac{i}{2}\left(z - z^{-1}\right)^2 \mathbf{I}_N - i\mathbf{A} & -iz\mathbf{Q}_n + iz^{-1}\mathbf{Q}_{n-1} \\ iz^{-1}\mathbf{R}_n - iz\mathbf{R}_{n-1} & -i\mathbf{R}_n\mathbf{Q}_{n-1} + \frac{i}{2}\left(z - z^{-1}\right)^2 \mathbf{I}_M + i\mathbf{B} \end{pmatrix} \mathbf{v}_n,$$

(5.1.2b)

where, as usual, $\mathbf{I}_N$ is the $N \times N$ identity matrix and $\mathbf{I}_M$ is the $M \times M$ identity matrix.

In [83], [84] the IST for an eigenvalue problem that is equivalent to (5.1.2a) was formulated. In particular, the completeness relation for the squared solutions was proved and the inverse problem was reduced to a Riemann–Hilbert problem with canonical normalization.

We observe that equation (5.1.2b) reduces to the scalar equation (3.2.5) when $N = M = 1$ and $\mathbf{A} = \mathbf{B} = \mathbf{0}$. Further, note that the matrices $\mathbf{A}$ and $\mathbf{B}$ in (5.1.2b) and in the system (5.1.1a)–(5.1.1b) can be absorbed by the gauge transformation

$$\hat{\mathbf{Q}}_n = e^{i\tau\mathbf{A}}\mathbf{Q}_n e^{i\tau\mathbf{B}}, \qquad \hat{\mathbf{R}}_n = e^{-i\tau\mathbf{B}}\mathbf{R}_n e^{-i\tau\mathbf{A}}, \qquad \hat{\mathbf{v}}_n = \begin{pmatrix} e^{i\tau\mathbf{A}} & \mathbf{0} \\ \mathbf{0} & e^{-i\tau\mathbf{B}} \end{pmatrix} \mathbf{v}_n.$$

$$(5.1.3)$$

That is, (5.1.2a)–(5.1.2b) are satisfied under the substitutions

$$\mathbf{Q}_n \to \hat{\mathbf{Q}}_n, \qquad \mathbf{R}_n \to \hat{\mathbf{R}}_n, \qquad \mathbf{v}_n \to \hat{\mathbf{v}}_n, \qquad \mathbf{A}, \mathbf{B} \to \mathbf{0}.$$

The system (5.1.1a)–(5.1.1b) does not, in general, admit the reduction $\mathbf{R}_n = \mp\mathbf{Q}_n^H$, and the asymmetry depends on the noncommutativity of matrix multiplication. However, if one takes $M = N$ and restricts $\mathbf{R}_n$ and $\mathbf{Q}_n$ to be such that

$$\mathbf{R}_n\mathbf{Q}_n = \mathbf{Q}_n\mathbf{R}_n = \alpha_n\mathbf{I}_N \qquad (5.1.4)$$

and α_n is real when $\mathbf{R}_n = \mp\mathbf{Q}_n^H$, then this symmetry is a consistent reduction of (5.1.1a)–(5.1.1b), which reduces to the single (matrix) equation

$$i\frac{d}{d\tau}\mathbf{Q}_n = \mathbf{Q}_{n+1} - 2\mathbf{Q}_n + \mathbf{A}\mathbf{Q}_n + \mathbf{Q}_n\mathbf{B} + \mathbf{Q}_{n-1} - \alpha_n\left(\mathbf{Q}_{n+1} + \mathbf{Q}_{n-1}\right).$$

$$(5.1.5)$$

In the case $N = 2$, the matrices

$$\mathbf{Q}_n = \begin{pmatrix} Q_n^{(1)} & Q_n^{(2)} \\ (-1)^n R_n^{(2)} & (-1)^{n+1} R_n^{(1)} \end{pmatrix}, \qquad \mathbf{R}_n = \begin{pmatrix} R_n^{(1)} & (-1)^n Q_n^{(2)} \\ R_n^{(2)} & (-1)^{n+1} Q_n^{(1)} \end{pmatrix} \quad (5.1.6)$$

satisfy the condition (5.1.4) with

$$\mathbf{R}_n\mathbf{Q}_n = \mathbf{Q}_n\mathbf{R}_n = \alpha_n\mathbf{I} = \left(R_n^{(1)} Q_n^{(1)} + R_n^{(2)} Q_n^{(2)}\right)\mathbf{I},$$

where, as usual, $\mathbf{I}$ denotes the 2×2 identity matrix. Note that (5.1.6) is equivalent to

$$\mathbf{R}_n = (-1)^n\mathbf{P}\mathbf{Q}_n^T\mathbf{P}, \qquad (5.1.7)$$

where

$$\mathbf{P} = \begin{pmatrix} 0 & 1 \\ -1 & 0 \end{pmatrix} \qquad (5.1.8)$$

and the superscript T denotes matrix transposition.

When $\mathbf{A} = \mathbf{B} = \mathbf{0}$, equations (5.1.1a)–(5.1.1b) with the symmetry condition (5.1.7) imply that the vectors $\mathbf{q}_n = h^{-1}\left(Q_n^{(1)}, Q_n^{(2)}\right)^T$, $\mathbf{r}_n = h^{-1}\left(R_n^{(1)}, R_n^{(2)}\right)^T$ satisfy

$$i\frac{d}{dt}\mathbf{q}_n = \frac{\mathbf{q}_{n+1} + \mathbf{q}_{n-1} - 2\mathbf{q}_n}{h^2} - (\mathbf{r}_n \cdot \mathbf{q}_n)(\mathbf{q}_{n+1} + \mathbf{q}_{n-1}) \quad (5.1.9a)$$

$$-i\frac{d}{dt}\mathbf{r}_n = \frac{\mathbf{r}_{n+1} + \mathbf{r}_{n-1} - 2\mathbf{r}_n}{h^2} - (\mathbf{r}_n \cdot \mathbf{q}_n)(\mathbf{r}_{n+1} + \mathbf{r}_{n-1}) \quad (5.1.9b)$$

with $t = h^2 \tau$. If, in addition, $\mathbf{R}_n = \mp \mathbf{Q}_n^H$, that is, $\mathbf{r}_n = \mp \mathbf{q}_n^*$, then

$$i\frac{d}{dt}\mathbf{q}_n = \frac{\mathbf{q}_{n+1} + \mathbf{q}_{n-1} - 2\mathbf{q}_n}{h^2} \pm \|\mathbf{q}_n\|^2 (\mathbf{q}_{n+1} + \mathbf{q}_{n-1}). \qquad (5.1.10)$$

Thus the vector extension of the IDNLS is amongst the class of equations solved by the IST associated with (5.1.2a)–(5.1.2b). Equation (5.1.10) is referred to here as the integrable discrete vector NLS (IDVNLS).

We also note that the symmetry (5.1.4) between $\mathbf{Q}_n$ and $\mathbf{R}_n$ imposes restrictions on the choice of the gauge matrices $\mathbf{A}$ and $\mathbf{B}$. Indeed, if the symmetry condition (5.1.7) is imposed at the initial time, say $\tau = 0$,

$$\mathbf{R}_n(0) = (-1)^n \mathbf{P} \mathbf{Q}_n^T(0) \mathbf{P}, \qquad (5.1.11)$$

it is preserved by the time evolution if, and only if,

$$\frac{d}{d\tau}\left[\mathbf{R}_n(\tau) - (-1)^n \mathbf{P} \mathbf{Q}_n^T(\tau) \mathbf{P}\right]\bigg|_{\tau=0} = \mathbf{0}, \qquad (5.1.12)$$

that is, according to (5.1.1a)–(5.1.1b), for $\mathbf{A}$ and $\mathbf{B}$ satisfying

$$\mathbf{B} = 2\mathbf{I} + \mathbf{P}\mathbf{B}^T\mathbf{P}, \qquad \mathbf{A} = 2\mathbf{I} + \mathbf{P}\mathbf{A}^T\mathbf{P}. \qquad (5.1.13)$$

The condition (5.1.13) is satisfied, for instance, if $\mathbf{A} = \mathbf{B} = \mathbf{I}$.

To obtain solutions of the system (5.1.9a)–(5.1.9b) one first determines the evolution of the potentials $\hat{\mathbf{Q}}_n$ and $\hat{\mathbf{R}}_n$ such that

$$\hat{\mathbf{Q}}_n(0) = \mathbf{Q}_n(0), \qquad \hat{\mathbf{R}}_n(0) = \mathbf{R}_n(0),$$

which solve the system (5.1.1a)–(5.1.1b) with gauge $\mathbf{A} = \mathbf{B} = \mathbf{I}$, that is,

$$i\frac{d}{d\tau}\hat{\mathbf{Q}}_n = \hat{\mathbf{Q}}_{n+1} + \hat{\mathbf{Q}}_{n-1} - \hat{\mathbf{Q}}_{n+1}\hat{\mathbf{R}}_n\hat{\mathbf{Q}}_n - \hat{\mathbf{Q}}_n\hat{\mathbf{R}}_n\hat{\mathbf{Q}}_{n-1} \qquad (5.1.14a)$$

$$-i\frac{d}{d\tau}\hat{\mathbf{R}}_n = \hat{\mathbf{R}}_{n+1} + \hat{\mathbf{R}}_{n-1} - \hat{\mathbf{R}}_{n+1}\hat{\mathbf{Q}}_n\hat{\mathbf{R}}_n - \hat{\mathbf{R}}_n\hat{\mathbf{Q}}_n\hat{\mathbf{R}}_{n-1}, \qquad (5.1.14b)$$

and then uses the transformations

$$\mathbf{Q}_n(\tau) = e^{i\tau\mathbf{A}}\hat{\mathbf{Q}}_n(\tau)e^{i\tau\mathbf{B}} = e^{2i\tau}\hat{\mathbf{Q}}_n(\tau) \qquad (5.1.15a)$$

$$\mathbf{R}_n(\tau) = e^{-i\tau\mathbf{B}}\hat{\mathbf{R}}_n(\tau)e^{-i\tau\mathbf{A}} = e^{-2i\tau}\hat{\mathbf{R}}_n(\tau) \qquad (5.1.15b)$$

to obtain the solution of (5.1.1a)–(5.1.1b) with gauge $\mathbf{A} = \mathbf{B} = \mathbf{0}$.

The symmetry (5.1.4) has no counterpart in the IST for the scalar IDNLS. The symmetry between the potentials induces the additional symmetry in the scattering data associated with the block-matrix scattering problem (see the discussion on symmetries in Section 5.2.1). This additional symmetry must be included in the formulation of the block-matrix inverse scattering problem in order to obtain a solution of the system (5.1.5) and the important special case (5.1.10) via the IST.

5.2 The inverse scattering transform for IDMNLS

5.2.1 Direct scattering problem

Eigenfunctions

The scattering problem (5.1.2a) associated with the integrable discrete matrix NLS can be written compactly as

$$\mathbf{v}_{n+1} = \mathbf{S}_n \mathbf{v}_n, \tag{5.2.16}$$

where

$$\mathbf{S}_n = \begin{pmatrix} z\mathbf{I}_N & \mathbf{Q}_n \\ \mathbf{R}_n & z^{-1}\mathbf{I}_M \end{pmatrix}. \tag{5.2.17}$$

For decaying potentials this difference equation admits the asymptotic solutions

$$\phi_n(z) \sim z^n \begin{pmatrix} \mathbf{I}_N \\ \mathbf{0} \end{pmatrix}, \qquad \bar{\phi}_n(z) \sim z^{-n} \begin{pmatrix} \mathbf{0} \\ \mathbf{I}_M \end{pmatrix} \qquad \text{as } n \to -\infty \tag{5.2.18a}$$

and

$$\psi_n(z) \sim z^{-n} \begin{pmatrix} \mathbf{0} \\ \mathbf{I}_M \end{pmatrix}, \qquad \bar{\psi}_n(z) \sim z^n \begin{pmatrix} \mathbf{I}_N \\ \mathbf{0} \end{pmatrix} \qquad \text{as } n \to +\infty, \tag{5.2.18b}$$

which are determined by the boundary condition as $n \to \pm\infty$, respectively. These solutions are matrix-valued functions with the following dimensions:

$$\phi_n(z) : (N + M) \times N, \qquad \bar{\phi}_n(z) : (N + M) \times M$$
$$\psi_n(z) : (N + M) \times M, \qquad \bar{\psi}_n(z) : (N + M) \times N.$$

We prove in the following (see the section on scattering data) that both $\phi, \bar{\phi}$ and $\psi, \bar{\psi}$ form a basis of solutions.

As in the IST for the 2×2 scattering problem we dealt with in Chapter 3, we introduce eigenfunctions with constant (i.e., independent of n and z) boundary conditions as $n \to \pm\infty$ via the following transformations:

$$\mathbf{M}_n(z) = z^{-n}\phi_n(z), \qquad \bar{\mathbf{M}}_n(z) = z^n \bar{\phi}_n(z), \tag{5.2.19a}$$
$$\mathbf{N}_n(z) = z^n \psi_n(z), \qquad \bar{\mathbf{N}}_n(z) = z^{-n}\bar{\psi}_n(z). \tag{5.2.19b}$$

These modified eigenfunctions satisfy the constant boundary conditions:

$$\mathbf{M}_n(z) \sim \begin{pmatrix} \mathbf{I}_N \\ \mathbf{0} \end{pmatrix}, \qquad \bar{\mathbf{M}}_n(z) \sim \begin{pmatrix} \mathbf{0} \\ \mathbf{I}_M \end{pmatrix} \qquad \text{as } n \to -\infty \tag{5.2.20a}$$

and

$$N_n(z) \sim \begin{pmatrix} 0 \\ I_M \end{pmatrix}, \qquad \bar{N}_n(z) \sim \begin{pmatrix} I_N \\ 0 \end{pmatrix} \qquad \text{as } n \to +\infty \qquad (5.2.20b)$$

and the difference equations:

$$M_{n+1}(z) - z^{-1}ZM_n(z) = z^{-1}\tilde{Q}_n M_n(z) \qquad (5.2.21a)$$

$$\bar{M}_{n+1}(z) - zZ\bar{M}_n(z) = z\tilde{Q}_n\bar{M}_n(z) \qquad (5.2.21b)$$

$$N_{n+1}(z) - zZN_n(z) = z\tilde{Q}_n N_n(z) \qquad (5.2.21c)$$

$$\bar{N}_{n+1}(z) - z^{-1}Z\bar{N}_n(z) = z^{-1}\tilde{Q}_n\bar{N}_n(z), \qquad (5.2.21d)$$

where

$$Z = \begin{pmatrix} zI_N & 0 \\ 0 & z^{-1}I_M \end{pmatrix}, \qquad \tilde{Q}_n = \begin{pmatrix} 0 & Q_n \\ R_n & 0 \end{pmatrix}.$$

The difference equations (5.2.21a)–(5.2.21d), along with their respective boundary conditions (5.2.20a)–(5.2.20b), can be converted to summation equations by the method of Green's functions. The derivation of the Green's functions is essentially the same as for the 2×2 scattering problem. The Green's function corresponding to (5.2.21a) or, equivalently, (5.2.21d) is a solution of the difference equation

$$G_{n+1} - z^{-1}ZG_n = z^{-1}\delta_{0,n}I_{N+M}. \qquad (5.2.22)$$

If v_n satisfies the summation equation

$$v_n = w + \sum_{k=-\infty}^{+\infty} G_{n-k}\tilde{Q}_k v_k, \qquad (5.2.23)$$

where G_n is a solution of (5.2.22) and w satisfies

$$w - z^{-1}Zw = 0, \qquad (5.2.24)$$

then v_n is a solution of the difference equation (5.2.21a) or, equivalently, (5.2.21d). The Green's function is not unique, and, as we show below, the choice of the Green's function and the choice of the inhomogeneous term w together determine the eigenfunction and its analytical properties.

To find the Green's function explicitly, we first note that one can set the off-diagonal blocks of G_n to zero and write

$$G_n = \begin{pmatrix} g_n^{(1)}I_N & 0 \\ 0 & g_n^{(2)}I_M \end{pmatrix}, \qquad (5.2.25)$$

where, according to (5.2.22), $g_n^{(j)}$ must satisfy

$$g_{n+1}^{(j)} - b^{(j)}g_n^{(j)} = z^{-1}\delta_{0,n} \qquad j = 1, 2 \qquad (5.2.26)$$

with

$$b^{(1)} = 1, \qquad b^{(2)} = z^{-2}. \qquad (5.2.27)$$

Next, we represent $g_n^{(j)}$ and $\delta_{0,n}$ as Fourier integrals

$$g_n^{(j)} = \frac{1}{2\pi i} \oint_{|p|=1} p^{n-1} \hat{g}^{(j)}(p) dp, \qquad \delta_{0,n} = \frac{1}{2\pi i} \oint_{|p|=1} p^{n-1} dp.$$

Substituting these integrals into the difference equations (5.2.26) yields

$$\hat{g}^{(j)}(p) = z^{-1} \frac{1}{p - b^{(j)}}$$

and

$$g_n^{(j)} = z^{-1} \frac{1}{2\pi i} \oint_{|p|=1} \frac{1}{p - b^{(j)}} p^{n-1} dp. \qquad (5.2.28)$$

The integral in (5.2.28) depends only on whether the pole $b^{(j)}$ is located inside or outside the contour of integration. However, when $|z| = 1$ we have $|b^{(1)}| = |b^{(2)}| = 1$, that is, the poles are on the contour $|p| = 1$. As usual, we consider contours that are perturbed away from $|p| = 1$ to avoid the singularities. Let C^{out} be a contour enclosing $p = 0$ and $p = b^{(j)}$ and let C^{in} be a contour enclosing $p = 0$ but not $p = b^{(j)}$ (see Figure 3.1). Consequently, we get

$$g_n^{(j),\text{out}} = z^{-1} \frac{1}{2\pi i} \oint_{C^{\text{out}}} \frac{1}{p - b^{(j)}} p^{n-1} dp = z^{-1} \begin{cases} (b^{(j)})^{n-1} & n \geq 1 \\ 0 & n \leq 0 \end{cases}$$

$$(5.2.29)$$

and

$$g_n^{(j),\text{in}} = z^{-1} \frac{1}{2\pi i} \oint_{C^{\text{in}}} \frac{1}{p - b^{(j)}} p^{n-1} dp = z^{-1} \begin{cases} 0 & n \geq 1 \\ -(b^{(j)})^{n-1} & n \leq 0. \end{cases}$$

$$(5.2.30)$$

By substituting one or the other of (5.2.29) or (5.2.30) into (5.2.25), with $b^{(j)}$ given by (5.2.27), we obtain two Green's functions satisfying (5.2.26), that is,

$$\mathbf{G}_n^{\ell}(z) = z^{-1} \theta(n - 1) \begin{pmatrix} \mathbf{I}_N & \mathbf{0} \\ \mathbf{0} & z^{-2(n-1)} \mathbf{I}_M \end{pmatrix} \qquad (5.2.31a)$$

$$\bar{\mathbf{G}}_n^r(z) = -z^{-1} \theta(-n) \begin{pmatrix} \mathbf{I}_N & \mathbf{0} \\ \mathbf{0} & z^{-2(n-1)} \mathbf{I}_M \end{pmatrix}, \qquad (5.2.31b)$$

where $\theta(n)$ is the discrete version of the step function introduced in (3.2.23), that is,

$$\theta(n) = \sum_{k=-\infty}^{n} \delta_{0,k} = \begin{cases} 1 & n \geq 0 \\ 0 & n < 0. \end{cases}$$

Taking into account the boundary conditions (5.2.20a)–(5.2.20b) and the relation (5.2.24) for the inhomogeneous term in (5.2.23), if $\mathbf{Q}_n, \mathbf{R}_n \to 0$ as $n \to \pm\infty$, we get the following summation equations for $\mathbf{M}_n$ and $\tilde{\mathbf{N}}_n$:

$$\mathbf{M}_n = \begin{pmatrix} \mathbf{I}_N \\ 0 \end{pmatrix} + \sum_{k=-\infty}^{+\infty} \mathbf{G}_{n-k}^{\ell} \tilde{\mathbf{Q}}_k \mathbf{M}_k \qquad (5.2.32a)$$

$$\tilde{\mathbf{N}}_n = \begin{pmatrix} \mathbf{I}_N \\ 0 \end{pmatrix} + \sum_{k=-\infty}^{+\infty} \bar{\mathbf{G}}_{n-k}^{r} \tilde{\mathbf{Q}}_k \bar{\mathbf{N}}_k. \qquad (5.2.32b)$$

A similar approach yields the summation equations for $\bar{\mathbf{M}}_n(z)$ and $\mathbf{N}_n(z)$:

$$\bar{\mathbf{M}}_n = \begin{pmatrix} 0 \\ \mathbf{I}_M \end{pmatrix} + \sum_{k=-\infty}^{+\infty} \bar{\mathbf{G}}_{n-k}^{\ell} \tilde{\mathbf{Q}}_k \bar{\mathbf{M}}_k \qquad (5.2.32c)$$

$$\mathbf{N}_n = \begin{pmatrix} 0 \\ \mathbf{I}_M \end{pmatrix} + \sum_{k=-\infty}^{+\infty} \mathbf{G}_{n-k}^{r} \tilde{\mathbf{Q}}_k \mathbf{N}_k, \qquad (5.2.32d)$$

where

$$\bar{\mathbf{G}}_n^{\ell}(z) = z\theta(n-1) \begin{pmatrix} z^{2(n-1)}\mathbf{I}_N & 0 \\ 0 & \mathbf{I}_M \end{pmatrix} \qquad (5.2.33a)$$

$$\mathbf{G}_n^{r}(z) = -z\theta(-n) \begin{pmatrix} z^{2(n-1)}\mathbf{I}_N & 0 \\ 0 & \mathbf{I}_M \end{pmatrix}. \qquad (5.2.33b)$$

Existence and analyticity of the eigenfunctions

The existence of the eigenfunctions can be proved by iteration (i.e., Neumann series) as long as $\|\mathbf{Q}\|_1, \|\mathbf{R}\|_1 < \infty$. Here $\|\mathbf{Q}\|_1 = \sum_{n=-\infty}^{+\infty} \|\mathbf{Q}_n\|_a$, where $\|\mathbf{Q}_n\|_a$ is any matrix norm. We also note that, if the potentials $\mathbf{Q}_n$ and $\mathbf{R}_n$ have finite support ($\mathbf{Q}_n, \mathbf{R}_n = 0$ if $n < n_{\min}$ and $n > n_{\max}$), each of the eigenfunctions can be generated by a finite number of applications of the appropriate differ- ence equation (5.2.21a)–(5.2.21d) with the proper choice of boundary condi- tion. To prove the existence and the sectional analyticity of the eigenfunctions in a more general case, we will make use of Lemmas A1 and A2 proved in Appendix A.

Lemma 5.1 *If* $\|\mathbf{Q}\|_1$, $\|\mathbf{R}\|_1$ *are bounded, then* $\mathbf{M}_n(z), \mathbf{N}_n(z)$ *defined by* (5.2.32a), (5.2.32d) *are analytic functions of z for* $|z| > 1$ *and continuous for*

$|z| \geq 1$, and $\bar{\mathbf{N}}_n(z)$, $\bar{\mathbf{M}}_n(z)$ defined by (5.2.32b), (5.2.32c) are analytic functions of z for $|z| < 1$ and continuous for $|z| \leq 1$. Moreover, these eigenfunctions are unique in the space of continuous functions.

Proof Consider the eigenfunction $\mathbf{M}_n(z)$. The Neumann series

$$\mathbf{M}_n(z) = \sum_{j=0}^{\infty} \mathbf{C}_n^j(z), \qquad (5.2.34)$$

where

$$\mathbf{C}_n^0(z) = \begin{pmatrix} \mathbf{I}_N \\ \mathbf{0} \end{pmatrix} \qquad (5.2.35a)$$

$$\mathbf{C}_n^{j+1}(z) = \sum_{k=-\infty}^{+\infty} \mathbf{G}_{n-k}^{\ell}(z)\tilde{\mathbf{Q}}_k\mathbf{C}_k^j(z), \qquad j \geq 0 \qquad (5.2.35b)$$

is, formally, a solution of the summation equation (5.2.32a). To make this rigorous, we establish a bound on the $\mathbf{C}_n^j$ such that the series representation (5.2.34) converges absolutely and uniformly in n and absolutely and uniformly in z in the region $|z| \geq 1$.

The summation equation for $\mathbf{M}_n(z)$ can be written as

$$\mathbf{M}_n^{(\mathrm{up})}(z) = \mathbf{I}_N + z^{-1} \sum_{k=-\infty}^{n-1} \mathbf{Q}_k\mathbf{M}_k^{(\mathrm{dn})}(z) \qquad (5.2.36a)$$

$$\mathbf{M}_n^{(\mathrm{dn})}(z) = \sum_{k=-\infty}^{n-1} z^{-2(n-k-1)-1}\mathbf{R}_k\mathbf{M}_k^{(\mathrm{up})}(z), \qquad (5.2.36b)$$

and equation (5.2.35b) in upper/lower component form is

$$\mathbf{C}_n^{j+1,(\mathrm{up})}(z) = z^{-1} \sum_{k=-\infty}^{n-1} \mathbf{Q}_k\mathbf{C}_k^{j,(\mathrm{dn})}(z),$$

$$\mathbf{C}_n^{j+1,(\mathrm{dn})}(z) = z^{-1} \sum_{k=-\infty}^{n-1} z^{-2(n-1-k)}\mathbf{R}_k\mathbf{C}_k^{j,(\mathrm{up})}(z). \qquad (5.2.37)$$

Then, since $\mathbf{C}_n^{0,(\mathrm{dn})} = \mathbf{0}$, it follows that also $\mathbf{C}_n^{2j+1,(\mathrm{up})} = \mathbf{0}$, $\mathbf{C}_n^{2j,(\mathrm{dn})} = \mathbf{0}$ for any $j \in \mathbb{N}$, and we only need to find bounds for $\mathbf{C}_n^{2j,(\mathrm{up})}$ and $\mathbf{C}_n^{2j+1,(\mathrm{dn})}$.

We prove by induction on j that for $|z| \geq 1$

$$\left\| \mathbf{C}_n^{2j+1,(\mathrm{dn})}(z) \right\|_a \leq \frac{\left(\sum_{k=-\infty}^{n-1} \|\mathbf{R}_k\|_a \right)^{j+1} \left(\sum_{k=-\infty}^{n-1} \|\mathbf{Q}_k\|_a \right)^{j}}{(j+1)!} \frac{}{j!} \qquad (5.2.38a)$$

$$\left\| \mathbf{C}_n^{2(j+1),(\mathrm{up})}(z) \right\|_a \leq \frac{\left(\sum_{k=-\infty}^{n-1} \|\mathbf{R}_k\|_a \right)^{j+1} \left(\sum_{k=-\infty}^{n-1} \|\mathbf{Q}_k\|_a \right)^{j+1}}{(j+1)!} \frac{}{(j+1)!}. \qquad (5.2.38b)$$

The proof follows from the inductive step through the application of (5.2.37) twice. Indeed, one iteration yields

$$\left\| \mathbf{C}_n^{2j+1,(\mathrm{dn})}(z) \right\|_a \leq \sum_{k=-\infty}^{n-1} \| \mathbf{R}_k \|_a \frac{\left(\sum_{\ell=-\infty}^{k-1} \| \mathbf{R}_\ell \|_a \right)^j}{j!} \frac{\left(\sum_{\ell=-\infty}^{k-1} \| \mathbf{Q}_\ell \|_a \right)^j}{j!}$$

$$\leq \frac{\left(\sum_{k=-\infty}^{n-1} \| \mathbf{Q}_k \|_a \right)^j}{j!} \sum_{k=-\infty}^{n-1} \| \mathbf{R}_k \|_a \frac{\left(\sum_{\ell=-\infty}^{k-1} \| \mathbf{R}_\ell \|_a \right)^j}{j!}.$$

Then, applying Lemma A.2 completes the induction for (5.2.38a).

The next iteration of (5.2.37) yields

$$\left\| \mathbf{C}_n^{2j+2,(\mathrm{up})}(z) \right\|_a \leq |z|^{-1} \sum_{k=-\infty}^{n-1} \| \mathbf{Q}_k \|_a \left\| \mathbf{C}_k^{2j+1,(\mathrm{dn})}(z) \right\|_a,$$

and for $|z| \geq 1$

$$\left\| \mathbf{C}_n^{2j+2,(\mathrm{up})}(z) \right\|_a \leq \sum_{k=-\infty}^{n-1} \| \mathbf{Q}_k \|_a \frac{\left(\sum_{\ell=-\infty}^{k-1} \| \mathbf{Q}_\ell \|_a \right)^j}{j!} \frac{\left(\sum_{\ell=-\infty}^{k-1} \| \mathbf{R}_\ell \|_a \right)^{j+1}}{(j+1)!}$$

$$\leq \frac{\left(\sum_{k=-\infty}^{n-1} \| \mathbf{R}_k \|_a \right)^{j+1}}{(j+1)!} \sum_{k=-\infty}^{n-1} \| \mathbf{Q}_k \|_a \frac{\left(\sum_{\ell=-\infty}^{k-1} \| \mathbf{Q}_\ell \|_a \right)^j}{j!}.$$

Then we again use Lemma A2 to complete the induction.

The bounds (5.2.38a)–(5.2.38b) are absolutely and uniformly (in n) summable if $\| \mathbf{Q} \|_1, \| \mathbf{R} \|_1 < \infty$, where $\| \cdot \|_1$ is the ℓ^1-norm. Moreover, in this case, the Neumann series (5.2.34) converges uniformly for all $|z| \geq 1$. Hence, the eigenfunction $\mathbf{M}_n(z)$ exists and is continuous in the region $|z| \geq 1$. Note that we originally constructed $\mathbf{M}_n(z)$ for $|z| = 1$, but the summation equation allows us to extend $\mathbf{M}_n(z)$ to the region $|z| \geq 1$.

Because the Neumann series (5.2.34) converges absolutely, this yields a convergent Laurent series in powers of z^{-1} for the solution $\mathbf{M}_n(z)$. Thus, the bounds (5.2.38a)–(5.2.38b) establish that $\mathbf{M}_n(z)$ is analytic in the region $|z| > 1$ when $\mathbf{Q}_n, \mathbf{R}_n \in \ell^1$. In an analogous manner, one can prove existence, continuity, and analyticity of the remaining eigenfunctions $\mathbf{N}_n(z), \bar{\mathbf{M}}_n(z), \bar{\mathbf{N}}_n(z)$.

Uniqueness of solution of the discrete integral equations (5.2.21a)–(5.2.21d) in the space of continuous functions can be proved in the same manner as in the scalar case.

Note that, to generate $\mathbf{N}_n(z)$ and $\bar{\mathbf{N}}_n(z)$, the maps

$$\mathbf{N}_n(z) = z^{-1} \mathbf{S}_n^{-1}(z) \mathbf{N}_{n+1}(z)$$

$$\bar{\mathbf{N}}_n(z) = z \mathbf{S}_n^{-1}(z) \bar{\mathbf{N}}_{n+1}(z),$$

where $\mathbf{S}_n(z)$ is given by (5.2.17), must be well defined because these eigenfunctions are determined by a boundary condition in the limit $n \to +\infty$. That is, the matrix $\mathbf{S}_n(z)$ must be invertible. Equivalently, it must be that

$$\det \mathbf{S}_n = z^{N-M} \det(\mathbf{I}_M - \mathbf{R}_n \mathbf{Q}_n) = z^{N-M} \det(\mathbf{I}_N - \mathbf{Q}_n \mathbf{R}_n) \neq 0, \quad (5.2.39)$$

where we used the Schur identities to compute the determinant of the block-partioned matrix $\mathbf{S}_n$ (see [81]). Specifically, for a matrix $\mathbf{\Delta}$ partitioned in four blocks

$$\mathbf{\Delta} = \begin{pmatrix} \mathbf{A} & \mathbf{B} \\ \mathbf{C} & \mathbf{D} \end{pmatrix}, \quad (5.2.40)$$

where $\mathbf{A}$ and $\mathbf{D}$ are square matrices, we have

$$\det \mathbf{\Delta} = \det \mathbf{A} \det\left(\mathbf{D} - \mathbf{C}\mathbf{A}^{-1}\mathbf{B}\right) \quad \text{if} \quad \det \mathbf{A} \neq 0 \quad (5.2.41\text{a})$$
$$\det \mathbf{\Delta} = \det \mathbf{D} \det\left(\mathbf{A} - \mathbf{B}\mathbf{D}^{-1}\mathbf{C}\right) \quad \text{if} \quad \det \mathbf{D} \neq 0. \quad (5.2.41\text{b})$$

Note that (5.2.39) is reminiscent of the condition $1 - R_n Q_n \neq 0$ that was required in the 2×2 scattering problem. This condition holds if, for example, $\mathbf{Q}_n$ and $\mathbf{R}_n$ are sufficiently small (in any matrix norm) or if all the eigenvalues of the product $\mathbf{R}_n \mathbf{Q}_n$ are negative. From here on, we assume that $\mathbf{Q}_n$, $\mathbf{R}_n$ satisfy (5.2.39).

Properties of the eigenfunctions in the complex z-plane

When the potentials have finite support, $\mathbf{M}_n(z)$ and $\mathbf{N}_n(z)$ are polynomials in z^{-1} while $\bar{\mathbf{M}}_n(z)$ and $\bar{\mathbf{N}}_n(z)$ are polynomials in z for each finite n. Therefore, all of the eigenfunctions are analytic everywhere in z, except possibly at $z = 0$ and $z = \infty$. Moreover, we showed that if the potentials have a finite ℓ^1-norm, the eigenfunctions $\mathbf{M}_n, \mathbf{N}_n$ are analytic in the region $|z| > 1$ while $\bar{\mathbf{M}}_n, \bar{\mathbf{N}}_n$ are in the region $|z| < 1$. Therefore $\mathbf{N}_n(z)$ and $\mathbf{M}_n(z)$ have Laurent series expansions in z^{-1} that converge in the region of analyticity. Similarly, $\bar{\mathbf{N}}_n(z)$ and $\bar{\mathbf{M}}_n(z)$ have power series expansions in z that converge in the region of analyticity.

Let us write the Laurent series expansion for $\mathbf{M}_n(z)$ as

$$\mathbf{M}_n^{(\mathrm{up})}(z) = \sum_{j=0}^{+\infty} z^{-j} \mathbf{M}_n^{(\mathrm{up}),-j} \quad (5.2.42\text{a})$$

$$\mathbf{M}_n^{(\mathrm{dn})}(z) = \sum_{j=0}^{+\infty} z^{-j} \mathbf{M}_n^{(\mathrm{dn}),-j} \quad (5.2.42\text{b})$$

and the summation equation (5.2.32a) for $\mathbf{M}_n(z)$ in the form

$$\mathbf{M}_n^{(\text{up})}(z) = \mathbf{I}_N + \sum_{k=-\infty}^{n-1} z^{-1} \mathbf{Q}_k \mathbf{M}_k^{(\text{dn})}(z) \tag{5.2.43a}$$

$$\mathbf{M}_n^{(\text{dn})}(z) = \sum_{k=-\infty}^{n-1} z^{-2(n-k)+1} \mathbf{R}_k \mathbf{M}_k^{(\text{up})}(z). \tag{5.2.43b}$$

If we substitute the expansions (5.2.42a)–(5.2.42b) into (5.2.43a)–(5.2.43b) and match the terms of $O(1)$, we get

$$\mathbf{M}_n^{(\text{up}),0} = \mathbf{I}_N, \qquad \mathbf{M}_n^{(\text{dn}),0} = \mathbf{0},$$

which anchors an induction on the coefficients of the expansion. Indeed, one can show by induction that for any integer $j \geq 0$

$$\mathbf{M}_n^{(\text{up}),-(2j+1)} = \mathbf{0}, \qquad \mathbf{M}_n^{(\text{dn}),-2j} = \mathbf{0}.$$

Moreover, matching the corresponding powers of z^{-1} into the summation equations (5.2.43a)–(5.2.43b) yields the following recursive relations:

$$\mathbf{M}_n^{(\text{dn}),-2j+1} = \sum_{k=n-j}^{n-1} \mathbf{R}_k \mathbf{M}_k^{(\text{up}),-2(j+k-n)} \tag{5.2.44}$$

$$\mathbf{M}_n^{(\text{up}),-2j} = \sum_{k=-\infty}^{n-1} \mathbf{Q}_k \mathbf{M}_k^{(\text{dn}),-2j+1}. \tag{5.2.45}$$

Analogously, for the coefficients of the power series expansion of $\bar{\mathbf{M}}_n(z)$ about $z = 0$, that is,

$$\bar{\mathbf{M}}_n^{(\text{up})}(z) = \sum_{j=0}^{+\infty} z^j \bar{\mathbf{M}}_n^{(\text{up}),j}, \tag{5.2.46a}$$

$$\bar{\mathbf{M}}_n^{(\text{dn})}(z) = \sum_{j=0}^{+\infty} z^j \bar{\mathbf{M}}_n^{(\text{dn}),j}, \tag{5.2.46b}$$

one obtains the following:

$$\bar{\mathbf{M}}_n^{(\text{up}),0} = \mathbf{0}, \qquad \bar{\mathbf{M}}_n^{(\text{dn}),0} = \mathbf{I}_M.$$

Then it is easy to prove by induction that, for any integer $j \geq 1$,

$$\bar{\mathbf{M}}_n^{(\text{up}),2j} = \mathbf{0}, \qquad \bar{\mathbf{M}}_n^{(\text{dn}),2j-1} = \mathbf{0}$$

and

$$\bar{\mathbf{M}}_n^{(\text{up}),2j-1} = \sum_{k=n-j}^{n-1} \mathbf{Q}_k \bar{\mathbf{M}}_k^{(\text{dn}),2(j-n+k)} \tag{5.2.47}$$

$$\bar{\mathbf{M}}_n^{(\mathrm{dn}),2j} = \sum_{k=-\infty}^{n-1} \mathbf{R}_k \bar{\mathbf{M}}_k^{(\mathrm{up}),2j-1}. \tag{5.2.48}$$

Therefore we can write

$$\mathbf{M}_n(z) = \begin{pmatrix} \mathbf{I}_N + O(z^{-2}, \text{even}) \\ z^{-1}\mathbf{R}_{n-1} + O(z^{-3}, \text{odd}) \end{pmatrix} \tag{5.2.49a}$$

$$\bar{\mathbf{M}}_n(z) = \begin{pmatrix} z\mathbf{Q}_{n-1} + O(z^3, \text{odd}) \\ \mathbf{I}_M + O(z^2, \text{even}) \end{pmatrix}, \tag{5.2.49b}$$

where "even" indicates that the remaining terms are even powers of z and "odd" indicates that the remaining terms are odd powers.

The expansions in z of $\mathbf{N}_n(z)$ and $\bar{\mathbf{N}}_n(z)$ are obtained from the summation equations (5.2.32d) and (5.2.32b), respectively. In upper/lower component form, equation (5.2.32d) is written as

$$\mathbf{N}_n^{(\mathrm{up})}(z) = -\sum_{k=n}^{\infty} z^{-2(k-n)-1} \mathbf{Q}_k \mathbf{N}_k^{(\mathrm{dn})}(z) \tag{5.2.50a}$$

$$\mathbf{N}_n^{(\mathrm{dn})}(z) = \mathbf{I}_M - \sum_{k=n}^{\infty} z \mathbf{R}_k \mathbf{N}_k^{(\mathrm{up})}(z), \tag{5.2.50b}$$

and we substitute the expansions

$$\mathbf{N}_n^{(\mathrm{up})}(z) = \sum_{j=0}^{+\infty} z^{-j} \mathbf{N}_n^{(\mathrm{up}),-j}$$

$$\mathbf{N}_n^{(\mathrm{dn})}(z) = \sum_{j=0}^{+\infty} z^{-j} \mathbf{N}_n^{(\mathrm{dn}),-j}$$

into (5.2.50a)–(5.2.50b) to obtain the equations

$$\mathbf{N}_n^{(\mathrm{up}),0} = 0 \tag{5.2.51a}$$

$$\mathbf{N}_n^{(\mathrm{up}),-1} = -\mathbf{Q}_n \mathbf{N}_n^{(\mathrm{dn}),0} \qquad \mathbf{N}_n^{(\mathrm{dn}),0} = \mathbf{I}_M - \sum_{k=n}^{+\infty} \mathbf{R}_k \mathbf{N}_k^{(\mathrm{up}),-1} \tag{5.2.51b}$$

$$\mathbf{N}_n^{(\mathrm{up}),-2} = -\mathbf{Q}_n \mathbf{N}_n^{(\mathrm{dn}),-1} \qquad \mathbf{N}_n^{(\mathrm{dn}),-1} = -\sum_{k=n}^{+\infty} \mathbf{R}_k \mathbf{N}_k^{(\mathrm{up}),-2}. \tag{5.2.51c}$$

Further manipulation yields

$$\mathbf{N}_{n+1}^{(\mathrm{dn}),0} = (\mathbf{I}_M - \mathbf{R}_n \mathbf{Q}_n) \mathbf{N}_n^{(\mathrm{dn}),0}$$

$$\mathbf{N}_{n+1}^{(\mathrm{dn}),-1} = (\mathbf{I}_M - \mathbf{R}_n \mathbf{Q}_n) \mathbf{N}_n^{(\mathrm{dn}),-1}.$$

Solving these equations with the boundary condition (5.2.20b) yields

$$\mathbf{N}_n^{(dn),0} = (\mathbf{I}_M - \mathbf{R}_n\mathbf{Q}_n)^{-1}(\mathbf{I}_M - \mathbf{R}_{n+1}\mathbf{Q}_{n+1})^{-1}\cdots = \prod_{\substack{k=n \\ \text{right}}}^{+\infty}(\mathbf{I}_M - \mathbf{R}_k\mathbf{Q}_k)^{-1},$$

where the term "right" in the product indicates that the matrix with index k occurs to the right of the matrix with index $k-1$. Also,

$$\mathbf{N}_n^{(dn),-1} = \mathbf{0}.$$

By substituting these expressions back into (5.2.51b)–(5.2.51c) we obtain

$$\mathbf{N}_n(z) = \begin{pmatrix} -z^{-1}\mathbf{Q}_n\mathbf{\Delta}_n^{-1} + O(z^{-3}) \\ \mathbf{\Delta}_n^{-1} + O(z^{-2}) \end{pmatrix}, \tag{5.2.52}$$

where

$$\mathbf{\Delta}_n = \ldots(\mathbf{I}_M - \mathbf{R}_{n+1}\mathbf{Q}_{n+1})(\mathbf{I}_M - \mathbf{R}_n\mathbf{Q}_n) \equiv \prod_{\substack{k=n \\ \text{left}}}^{+\infty}(\mathbf{I}_M - \mathbf{R}_k\mathbf{Q}_k) \tag{5.2.53}$$

and "left" indicates that the matrix with index k occurs to the left of the matrix with index $k-1$.

Analogously, we derive the expansion for $\bar{\mathbf{N}}_n(z)$ in powers of z,

$$\bar{\mathbf{N}}_n(z) = \begin{pmatrix} \mathbf{\Omega}_n^{-1} + O(z^2) \\ -z\mathbf{R}_n\mathbf{\Omega}_n^{-1} + O(z^3) \end{pmatrix}, \tag{5.2.54}$$

where

$$\mathbf{\Omega}_n = \ldots(\mathbf{I}_M - \mathbf{Q}_{n+1}\mathbf{R}_{n+1})(\mathbf{I}_M - \mathbf{Q}_n\mathbf{R}_n) \equiv \prod_{\substack{k=n \\ \text{left}}}^{+\infty}(\mathbf{I}_N - \mathbf{Q}_k\mathbf{R}_k) \tag{5.2.55}$$

and, as before, "left" indicates that the matrix with index k occurs to the left of the matrix with index $k-1$.

Scattering data

The eigenfunctions $\bar{\psi}_n(z)$ and $\psi_n(z)$ together constitute $N + M$ solutions of the difference equation (5.2.16). To show that these solutions are linearly independent for all n, we calculate their Wronskian. The matrix

$$\mathbf{\Psi}_n(z) = \left(\bar{\psi}_n(z), \psi_n(z)\right)$$

is an $(N + M) \times (N + M)$ square matrix. Hence, we define

$$W\left(\bar{\psi}_n(z), \psi_n(z)\right) = \det\mathbf{\Psi}_n(z) \tag{5.2.56}$$

and similarly for any collection of matrices that, all together, have the same number of rows and columns. In particular,

$$
\begin{aligned}
W\left(\bar{\psi}_{n+1}(z),\psi_{n+1}(z)\right) &= W\left(\mathbf{S}_n(z)\bar{\psi}_n(z),\mathbf{S}_n(z)\psi_n(z)\right) \\
&= \det \mathbf{S}_n(z) W\left(\bar{\psi}_n(z),\psi_n(z)\right) \\
&= z^{N-M}\det\left(\mathbf{I}_M - \mathbf{R}_n\mathbf{Q}_n\right)W\left(\bar{\psi}_n(z),\psi_n(z)\right),
\end{aligned}
$$

and since we assumed $\det\left(\mathbf{I}_M - \mathbf{R}_n\mathbf{Q}_n\right) \neq 0$ for any $n \in \mathbb{Z}$, it follows that for any $j \in \mathbb{N}$

$$
\begin{aligned}
W\left(\bar{\psi}_n(z),\psi_n(z)\right) &= \frac{z^{j(M-N)}}{\prod_{l=0}^{j-1}\det\left(\mathbf{I}_M - \mathbf{R}_{n+l}\mathbf{Q}_{n+l}\right)} W\left(\bar{\psi}_{n+j}(z),\psi_{n+j}(z)\right) \\
&= \frac{z^{n(N-M)}}{\prod_{l=0}^{j-1}\det\left(\mathbf{I}_M - \mathbf{R}_{n+l}\mathbf{Q}_{n+l}\right)} W\left(\bar{\mathbf{N}}_{n+j}(z),\mathbf{N}_{n+j}(z)\right).
\end{aligned}
$$

Taking the limit as $j \to \infty$, we obtain

$$
W\left(\bar{\psi}_n(z),\psi_n(z)\right) = \frac{z^{n(N-M)}}{\prod_{l=n}^{\infty}\det\left(\mathbf{I}_M - \mathbf{R}_l\mathbf{Q}_l\right)}, \tag{5.2.57}
$$

which is nonzero, showing that ψ_n and $\bar{\psi}_n$ are linearly independent.

The functions $\phi_n(z)$ and $\bar{\phi}_n(z)$ constitute a second set of $M + N$ solutions of the scattering problem (5.2.16) and

$$
W\left(\phi_n(z),\bar{\phi}_n(z)\right) = z^{n(N-M)}\prod_{j=-\infty}^{n-1}\det\left(\mathbf{I}_M - \mathbf{R}_j\mathbf{Q}_j\right). \tag{5.2.58}
$$

Hence, the solutions in $\phi_n(z)$ and $\bar{\phi}_n(z)$ are linearly dependent on the set of solutions $\psi_n(z)$ and $\bar{\psi}_n(z)$. This dependence can be expressed as

$$
\phi_n(z) = \psi_n(z)\mathbf{b}(z) + \bar{\psi}_n(z)\mathbf{a}(z) \tag{5.2.59a}
$$

$$
\bar{\phi}_n(z) = \psi_n(z)\bar{\mathbf{a}}(z) + \bar{\psi}_n(z)\bar{\mathbf{b}}(z), \tag{5.2.59b}
$$

where $\mathbf{b}(z)$ is an $M \times N$ matrix, $\mathbf{a}(z)$ is an $N \times N$ matrix, $\bar{\mathbf{a}}(z)$ is an $M \times M$ matrix, and $\bar{\mathbf{b}}(z)$ is an $N \times M$ matrix. In block-matrix form, (5.2.59a)–(5.2.59b) are

$$
\left(\phi_n, \bar{\phi}_n\right) = \left(\bar{\psi}_n, \psi_n\right)\begin{pmatrix} \mathbf{a} & \bar{\mathbf{b}} \\ \mathbf{b} & \bar{\mathbf{a}} \end{pmatrix}, \tag{5.2.60}
$$

and in terms of the eigenfunctions (5.2.19a)–(5.2.19b), equations (5.2.59a)–(5.2.59b) are

$$
\mathbf{M}_n(z) = z^{-2n}\mathbf{N}_n(z)\mathbf{b}(z) + \bar{\mathbf{N}}_n(z)\mathbf{a}(z) \tag{5.2.61a}
$$

$$
\bar{\mathbf{M}}_n(z) = \mathbf{N}_n(z)\bar{\mathbf{a}}(z) + z^{2n}\bar{\mathbf{N}}_n(z)\bar{\mathbf{b}}(z). \tag{5.2.61b}
$$

These equations define the coefficients $\mathbf{a}(z)$, $\bar{\mathbf{a}}(z)$, $\mathbf{b}(z)$, and $\bar{\mathbf{b}}(z)$ for any z such that all four eigenfunctions exist.

The scattering equations (5.2.59a)–(5.2.59b) have the counterparts

$$\psi_n(z) = \phi_n(z)\mathbf{d}(z) + \bar{\phi}_n(z)\mathbf{c}(z) \tag{5.2.62a}$$

$$\bar{\psi}_n(z) = \phi_n(z)\bar{\mathbf{c}}(z) + \bar{\phi}_n(z)\bar{\mathbf{d}}(z), \tag{5.2.62b}$$

which express the eigenfunctions with boundary conditions defined as $n \to +\infty$ (i.e., the "right" eigenfunctions $\psi_n(z)$ and $\bar{\psi}_n(z)$) in terms of eigenfunctions with boundary conditions defined as $n \to -\infty$ (i.e., the "left" eigenfunctions $\phi_n(z)$, $\bar{\phi}_n(z)$). In block-matrix form,

$$(\bar{\psi}_n, \psi_n) = (\phi_n, \bar{\phi}_n)\begin{pmatrix} \bar{\mathbf{c}} & \mathbf{d} \\ \bar{\mathbf{d}} & \mathbf{c} \end{pmatrix}, \tag{5.2.63}$$

and by comparing (5.2.60) and (5.2.63) we get the relation between "left" and "right" scattering data,

$$\begin{pmatrix} \mathbf{a} & \bar{\mathbf{b}} \\ \mathbf{b} & \bar{\mathbf{a}} \end{pmatrix}^{-1} = \begin{pmatrix} \bar{\mathbf{c}} & \mathbf{d} \\ \bar{\mathbf{d}} & \mathbf{c} \end{pmatrix}, \tag{5.2.64}$$

which is valid for any z such that all the scattering coefficients are well defined. In particular, this relation holds for $|z| = 1$, that is, for all z such that $z = 1/z^*$.

The scattering coefficients $\mathbf{a}(z)$ and $\mathbf{b}(z)$ can be written as explicit sums of the eigenfunction $\mathbf{M}_n(z)$ and the potential $\tilde{\mathbf{Q}}_n$. The formula is derived as follows. First, we obtain the relation

$$\mathbf{M}_n(z) - \bar{\mathbf{N}}_n(z)\mathbf{a}(z)$$
$$= \begin{pmatrix} \mathbf{I}_N - \mathbf{a}(z) \\ \mathbf{0} \end{pmatrix} + \sum_{k=-\infty}^{+\infty} \left\{ \mathbf{G}_{n-k}^\ell(z)\tilde{\mathbf{Q}}_k\mathbf{M}_k(z) - \bar{\mathbf{G}}_{n-k}^r(z)\tilde{\mathbf{Q}}_k\bar{\mathbf{N}}_k(z)\mathbf{a}(z) \right\}$$

by substituting the right-hand sides of the summation equations (5.2.32a) and (5.2.32b) for $\mathbf{M}_n(z)$ and $\bar{\mathbf{N}}_n(z)$, respectively. Then, we use the identity

$$\mathbf{G}_n^\ell(z) = \bar{\mathbf{G}}_n^r(z) + z^{-1}\begin{pmatrix} \mathbf{I}_N & \mathbf{0} \\ \mathbf{0} & z^{-2(n-1)}\mathbf{I}_M \end{pmatrix}$$

and the relation (5.2.61a) to replace $\mathbf{M}_n(z) - \bar{\mathbf{N}}_n(z)\mathbf{a}(z)$ with $z^{-2n}\mathbf{N}_n(z)\mathbf{b}(z)$, so that

$$z^{-2n}\mathbf{N}_n(z)\mathbf{b}(z) = \begin{pmatrix} \mathbf{I}_N - \mathbf{a}(z) \\ \mathbf{0} \end{pmatrix} + \sum_{k=-\infty}^{+\infty} z^{-2k}\bar{\mathbf{G}}_{n-k}^r(z)\tilde{\mathbf{Q}}_k\mathbf{N}_k(z)\mathbf{b}(z)$$
$$+ \sum_{k=-\infty}^{+\infty}\begin{pmatrix} z^{-1}\mathbf{I}_N & \mathbf{0} \\ \mathbf{0} & z^{-2(n-k)+1}\mathbf{I}_M \end{pmatrix}\tilde{\mathbf{Q}}_k\mathbf{M}_k(z).$$

Finally, taking into account that $\bar{\mathbf{G}}_n^r(z) = z^{-2n}\mathbf{G}_n^r(z)$, we obtain

$$z^{-2n}\left\{\mathbf{N}_n(z) - \sum_{k=-\infty}^{+\infty}\mathbf{G}_{n-k}^r(z)\tilde{\mathbf{Q}}_k\mathbf{N}_k(z)\right\}\mathbf{b}(z)$$

$$= \begin{pmatrix}\mathbf{I}_N - \mathbf{a}(z)\\ \mathbf{0}\end{pmatrix} + \sum_{k=-\infty}^{+\infty}\begin{pmatrix}z^{-1}\mathbf{I}_N & \mathbf{0}\\ \mathbf{0} & z^{-2(n-k)+1}\mathbf{I}_M\end{pmatrix}\tilde{\mathbf{Q}}_k\mathbf{M}_k(z).$$

Due to the summation equation (5.2.32d) for $\mathbf{N}_n(z)$, the term in curly braces is $(\mathbf{0}, \mathbf{I}_M)^T$, and we have

$$\begin{pmatrix}\mathbf{a}(z)\\ z^{-2n}\mathbf{b}(z)\end{pmatrix} = \begin{pmatrix}\mathbf{I}_N\\ \mathbf{0}\end{pmatrix} + \sum_{k=-\infty}^{+\infty}\begin{pmatrix}z^{-1}\mathbf{I}_N & \mathbf{0}\\ \mathbf{0} & z^{-2(n-k)+1}\mathbf{I}_M\end{pmatrix}\tilde{\mathbf{Q}}_k\mathbf{M}_k(z).$$

Finally, we conclude that

$$\mathbf{a}(z) = \mathbf{I}_N + \sum_{k=-\infty}^{+\infty}z^{-1}\mathbf{Q}_k\mathbf{M}_k^{(\mathrm{dn})}(z) \tag{5.2.65a}$$

$$\mathbf{b}(z) = \sum_{k=-\infty}^{+\infty}z^{2k+1}\mathbf{R}_k\mathbf{M}_k^{(\mathrm{up})}(z). \tag{5.2.65b}$$

The same approach works for $\bar{\mathbf{a}}(z)$ and $\bar{\mathbf{b}}(z)$ and for $\mathbf{c}(z)$, $\mathbf{d}(z)$, $\bar{\mathbf{c}}(z)$, and $\bar{\mathbf{d}}(z)$. The corresponding expressions are

$$\bar{\mathbf{a}}(z) = \mathbf{I}_M + \sum_{k=-\infty}^{+\infty}z\mathbf{R}_k\bar{\mathbf{M}}_k^{(\mathrm{up})}(z) \tag{5.2.65c}$$

$$\bar{\mathbf{b}}(z) = \sum_{k=-\infty}^{+\infty}z^{-2k-1}\mathbf{Q}_k\bar{\mathbf{M}}_k^{(\mathrm{dn})}(z) \tag{5.2.65d}$$

$$\mathbf{c}(z) = \mathbf{I}_M - \sum_{k=-\infty}^{+\infty}z\mathbf{R}_k\mathbf{N}_k^{(\mathrm{up})}(z) \tag{5.2.65e}$$

$$\mathbf{d}(z) = -\sum_{k=-\infty}^{+\infty}z^{-2k-1}\mathbf{Q}_k\mathbf{N}_k^{(\mathrm{dn})}(z) \tag{5.2.65f}$$

$$\bar{\mathbf{c}}(z) = \mathbf{I}_N - \sum_{k=-\infty}^{+\infty}z^{-1}\mathbf{Q}_k\bar{\mathbf{N}}_k^{(\mathrm{dn})}(z) \tag{5.2.65g}$$

$$\bar{\mathbf{d}}(z) = -\sum_{k=-\infty}^{+\infty}z^{2k+1}\mathbf{R}_k\bar{\mathbf{N}}_k^{(\mathrm{up})}(z). \tag{5.2.65h}$$

The expressions (5.2.65a) and (5.2.65c) imply that $\mathbf{a}(z)$ is analytic in the same region as $\mathbf{M}_n(z)$, that is, $|z| > 1$, and $\bar{\mathbf{a}}(z)$ is analytic in the same region as $\bar{\mathbf{M}}_n(z)$, that is, $|z| < 1$. Similarly, (5.2.65e) and (5.2.65g), together with (5.2.52) and

(5.2.54), imply that $\mathbf{c}(z)$ is analytic in the same region as $\mathbf{N}_n(z)$ (and, consequently, as $\mathbf{a}(z)$) and $\bar{\mathbf{c}}(z)$ is analytic in the same region as $\bar{\mathbf{N}}_n(z)$. By inserting the expansions (5.2.49a)–(5.2.49b), (5.2.52), and (5.2.54) for the eigenfunctions into the summation representations (5.2.65a)–(5.2.65g), we obtain the expansions for the scattering coefficients $\mathbf{a}(z)$, $\bar{\mathbf{a}}(z)$, $\mathbf{c}(z)$, and $\bar{\mathbf{c}}(z)$:

$$\mathbf{a}(z) = \mathbf{I}_N + O(z^{-2}, \text{even}),$$

$$\bar{\mathbf{c}}(z) = \mathbf{I}_N + \sum_{k=-\infty}^{+\infty} \mathbf{R}_k \mathbf{Q}_k \Delta_k^{-1} + O(z^2, \text{even}) \qquad (5.2.66\text{a})$$

$$\bar{\mathbf{a}}(z) = \mathbf{I}_M + O(z^2, \text{even}),$$

$$\mathbf{c}(z) = \mathbf{I}_M + \sum_{k=-\infty}^{+\infty} \mathbf{Q}_k \mathbf{R}_k \Omega_k^{-1} + O(z^{-2}, \text{even}), \qquad (5.2.66\text{b})$$

where Δ_n and Ω_n are given by (5.2.53) and (5.2.55). Hence $\mathbf{a}(z)$, $\bar{\mathbf{a}}(z)$ and $\mathbf{c}(z)$, $\bar{\mathbf{c}}(z)$ are even functions of the spectral parameter z. Moreover, the expressions (5.2.65b), (5.2.65d), (5.2.65f), and (5.2.65h) imply that $\mathbf{b}(z)$, $\bar{\mathbf{b}}(z)$ and $\mathbf{d}(z)$, $\bar{\mathbf{d}}(z)$ are odd functions of z.

Together with the eigenfunctions $\mathbf{M}_n(z)$ and $\bar{\mathbf{M}}_n(z)$, it is convenient to define

$$\boldsymbol{\mu}_n(z) = \mathbf{M}_n(z)\mathbf{a}^{-1}(z), \qquad \bar{\boldsymbol{\mu}}_n(z) = \bar{\mathbf{M}}_n(z)\bar{\mathbf{a}}^{-1}(z). \qquad (5.2.67)$$

Note that $\boldsymbol{\mu}_n(z)$ and $\bar{\boldsymbol{\mu}}_n(z)$ are meromorphic in z and have poles where, respectively, $\det \mathbf{a}(z) = 0$ and $\det \bar{\mathbf{a}}(z) = 0$. In terms of these functions, the relations (5.2.61a)–(5.2.61b) are

$$\boldsymbol{\mu}_n(z) - \bar{\mathbf{N}}_n(z) = z^{-2n}\mathbf{N}_n(z)\rho(z) \qquad (5.2.68\text{a})$$

$$\bar{\boldsymbol{\mu}}_n(z) - \mathbf{N}_n(z) = z^{2n}\bar{\mathbf{N}}_n(z)\bar{\rho}(z), \qquad (5.2.68\text{b})$$

where the reflection coefficients

$$\rho(z) = \mathbf{b}(z)\mathbf{a}^{-1}(z), \qquad \bar{\rho}(z) = \bar{\mathbf{b}}(z)\bar{\mathbf{a}}^{-1}(z) \qquad (5.2.69)$$

are part of the (n-independent) scattering data.

The system (5.2.68a)–(5.2.68b) is the starting point for the inverse problem. Indeed, if $\rho(z)$ and $\bar{\rho}(z)$ are known functions on the unit circle $|z| = 1$, then (5.2.68a)–(5.2.68b) is the boundary condition of a Riemann–Hilbert problem that we use to compute the functions $\boldsymbol{\mu}_n(z)$, $\mathbf{N}_n(z)$ in the region $|z| > 1$ and $\bar{\boldsymbol{\mu}}_n(z)$ and $\bar{\mathbf{N}}_n(z)$ in the region $|z| < 1$.

Proper eigenvalues and norming constants

Just as for the scalar case, we define a proper eigenvalue to be a value of z such that the scattering problem (5.2.16) has a solution that decays as $n \to \pm\infty$. If

$|z| > 1$, then the solutions $\phi_n(z)$ decay as $n \to -\infty$ while the solutions $\psi_n(z)$ decay as $n \to +\infty$. On the other hand, the solutions $\bar{\phi}_n(z)$ blow up as $n \to +\infty$ and the solutions $\bar{\psi}_n(z)$ blow up as $n \to -\infty$. Recall that $\phi_n(z), \bar{\phi}_n(z)$ and $\psi_n(z), \bar{\psi}_n(z)$ are both bases of solutions of the scattering problem. Therefore, if $|z_j| > 1$ is an eigenvalue, it must be that one of the solutions in the span of $\phi_n(z)$ is in the span of $\psi_n(z)$. That is, z_j is an eigenvalue if, and only if,

$$W\left(\phi_n(z_j), \psi_n(z_j)\right) = 0. \tag{5.2.70}$$

Similarly, $\bar{z}_\ell, |\bar{z}_\ell| < 1$, is an eigenvalue if and only if

$$W\left(\bar{\phi}_n(\bar{z}_\ell), \bar{\psi}_n(\bar{z}_\ell)\right) = 0. \tag{5.2.71}$$

From the other side, the Wronskian (5.2.56) of the eigenfunctions can be related to the scattering coefficient $\mathbf{a}(z)$ as follows:

$$
\begin{aligned}
W\left(\phi_n(z), \psi_n(z)\right) &= W\left(\left(\bar{\psi}_n(z), \psi_n(z)\right)\begin{pmatrix}\mathbf{a}(z)\\ \mathbf{b}(z)\end{pmatrix}, \left(\bar{\psi}_n(z), \psi_n(z)\right)\begin{pmatrix}\mathbf{0}\\ \mathbf{I}_M\end{pmatrix}\right)\\
&= W(\bar{\psi}_n(z), \psi_n(z)) \det\begin{pmatrix}\mathbf{a}(z) & \mathbf{0}\\ \mathbf{b}(z) & \mathbf{I}_M\end{pmatrix}\\
&= \frac{z^{n(N-M)}}{\prod_{j=n}^{+\infty} \det\left(\mathbf{I}_M - \mathbf{R}_j\mathbf{Q}_j\right)} \det\mathbf{a}(z).
\end{aligned}
\tag{5.2.72}
$$

Therefore, the eigenvalues in the region $|z| > 1$ are the points $z = z_j$ such that $\det\mathbf{a}(z_j) = 0$. Similarly, one can show that

$$W\left(\bar{\psi}_n(z), \bar{\phi}_n(z)\right) = \frac{z^{n(N-M)}}{\prod_{j=n}^{+\infty} \det\left(\mathbf{I}_M - \mathbf{R}_j\mathbf{Q}_j\right)} \det\bar{\mathbf{a}}(z), \tag{5.2.73}$$

and, therefore, the eigenvalues in the region $|z| < 1$ are the points $z = \bar{z}_\ell$ such that $\det\bar{\mathbf{a}}(\bar{z}_\ell) = 0$. There are no eigenvalues on the circle $|z| = 1$ because, on this circle, none of the basis eigenfunctions vanish as $n \to \pm\infty$. We also assume that $\det\mathbf{a}(z), \det\bar{\mathbf{a}}(z) \neq 0$ for $|z| = 1$. Moreover, we assume that there is a finite number of eigenvalues in the regions $|z| > 1$ and $|z| < 1$.

The function $\mu_n(z)$ has poles precisely at the points $z = z_j$ such that $\det\mathbf{a}(z_j) = 0$. Similarly, the function $\bar{\mu}_n(z)$ has poles precisely at the points $z = \bar{z}_\ell$ such that $\det\mathbf{a}(\bar{z}_\ell) = 0$. That is, the eigenvalues of the scattering problem are the poles of the meromorphic functions $\mu_n(z)$ and $\bar{\mu}_n(z)$ in the regions $|z| > 1$ and $|z| < 1$, respectively.

We now obtain the expressions for the residues of these poles. If the pole at $z = z_j$ is simple (equivalently, the zero of $\det\mathbf{a}(z_j)$ is simple), then

$$\operatorname{Res}\left(\mu_n; z_j\right) = \lim_{z \to z_j}\left(z - z_j\right)\mu_n(z).$$

The relation (5.2.68a) holds for any z such that all the eigenfunctions exist; therefore, if we write

$$\mathbf{a}^{-1}(z) = \frac{1}{a(z)}\boldsymbol{\alpha}(z),$$

where $a(z) = \det \mathbf{a}(z)$ and $\boldsymbol{\alpha}(z)$ is the cofactor matrix of $\mathbf{a}(z)$, then

$$\begin{aligned}
\mathrm{Res}\left(\boldsymbol{\mu}_n; z_j\right) &= \lim_{z \to z_j} \left\{ (z - z_j)\bar{\mathbf{N}}_n(z) + (z - z_j)z^{-2n}\mathbf{N}_n(z)\rho(z) \right\} \\
&= z_j^{-2n}\mathbf{N}_n(z_j) \lim_{z \to z_j}(z - z_j)\rho(z) \\
&= z_j^{-2n}\mathbf{N}_n(z_j)\mathbf{C}_j,
\end{aligned}$$
(5.2.74)

where the $M \times N$ matrix

$$\mathbf{C}_j = \lim_{z \to z_j}(z - z_j)\rho(z) = \frac{1}{a'(z_j)}\mathbf{b}(z_j)\boldsymbol{\alpha}(z_j) \qquad (5.2.75)$$

is the *norming constant* that relates $\mathrm{Res}\left(\boldsymbol{\mu}_n; z_j\right)$ and $\mathbf{N}_n(z_j)$ and $'$ denotes the derivative with respect to z.

By a similar procedure, we can find an expression for the residues of the poles of $\bar{\boldsymbol{\mu}}_n(z)$ in the region $|z| < 1$. We write

$$\bar{\mathbf{a}}^{-1}(z) = \frac{1}{\bar{a}(z)}\bar{\boldsymbol{\alpha}}(z),$$

where $\bar{a}(z) = \det \bar{\mathbf{a}}(z)$ and $\bar{\boldsymbol{\alpha}}(z)$ is the cofactor matrix of $\bar{\mathbf{a}}(z)$. If we assume $\bar{z}_\ell$ to be a simple zero of $\bar{a}(z)$, then

$$\mathrm{Res}\left(\bar{\boldsymbol{\mu}}_n; \bar{z}_\ell\right) = \bar{z}_\ell^{2n}\bar{\mathbf{N}}_n(\bar{z}_\ell)\bar{\mathbf{C}}_\ell, \qquad (5.2.76)$$

where, as above, the $N \times M$ matrix

$$\bar{\mathbf{C}}_\ell = \lim_{z \to \bar{z}_\ell}(z - \bar{z}_\ell)\bar{\rho}(z) = \frac{1}{\bar{a}'(\bar{z}_\ell)}\bar{\mathbf{b}}(\bar{z}_\ell)\bar{\boldsymbol{\alpha}}(\bar{z}_\ell) \qquad (5.2.77)$$

is the norming constant that relates the residue of $\bar{\boldsymbol{\mu}}_n(z)$ to the function $\bar{\mathbf{N}}_n(z)$.

Symmetries

In this subsection we compute symmetries of the scattering data that are induced by the symmetries in the potentials required to reduce the discrete matrix NLS (5.1.1a)–(5.1.1b) to the system (5.1.9a)–(5.1.9b), as indicated at the beginning of this section. Also, these symmetries in the scattering data must be included in the formulation of the inverse problem in order to construct solutions.

Previously, we showed that $\mathbf{a}(z)$ and $\bar{\mathbf{a}}(z)$ are even functions of z. The determinants of these functions are therefore also even functions of z. In particular, if $\det \mathbf{a}(z_j) = 0$, then $\det \mathbf{a}(-z_j) = 0$ and the eigenvalues appear in pairs (the same holds for the eigenvalues $|\bar{z}_j| < 1$). Without loss of generality, we will denote by z_j the eigenvalues such that $|\arg z_j| < \frac{\pi}{2}$ and $-z_j$ will be understood to be the eigenvalue such that $|\arg(-z_j)| > \frac{\pi}{2}$. If $|\arg z_j| = \frac{\pi}{2}$, we consider z_j to be the eigenvalue such that $\arg z_j = \frac{\pi}{2}$ and therefore $\arg(-z_j) = -\frac{\pi}{2}$. The same convention holds for the eigenvalues $\bar{z}_l$.

Let $\mathbf{C}_j^{\pm}$ denote the norming constants at $\pm z_j$. Then, recalling that $\mathbf{a}(z)$ is an even function of z while $\mathbf{b}(z)$ is odd, we have

$$\mathbf{C}_j^{-} = \frac{1}{a'(-z_j)}\mathbf{b}(-z_j)\alpha(-z_j) = \frac{1}{-a'(z_j)}(-\mathbf{b}(z_j))\alpha(z_j) = \mathbf{C}_j^{+}. \qquad (5.2.78)$$

Hence, we can drop the superscprits and denote both norming constants as $\mathbf{C}_j$. Similarly, the norming constant associated with $-\bar{z}_\ell$ is equal to the norming constant associated with the eigenvalue $\bar{z}_\ell$. Hence, we have established the following result:

Symmetry 5.1 For any potentials $\mathbf{Q}_n$, $\mathbf{R}_n$ such that the eigenfunctions are well defined, all of the eigenvalues appear in pairs $\pm z_j$ ($\pm\bar{z}_\ell$). Moreover, the norming constant associated with $-z_j$ ($-\bar{z}_\ell$) is equal to the norming constant associated with $+z_j$ ($+\bar{z}_\ell$), and we denote both by $\mathbf{C}_j$ ($\bar{\mathbf{C}}_\ell$).

Recall that, to reduce the discrete matrix NLS (5.1.1a)–(5.1.1b) to the discrete VNLS, we required two symmetries. First we impose the reduction (5.1.4) on the potentials, that is, we take $N = M$ and restrict ourselves to potential matrices $\mathbf{Q}_n$, $\mathbf{R}_n$ such that

$$\mathbf{Q}_n\mathbf{R}_n = \mathbf{R}_n\mathbf{Q}_n = \alpha_n\mathbf{I}. \qquad (5.2.79a)$$

For $N = 2$, we obtain potentials that satisfy this symmetry by imposing the condition (5.1.6) or, equivalently, (5.1.7). Second, to reduce the system (5.1.9a)–(5.1.9b) to the single-vector equation (5.1.10), we impose the further symmetry

$$\mathbf{R}_n = \mp\mathbf{Q}_n^{H}. \qquad (5.2.79b)$$

Each of the symmetries (5.2.79a) and (5.2.79b) induces a symmetry in the scattering data.

We first consider the case when $N = 2$ and the potentials satisfy the symmetry (5.1.6) or, equivalently, (5.1.7). We show how to compute the corresponding symmetries of the norming constants and reflection coefficients.

To compute the symmetry of the reflection coefficients, we define the functional

$$\mathbf{f}_n(z) = \left[\hat{\mathbf{P}}_{n-1}\bar{\phi}_n(i/z)\right]^{T}\phi_n(z),$$

where, as usual, the superscript T denotes standard matrix transposition and

$$\hat{\mathbf{P}}_n = \begin{pmatrix} (-i)^n \mathbf{P} & \mathbf{0} \\ \mathbf{0} & i^{n+1} \mathbf{P} \end{pmatrix}.$$

$\mathbf{P}$ is given by (5.1.8), and $\phi_n(z)$ and $\bar{\phi}_n(z)$ are the solutions of the block-matrix scattering problem (5.2.16) defined by the boundary conditions (5.2.18a). Note that this functional is well defined for $|z| \geq 1$ when (i) $\phi_n(z)$ exists for $|z| \geq 1$ and (ii) $\bar{\phi}_n(z)$ exists for $|z| \leq 1$. We relate this functional to the reflection coefficients by comparing the limits $\lim_{n \to +\infty} \mathbf{f}_n$ and $\lim_{n \to -\infty} \mathbf{f}_n$.

First, we calculate a recursion formula for $\mathbf{f}_n$ by making use of the relations

$$\mathbf{f}_{n+1}(z) = \left[\hat{\mathbf{P}}_n \bar{\phi}_{n+1}(i/z) \right]^T \phi_{n+1}(z)$$

$$= \left[\hat{\mathbf{P}}_n \mathbf{S}_n(i/z) \bar{\phi}_n(i/z) \right]^T \mathbf{S}_n(z) \phi_n(z)$$

$$= \left[\hat{\mathbf{P}}_{n-1} \bar{\phi}_n(i/z) \right]^T \left[\hat{\mathbf{P}}_n \mathbf{S}_n(i/z) (\hat{\mathbf{P}}_{n-1})^{-1} \right]^T \mathbf{S}_n(z) \phi_n(z),$$

where the second equality holds because, by definition, $\phi_n(z)$ and $\bar{\phi}_n(z)$ satisfy the scattering problem (5.2.16). The last expression can be simplified further. A direct calculation shows that

$$\left[\hat{\mathbf{P}}_n \mathbf{S}_n(i/z) \left(\hat{\mathbf{P}}_{n-1} \right)^{-1} \right]^T \mathbf{S}_n(z) = (1 - \alpha_n) \mathbf{I},$$

where, as we have assumed, $\mathbf{Q}_n \mathbf{R}_n = \mathbf{R}_n \mathbf{Q}_n = \alpha_n \mathbf{I}$. Hence,

$$\mathbf{f}_{n+1}(z) = (1 - \alpha_n) \left[\hat{\mathbf{P}}_{n-1} \bar{\phi}_n(i/z) \right]^T \phi_n(z) = (1 - \alpha_n) \mathbf{f}_n(z), \qquad (5.2.80)$$

and for any $j \in \mathbb{N}$

$$\mathbf{f}_n(z) = \left[\prod_{l=n}^{n+j-1} (1 - \alpha_l) \right]^{-1} \mathbf{f}_{n+j}(z) \qquad (5.2.81a)$$

$$\mathbf{f}_n(z) = \left[\prod_{l=n-j}^{n-1} (1 - \alpha_l) \right] \mathbf{f}_{n-j}(z). \qquad (5.2.81b)$$

Now we evaluate the right-hand sides of (5.2.81a)–(5.2.81b) in the limit $j \to \infty$. First we rewrite the right-hand side of (5.2.81a) by using the scattering equations (5.2.59a)–(5.2.59b) to obtain

$$\mathbf{f}_{n+j}(z) = i^{n+j} \left(\bar{\mathbf{b}}(i/z) \right)^T \left(\bar{\mathbf{N}}_{n+j}(i/z) \right)^T \left(\hat{\mathbf{P}}_{n+j-1} \right)^T \bar{\mathbf{N}}_{n+j}(z) \mathbf{a}(z)$$

$$+ i^{n+j} z^{-2(n+j)} \left(\bar{\mathbf{b}}(i/z) \right)^T \left(\bar{\mathbf{N}}_{n+j}(i/z) \right)^T \left(\hat{\mathbf{P}}_{n+j-1} \right)^T \mathbf{N}_{n+j}(z) \mathbf{b}(z)$$

$$+ i^{-(n+j)} z^{2(n+j)} (\bar{\mathbf{a}}(i/z))^T \left(\mathbf{N}_{n+j}(i/z)\right)^T \left(\hat{\mathbf{P}}_{n+j-1}\right)^T \bar{\mathbf{N}}_{n+j}(z)\mathbf{a}(z)$$

$$+ i^{-(n+j)} (\bar{\mathbf{a}}(i/z))^T \left(\mathbf{N}_{n+j}(i/z)\right)^T \left(\hat{\mathbf{P}}_{n+j-1}\right)^T \mathbf{N}_{n+j}(z)\mathbf{b}(z),$$

which, for $|z| = 1$, yields

$$\lim_{j \to +\infty} \mathbf{f}_{n+j}(z) = -i \left(\bar{\mathbf{b}}(i/z)\right)^T \mathbf{P}\, \mathbf{a}(z) - (\bar{\mathbf{a}}(i/z))^T \mathbf{P}\, \mathbf{b}(z).$$

Hence, by (5.2.81a), we have

$$\mathbf{f}_n(z) = \left[\prod_{l=n}^{+\infty}(1 - \alpha_l)\right]^{-1} \left[-i \left(\bar{\mathbf{b}}(i/z)\right)^T \mathbf{P}\, \mathbf{a}(z) - (\bar{\mathbf{a}}(i/z))^T \mathbf{P}\, \mathbf{b}(z)\right].$$

$$(5.2.82)$$

On the other hand,

$$\lim_{j \to +\infty} \mathbf{f}_{n-j}(z) = \lim_{j \to +\infty} \left(\bar{\phi}_{n-j}(i/z)\right)^T \left(\hat{\mathbf{P}}_{n-j-1}\right)^T \phi_{n-j}(z)$$

$$= \lim_{j \to +\infty} i^{j-n} z^{2(n-j)} \left(\bar{\mathbf{M}}_{n-j}(i/z)\right)^T \left(\hat{\mathbf{P}}_{n-j-1}\right)^T \mathbf{M}_{n-j}(z)$$

$$= \mathbf{0};$$

so, by (5.2.81b), we get

$$\mathbf{f}_n(z) = \left[\prod_{l=-\infty}^{n-1}(1 - \alpha_l)\right]\mathbf{0} = \mathbf{0}. \qquad (5.2.83)$$

Comparing (5.2.82) and (5.2.83), we obtain

$$-i(\bar{\mathbf{b}}(i/z))^T \mathbf{P}\, \mathbf{a}(z) - (\bar{\mathbf{a}}(i/z))^T \mathbf{P}\, \mathbf{b}(z) = \mathbf{0},$$

which, by definition (5.2.69) of the reflection coefficients, is equivalent to

$$\bar{\rho}(i/z) = -i\mathbf{P}\, (\rho(z))^T \mathbf{P}; \qquad (5.2.84)$$

that is, if the potentials satisfy the symmetry (5.1.4), then the reflection coefficients satisfy (5.2.84).

The symmetry (5.2.79a) in the potentials also induces a symmetry in the eigenvalues. We have shown that the eigenvalues in $|z| > 1$ are the zeros of $\det \mathbf{a}(z)$ and the eigenvalues in $|z| < 1$ are the zeros of $\det \bar{\mathbf{a}}(z)$. To establish the relation between the eigenvalues in $|z| > 1$ and the eigenvalues in $|z| < 1$, we first establish a relationship between $\det \mathbf{a}(z)$ and $\det \mathbf{c}(z)$. Then we establish a relation between $\mathbf{c}(z)$ and $\bar{\mathbf{a}}(i/z)$. Note that if $|z| > 1$, then $|i/z| < 1$.

Equation (5.2.72) gives the relation between $\det \mathbf{a}(z)$ and the Wronskian of $\phi_n(z)$, $\psi_n(z)$. With $M = N$ and the additional symmetry (5.2.79a), equation

(5.2.72) simplifies to

$$W\left(\phi_n(z), \psi_n(z)\right) = \left[\prod_{j=n}^{+\infty} (1 - \alpha_j)\right]^{-N} \det \mathbf{a}(z). \qquad (5.2.85)$$

On the other hand,

$$W\left(\phi_n(z), \psi_n(z)\right) = W\left(\left(\phi_n(z), \bar{\phi}_n(z)\right)\begin{pmatrix} \mathbf{I}_N \\ \mathbf{0} \end{pmatrix}, \left(\phi_n(z), \bar{\phi}_n(z)\right)\begin{pmatrix} \mathbf{d}(z) \\ \mathbf{c}(z) \end{pmatrix}\right)$$

$$= \left[\prod_{j=-\infty}^{n-1} (1 - \alpha_j)\right]^N \det \mathbf{c}(z), \qquad (5.2.86)$$

where we used (5.2.58). Comparing (5.2.85) and (5.2.86), we obtain

$$\det \mathbf{c}(z) = \left[\prod_{j=-\infty}^{+\infty} (1 - \alpha_j)\right]^{-N} \det \mathbf{a}(z). \qquad (5.2.87)$$

Analogously, the functional

$$\hat{\mathbf{f}}_n(z) = \left[\hat{\mathbf{P}}_{n-1}\bar{\phi}_n(i/z)\right]^T \psi_n(z)$$

is well defined for $|z| \geq 1$ when (i) $\psi_n(z)$ exists for $|z| \geq 1$ and (ii) $\bar{\phi}_n(z)$ exists for $|z| \leq 1$. By the same argument that established (5.2.80) for the functional $\mathbf{f}_n(z)$, we obtain

$$\hat{\mathbf{f}}_{+\infty}(z) = \left[\prod_{j=-\infty}^{+\infty} (1 - \alpha_j)\right] \hat{\mathbf{f}}_{-\infty}(z),$$

where the subscript $\pm\infty$ is understood to mean the limit as $n \to \pm\infty$. By evaluating the limits we obtain

$$\bar{\mathbf{a}}^T(i/z)\,\mathbf{P} = \left[\prod_{j=-\infty}^{+\infty} (1 - \alpha_j)\right] \mathbf{P}\,\mathbf{c}(z). \qquad (5.2.88)$$

Analogously, if we consider

$$\hat{\mathbf{h}}_n(z) = \left[\hat{\mathbf{P}}_{n-1}\bar{\psi}_n(i/z)\right]^T \phi_n(z),$$

which is well defined for $|z| \geq 1$, exactly the same arguments lead to the relation

$$\mathbf{P}\,\mathbf{a}(z) = \left[\prod_{j=-\infty}^{+\infty} (1 - \alpha_j)\right] \bar{\mathbf{c}}^T(i/z)\,\mathbf{P}; \qquad (5.2.89)$$

therefore

$$\det \bar{\mathbf{a}}(i/z) = \left[\prod_{j=-\infty}^{+\infty} (1 - \alpha_j) \right]^M \det \mathbf{c}(z) \qquad (5.2.90)$$

$$\det \mathbf{a}(z) = \left[\prod_{j=-\infty}^{+\infty} (1 - \alpha_j) \right]^N \det \bar{\mathbf{c}}(i/z). \qquad (5.2.91)$$

Comparing (5.2.87) and (5.2.90) we have

$$\det \bar{\mathbf{a}}(i/z) = \det \mathbf{a}(z),$$

which shows that $\hat{z}_j = i/z_j$ is an eigenvalue if, and only if, z_j is an eigenvalue.

Now we compute the symmetry of the norming constants associated with the pair of poles z_j, $\hat{z}_j = i/z_j$. We assume, as we have previously, that all the eigenfunctions are defined in the neighborhood of each eigenvalue. Let $\mathbf{C}_j$ be the norming constant associated with the eigenvalue z_j and $\hat{\mathbf{C}}_j$ be the norming constant associated with the eigenvalue $\hat{z}_j$. The norming constants are given by (5.2.75), so

$$\hat{\mathbf{C}}_j = \lim_{z \to \hat{z}_j} (z - \hat{z}_j)\bar{\rho}(z) = -iz_j^{-2} \lim_{w \to z_j} \left(w - z_j\right) \bar{\rho}(i/w),$$

and, applying the symmetry (5.2.84), we obtain

$$\hat{\mathbf{C}}_j = -z_j^{-2}\mathbf{P}\,\mathbf{C}_j^T\mathbf{P}.$$

Hence, we have shown that:

Symmetry 5.2 If the potentials $\mathbf{Q}_n$, $\mathbf{R}_n$ with $N = 2$ satisfy the symmetry (5.1.6) or, equivalently, (5.1.7), then:

(i) The reflection coefficients satisfy the symmetry

$$\bar{\rho}(i/z) = -i\mathbf{P}\,\rho^T(z)\,\mathbf{P}. \qquad (5.2.92a)$$

(ii) $\hat{z}_j = i/z_j$ is an eigenvalue such that $|\hat{z}_j| < 1$ if and only if z_j is an eigenvalue such that $|z_j| > 1$.

(iii) The norming constants associated with these poles have the symmetry

$$\hat{\mathbf{C}}_j = -z_j^{-2}\mathbf{P}\,\mathbf{C}_j^T\mathbf{P}, \qquad (5.2.92b)$$

where $\mathbf{C}_j$ is the norming constant associated with z_j and $\hat{\mathbf{C}}_j$ is the norming constant associated with $\hat{z}_j$. As a consequence, $\bar{z}_j = i/\bar{z}_j$ is an eigenvalue such that $|\bar{z}_j| > 1$ if, and only if, $\bar{z}_j$ is an eigenvalue such that

$|\bar{z}_j| < 1$ and the norming constants associated with these poles have the symmetry

$$\tilde{\mathbf{C}}_j = \bar{z}_j^{-2} \mathbf{P} \, \bar{\mathbf{C}}_j^T \mathbf{P}, \qquad (5.2.92\text{c})$$

where $\bar{\mathbf{C}}_j$ is the norming constant associated with $\bar{z}_j$ and $\tilde{\mathbf{C}}_j$ is the norming constant associated with $\tilde{z}_j$. As a consequence, $J = \bar{J}$, that is, the number of eigenvalues inside the unit circle is equal to the number of eigenvalues outside.

Now we turn to the symmetry in the scattering data induced by the symmetry (5.2.79b), that is, $\mathbf{R}_n = \mp \mathbf{Q}_n^H$, when the symmetry (5.2.79a), $\mathbf{Q}_n \mathbf{R}_n = \mathbf{R}_n \mathbf{Q}_n = \alpha_n \mathbf{I}_N$, also holds. This symmetry in the scattering data is determined by methods similar to those discussed previously, but the calculation holds for any N. First, we consider the functional

$$\mathbf{g}_n^{\pm}(z) = \left[\sigma_{\pm} \bar{\phi}_n(1/z^*) \right]^H \phi_n(z),$$

where

$$\sigma_{\pm} = \begin{pmatrix} \mathbf{I}_N & \mathbf{0} \\ \mathbf{0} & \pm \mathbf{I}_M \end{pmatrix}. \qquad (5.2.93)$$

Using the scattering problem (5.2.16), it is easy to show that when symmetries (5.2.79a) and (5.2.79b) hold, these functionals satisfy the recursion relation

$$\mathbf{g}_{n+1}^{\pm}(z) = (1 - \alpha_n)\mathbf{g}_n^{\pm}(z). \qquad (5.2.94)$$

Hence,

$$\mathbf{g}_{+\infty}^{\pm}(z) = \left[\prod_{j=-\infty}^{+\infty} (1 - \alpha_j) \right] \mathbf{g}_{-\infty}^{\pm}(z),$$

where the subscripts $\pm\infty$ are understood to denote the limits of $\mathbf{g}_n^{\pm}$ as $n \to \pm\infty$. By evaluating these asymptotics by means of equations (5.2.18a)–(5.2.18b) and (5.2.59a)–(5.2.59b), we conclude that (for $|z| = 1$, or also $|z| \geq 1$ provided the decay of the eigenfunctions is fast enough), the following relation holds:

$$\pm \bar{\mathbf{a}}^H (1/z^*) \mathbf{b}(z) + \bar{\mathbf{b}}^H (1/z^*) \mathbf{a}(z) = \mathbf{0}$$

or, equivalently, in terms of the reflection coefficients,

$$\bar{\rho}(z) = \mp \rho^H (1/z^*) \qquad (5.2.95)$$

if symmetries (5.2.79a) and (5.2.79b) hold for the potentials.

These symmetries in the potentials also induce a symmetry in the eigenvalues. Indeed, the functionals $\hat{\mathbf{g}}_n^{\pm}(z) = \left[\sigma_{\pm}\bar{\phi}_n(1/z^*)\right]^H \psi_n(z)$ and $\hat{\mathbf{l}}_n^{\pm}(z) = \left[\sigma_{\pm}\bar{\psi}_n(1/z^*)\right]^H \phi_n(z)$ satisfy the recursion relation (5.2.94), and the comparison between their asymptotic behaviors as $n \to \pm\infty$ yields

$$\bar{\mathbf{a}}^H(1/z^*) = \left[\prod_{j=-\infty}^{+\infty} (1 - \alpha_j)\right] \mathbf{c}(z) \qquad (5.2.96a)$$

$$\mathbf{a}(z) = \left[\prod_{j=-\infty}^{+\infty} (1 - \alpha_j)\right] \bar{\mathbf{c}}^H(1/z^*). \qquad (5.2.96b)$$

Finally, similar calculations show that

$$\bar{\mathbf{b}}^H(1/z^*) = \left[\prod_{j=-\infty}^{+\infty} (1 - \alpha_j)\right] \bar{\mathbf{d}}(z) \qquad (5.2.96c)$$

$$\mathbf{b}^H(1/z^*) = \left[\prod_{j=-\infty}^{+\infty} (1 - \alpha_j)\right] \mathbf{d}(z). \qquad (5.2.96d)$$

Note that (5.2.96a) and (5.2.96b) also imply

$$\det \bar{\mathbf{a}}(z) = \left[\prod_{j=-\infty}^{+\infty} (1 - \alpha_j)\right]^M \left(\det \mathbf{c}(1/z^*)\right)^*$$

$$\det \mathbf{a}(z) = \left[\prod_{j=-\infty}^{+\infty} (1 - \alpha_j)\right]^N \left(\det \bar{\mathbf{c}}(1/z^*)\right)^*,$$

which, taking into account (5.2.87), give

$$\det \bar{\mathbf{a}}(1/z^*) = \left[\prod_{j=-\infty}^{+\infty} (1 - \alpha_j)\right]^{M-N} \left(\det \mathbf{a}(z)\right)^*. \qquad (5.2.97)$$

We therefore conclude that $\bar{z}_j = 1/z_j^*$ is an eigenvalue if, and only if, z_j is an eigenvalue. The norming constant associated with the eigenvalue $\bar{z}_j$ can be related to the norming constant associated with z_j as follows:

$$\bar{\mathbf{C}}_j = \lim_{z \to \bar{z}_j} (z - \bar{z}_j)\, \bar{\rho}(z) = \pm \lim_{w \to z_j} \frac{w^* - z_j^*}{z_j^* w^*} \rho^H(w) = \pm(z_j^*)^{-2} \mathbf{C}_j^H.$$

Note that only the case corresponding to the upper sign, that is, for $\mathbf{R}_n = -\mathbf{Q}_n^H$, is relevant since no proper eigenvalues exist for the scattering problem with the symmetry reduction $\mathbf{R}_n = \mathbf{Q}_n^H$.

We have therefore shown that:

Symmetry 5.3 If the potentials satisfy the symmetry (5.2.79b), that is, $\mathbf{R}_n = \mp\mathbf{Q}_n^H$, in addition to the symmetry (5.2.79a), that is, $\mathbf{Q}_n\mathbf{R}_n = \mathbf{R}_n\mathbf{Q}_n = \alpha_n\mathbf{I}$, then:

(i) the reflection coefficients satisfy the symmetry

$$\bar{\rho}(z) = \mp\rho^H(1/z^*); \tag{5.2.98a}$$

(ii) $\bar{z}_j = 1/z_j^*$ is an eigenvalue such that $\left|\bar{z}_j\right| < 1$ if, and only if, z_j is an eigenvalue such that $\left|z_j\right| > 1$ and, consequently, the number of eigenvalues inside the unit circle is equal to the number of eigenvalues outside, that is, $J = \bar{J}$;

(iii) the norming constants associated with these paired eigenvalues satisfy the symmetry

$$\bar{\mathbf{C}}_j = \pm(z_j^*)^{-2}\mathbf{C}_j^H, \tag{5.2.98b}$$

where $\mathbf{C}_j$ is the norming constant associated with z_j and $\bar{\mathbf{C}}_j$ is the norming constant associated with $\bar{z}_j$.

As a consequence, if Symmetry 5.3 holds, then the eigenvalues appear in sets of eight,

$$\left\{\pm z_j, \pm i/z_j, \pm 1/z_j^*, \pm iz_j^*\right\}_{j=1}^{J}$$

as shown in Figure 5.1.

Note that, since we have assumed that on the unit circle $\det\mathbf{a}(z) \neq 0$ and $\det\bar{\mathbf{a}}(z) \neq 0$, from (5.2.64) we obtain the following relations:

$$\mathbf{a}(z) - \bar{\mathbf{b}}(z)\bar{\mathbf{a}}^{-1}(z)\mathbf{b}(z) = \bar{\mathbf{c}}^{-1}(z) \tag{5.2.99a}$$

$$\bar{\mathbf{a}}(z) - \mathbf{b}(z)\mathbf{a}^{-1}(z)\bar{\mathbf{b}}(z) = \mathbf{c}^{-1}(z). \tag{5.2.99b}$$

Then, taking into account the symmetries (5.2.96a)–(5.2.96b) and (5.2.96c)–(5.2.96d), we also have

$$\left[\mathbf{a}(z) - \bar{\mathbf{b}}(z)\bar{\mathbf{a}}^{-1}(z)\mathbf{b}(z)\right]\mathbf{a}^H(z) = \prod_{j=-\infty}^{+\infty}(1 - \alpha_j)\mathbf{I}_N \qquad \text{for } |z| = 1. \tag{5.2.100}$$

Equation (5.2.100) yields, from one side,

$$|\det\mathbf{a}(z)|^2 = \left[\prod_{j=-\infty}^{+\infty}(1 - \alpha_j)\right]^N [\det(\mathbf{I}_N - \bar{\rho}(z)\rho(z))]^{-1} \qquad \text{for } |z| = 1 \tag{5.2.101}$$

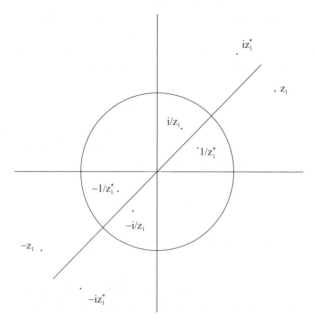

Figure 5.1: Symmetry of the eigenvalues for the block-matrix scattering problem (cf. Symmetry 5.3)

and, from the other side, taking into account (5.2.96b),

$$\det \begin{pmatrix} \mathbf{a} & \bar{\mathbf{b}} \\ \mathbf{b} & \bar{\mathbf{a}} \end{pmatrix} = \left[\prod_{j=-\infty}^{+\infty} (1 - \alpha_j) \right]^N. \qquad (5.2.102)$$

We can summarize the symmetries of the scattering data induced by the symmetries in the potentials as follows. In the general case, only Symmetry 5.1 holds, that is,

Symmetry 5.1 All the eigenvalues appear in pairs $\pm z_j$ ($\pm \bar{z}_\ell$) outside and inside the unit circle, respectively. Moreover, the norming constant associated with $-z_j$ (resp. $-\bar{z}_\ell$) is equal to the norming constant associated with $+z_j$ (resp. $+\bar{z}_\ell$).

If the potentials $\mathbf{Q}_n$, $\mathbf{R}_n$ with $N = 2$ satisfy the symmetry (5.1.6)–(5.1.8), that is, $\mathbf{R}_n = (-1)^n \mathbf{P} \mathbf{Q}_n^T \mathbf{P}$, then Symmetry 5.2 holds, that is,

Symmetry 5.2

(i) The reflection coefficients satisfy the symmetry

$$\bar{\rho}(i/z) = -i \mathbf{P} \, \rho^T(z) \, \mathbf{P}.$$

(ii) $\hat{z}_j = i/z_j$ is an eigenvalue such that $|\hat{z}_j| < 1$ if, and only if, z_j is an eigenvalue such that $|z_j| > 1$. As a consequence, $\bar{\tilde{z}}_j = i/\bar{z}_j$ is an eigenvalue such that $|\bar{\tilde{z}}_j| > 1$ if, and only if, $\bar{z}_j$ is an eigenvalue such that $|\bar{z}_j| < 1$.

(iii) The norming constants associated with these poles have the symmetry

$$\hat{\mathbf{C}}_j = -z_j^{-2}\mathbf{P}\,\mathbf{C}_j^T\mathbf{P},$$

where $\mathbf{C}_j$ is the norming constant associated with z_j and $\hat{\mathbf{C}}_j$ is the norming constant associated with $\hat{z}_j$. Moreover,

$$\tilde{\mathbf{C}}_j = \bar{z}_j^{-2}\mathbf{P}\,\bar{\mathbf{C}}_j^T\mathbf{P},$$

where $\bar{\mathbf{C}}_j$ is the norming constant associated with $\bar{z}_j$ and $\tilde{\mathbf{C}}_j$ is the norming constant associated with $\tilde{z}_j$.

Finally, if the potentials satisfy the symmetry $\mathbf{R}_n = \mp\mathbf{Q}_n^H$, in addition to the symmetry $\mathbf{Q}_n\mathbf{R}_n = \mathbf{R}_n\mathbf{Q}_n = \alpha_n\mathbf{I}$, then Symmetry 5.3 holds, that is,

Symmetry 5.3

(i) The reflection coefficients satisfy the symmetry

$$\bar{\rho}(z) = \mp\rho^H(1/z^*).$$

(ii) $\bar{z}_j = 1/z_j^*$ is an eigenvalue such that $|\bar{z}_j| < 1$ if, and only if, z_j is an eigenvalue such that $|z_j| > 1$, and therefore, taking into account Symmetry 5.2, discrete eigenvalues appear in octets $\left\{\pm z_j, \pm i/z_j, \pm 1/z_j^*, \pm iz_j^*\right\}_{j=1}^{j}$, with four inside the unit circle and four outside (cf. Figure 5.1). Note that this is a generalization with respect to the scalar IDNLS scattering problem in which the eigenvalues appear in quartets (cf. Chapter 3).

(iii) The norming constants associated with these paired eigenvalues satisfy the symmetry

$$\bar{\mathbf{C}}_j = \pm(z_j^*)^{-2}\mathbf{C}_j^H,$$

where $\mathbf{C}_j$ is the norming constant associated with z_j and $\bar{\mathbf{C}}_j$ is the norming constant associated with $\bar{z}_j$.

Trace formula

Assume $a(z) = \det \mathbf{a}(z)$ and $\bar{a}(z) = \det \bar{\mathbf{a}}(z)$ to have the simple zeros $\{\pm z_j : |z_j| > 1\}_{j=1}^J$ and $\{\pm \bar{z}_j : |\bar{z}_j| < 1\}_{j=1}^{\bar{J}}$, respectively, and define

$$\alpha(z) = \prod_{m=1}^{J} \frac{(z^2 - (z_m^*)^{-2})}{(z^2 - z_m^2)} a(z), \qquad \bar{\alpha}(z) = \prod_{m=1}^{\bar{J}} \frac{(z^2 - (\bar{z}_m^*)^{-2})}{(z^2 - \bar{z}_m^2)} \bar{a}(z).$$

$$(5.2.103)$$

According to (5.2.103), $\alpha(z)$ is analytic outside the unit circle, where it has no zeros, while $\bar{\alpha}(z)$ is analytic inside, where it has no zeros; moreover, due to (5.2.66a), $\alpha(z) \to 1$ as $|z| \to \infty$. Therefore, by taking into account that both a and $\bar{a}$ are even functions of z, we can write

$$\log a(z) = \sum_{m=1}^{J} \log \left(\frac{z^2 - z_m^2}{z^2 - (z_m^*)^2} \right)$$
$$- \frac{1}{2\pi i} \oint_{|w|=1} \frac{w \log (\bar{\alpha}(w)\alpha(w))}{w^2 - z^2} dw, \qquad |z| > 1 \quad (5.2.104a)$$

$$\log \bar{a}(z) = \sum_{m=1}^{\bar{J}} \log \left(\frac{z^2 - \bar{z}_m^2}{z^2 - (\bar{z}_m^*)^2} \right)$$
$$+ \frac{1}{2\pi i} \oint_{|w|=1} \frac{w \log (\alpha(w)\bar{\alpha}(w))}{w^2 - z^2} dw \qquad |z| < 1. \quad (5.2.104b)$$

If Symmetry 5.3 holds, the eigenvalues appear in sets of eight, $\{\pm z_m, \pm i/z_m, \pm 1/z_m^*, \pm i z_m^*\}$, that is, in particular $\bar{J} = J$ and $\bar{z}_j = 1/z_j^*$. Then, taking into account (5.2.97) and (5.2.101), from (5.2.104a)–(5.2.104b) we obtain

$$\log a(z) = \sum_{m=1}^{J} \log \frac{(z^2 - z_m^2)(z^2 + \bar{z}_m^{-2})}{(z^2 - \bar{z}_m^2)(z^2 + z_m^{-2})}$$
$$+ \frac{1}{2\pi i} \oint_{|w|=1} \frac{w \log [\det (\mathbf{I}_N - \bar{\rho}(w)\rho(w))]}{w^2 - z^2} dw, \quad (5.2.105)$$

which allows us to recover $a(z)$, $\bar{a}(z)$ for any z in the proper region from knowledge of the discrete eigenvalues and of the reflection coefficient on the unit circle.

The problem of reconstructing the matrices $\mathbf{a}(z)$ and $\bar{\mathbf{a}}(z)$ is more complicated. However, from inverse scattering one can reconstruct potentials and eigenfunctions and then obtain the matrix transmission coefficients by using, for instance, formulas (5.2.65a) and (5.2.65c).

5.2.2 *Inverse scattering problem*

Boundary conditions and residues

In this subsection, we reconstruct the potentials from the scattering data. As in the 2×2 scattering problem, we first reconstruct the eigenfunctions from the scattering data and then recover the potentials from the eigenfunctions.

To formulate the inverse problem as a Riemann–Hilbert problem in the complex variable z, we must specify (i) the boundary conditions of the eigenfunctions as $|z| \to \infty$ and (ii) the equations that determine the residues of the poles of the meromorphic functions $\mu_n(z)$ and $\bar{\mu}_n(z)$. According to the expansions (5.2.49a), (5.2.52), and (5.2.66a) and definition (5.2.67), we have the boundary conditions

$$\mathbf{N}_n(z) \to \begin{pmatrix} \mathbf{0} \\ \mathbf{\Delta}_n^{-1} \end{pmatrix}, \qquad \bar{\mu}_n(z) \to \begin{pmatrix} \mathbf{0} \\ \mathbf{I}_M \end{pmatrix}$$

as $|z| \to \infty$, with $\mathbf{\Delta}_n$ given by (5.2.53). Hence, the boundary condition for $\mathbf{N}_n(z)$ depends on $\mathbf{Q}_n$ and $\mathbf{R}_n$, which are unknowns in the inverse problem. Therefore, we introduce the following modified functions:

$$\mathbf{N}_n' = \begin{pmatrix} \mathbf{I}_N & \mathbf{0} \\ \mathbf{0} & \mathbf{\Delta}_n \end{pmatrix} \mathbf{N}_n = \begin{pmatrix} -z^{-1}\mathbf{Q}_n\mathbf{\Delta}_n^{-1} + O(z^{-3}) \\ \mathbf{I}_M + O(z^{-2}) \end{pmatrix} \qquad (5.2.106a)$$

$$\mu_n' = \begin{pmatrix} \mathbf{I}_N & \mathbf{0} \\ \mathbf{0} & \mathbf{\Delta}_n \end{pmatrix} \mu_n = \begin{pmatrix} \mathbf{I}_N + O(z^{-2}) \\ z^{-1}\mathbf{\Delta}_n\mathbf{R}_{n-1} + O(z^{-3}) \end{pmatrix} \qquad (5.2.106b)$$

$$\bar{\mathbf{N}}_n' = \begin{pmatrix} \mathbf{I}_N & \mathbf{0} \\ \mathbf{0} & \mathbf{\Delta}_n \end{pmatrix} \bar{\mathbf{N}}_n = \begin{pmatrix} \mathbf{\Omega}_n^{-1} + O(z^2) \\ -z\mathbf{\Delta}_n\mathbf{R}_n\mathbf{\Omega}_n^{-1} + O(z^3) \end{pmatrix} \qquad (5.2.106c)$$

$$\bar{\mu}_n' = \begin{pmatrix} \mathbf{I}_N & \mathbf{0} \\ \mathbf{0} & \mathbf{\Delta}_n \end{pmatrix} \bar{\mu}_n = \begin{pmatrix} z\mathbf{Q}_{n-1} + O(z^3) \\ \mathbf{\Delta}_n + O(z^2) \end{pmatrix}, \qquad (5.2.106d)$$

where $\mathbf{\Omega}_n$ is given by (5.2.55). Hence, by this modification of the eigenfunctions, we eliminate the dependence of the boundary conditions on the potentials. We emphasize that $'$ does not indicate a derivative with respect to z, but rather the modification of the eigenfunctions by the matrix prefactor.

From (5.2.68a)–(5.2.68b) it follows that these modified functions satisfy the jump conditions

$$\mu_n'(z) - \bar{\mathbf{N}}_n'(z) = z^{-2n}\mathbf{N}_n'(z)\rho(z) \qquad (5.2.107a)$$

$$\bar{\mu}_n'(z) - \mathbf{N}_n'(z) = z^{2n}\bar{\mathbf{N}}_n'(z)\bar{\rho}(z). \qquad (5.2.107b)$$

Also, the poles of $\mu_n'(z)$ and $\bar{\mu}_n'(z)$ are the same as the poles of $\mu_n(z)$ and $\bar{\mu}_n(z)$, respectively. The residues of these poles are determined by the relations:

$$\text{Res}\left(\mu'_n; z_j\right) = z_j^{-2n} \mathbf{N}'_n(z_j) \mathbf{C}_j \qquad (5.2.108a)$$

$$\text{Res}\left(\bar{\mu}'_n; \bar{z}_j\right) = \bar{z}_j^{2n} \bar{\mathbf{N}}'_n(\bar{z}_j) \bar{\mathbf{C}}_j, \qquad (5.2.108b)$$

which follow from (5.2.74) and (5.2.76). We therefore solve the Riemann–Hilbert problem for the modified eigenfunctions (5.2.106a)–(5.2.106d).

Recovery of the eigenfunctions

The first step of the inverse problem is to recover the functions $\mathbf{N}'_n(z)$ and $\bar{\mathbf{N}}'_n(z)$ from the scattering data. Let us consider first the case when there are no discrete eigenvalues, that is, μ'_n and $\bar{\mu}'_n$ have no poles. Introducing the $(N + M) \times (N + M)$ matrices

$$\mathbf{m}_n(z) = \left(\mu'_n(z), \mathbf{N}'_n(z)\right), \qquad \bar{\mathbf{m}}_n(z) = \left(\bar{\mathbf{N}}'_n(z), \bar{\mu}'_n(z)\right) \qquad (5.2.109a)$$

with $\mathbf{m}_n(z)$ analytic outside the unit circle $|z| = 1$ and $\bar{\mathbf{m}}_n(z)$ analytic inside, we can write the "jump" conditions (5.2.107a)–(5.2.107b) as

$$\mathbf{m}_n(z) - \bar{\mathbf{m}}_n(z) = \bar{\mathbf{m}}_n(z)\mathbf{V}_n(z) \qquad |z| = 1, \qquad (5.2.109b)$$

where

$$\mathbf{V}_n(z) = \begin{pmatrix} -\bar{\rho}(z)\rho(z) & -z^{2n}\bar{\rho}(z) \\ z^{-2n}\rho(z) & 0 \end{pmatrix} \qquad (5.2.109c)$$

and

$$\mathbf{m}_n(z) \to \mathbf{I}_{N+M} \qquad \text{as } |z| \to \infty. \qquad (5.2.109d)$$

Therefore (5.2.109b) can be regarded as a matrix Riemann–Hilbert boundary-value problem on $|z| = 1$ with boundary conditions given by (5.2.109d). In analogy with what was done for the scalar case, we introduce the integral operators

$$\bar{P}(\mathbf{F})(z) = \lim_{\substack{\zeta \to z \\ |\zeta| < 1}} \frac{1}{2\pi i} \oint_{|w|=1} \frac{\mathbf{F}(w)}{w - \zeta} dw \qquad (5.2.110a)$$

$$P(\mathbf{F})(z) = \lim_{\substack{\zeta \to z \\ |\zeta| > 1}} \frac{1}{2\pi i} \oint_{|w|=1} \frac{\mathbf{F}(w)}{w - \zeta} dw \qquad (5.2.110b)$$

defined for $|z| < 1$ and $|z| > 1$, respectively, for any matrix-valued function $\mathbf{F}(w)$ continuous on $|w| = 1$. Applying $\bar{P}$ to both sides of equations (5.2.109b) yields

$$\bar{\mathbf{m}}_n(z) = \mathbf{I}_{N+M} - \lim_{\substack{\zeta \to z \\ |\zeta| < 1}} \frac{1}{2\pi i} \oint_{|w|=1} \frac{\bar{\mathbf{m}}_n(w)\mathbf{V}_n(w)}{w - \zeta} dw, \qquad (5.2.111)$$

which allows us, in principle, to find $\bar{\mathbf{m}}_n(z)$ from the scattering data. In component form, (5.2.111) yields

$$\bar{\mathbf{N}}'_n(z) = \begin{pmatrix} \mathbf{I}_N \\ \mathbf{0} \end{pmatrix} - \lim_{\substack{\zeta \to z \\ |\zeta|<1}} \frac{1}{2\pi i} \oint_{|w|=1} \frac{w^{-2n}\bar{\mu}'_n(w)\rho(w) - \bar{\mathbf{N}}'_n(w)\bar{\rho}(w)\rho(w)}{w - \zeta} dw$$

(5.2.112a)

$$\bar{\mu}'_n(z) = \begin{pmatrix} \mathbf{0} \\ \mathbf{I}_M \end{pmatrix} + \lim_{\substack{\zeta \to z \\ |\zeta|<1}} \frac{1}{2\pi i} \oint_{|w|=1} \frac{w^{2n}\bar{\mathbf{N}}'_n(w)\bar{\rho}(w)}{w - \zeta} dw, \qquad (5.2.112b)$$

which is a system of linear integral equations, on $|z| = 1$, for $\bar{\mathbf{N}}'_n(z)$ and $\bar{\mu}'_n(z)$ in terms of the scattering data. The solutions of these integral equations can be analytically continued into the region $|z| < 1$.

Equivalently, by applying the outside projector P to both sides of equations (5.2.107a)–(5.2.107b), we get

$$\mathbf{N}'_n(z) = \begin{pmatrix} \mathbf{0} \\ \mathbf{I}_M \end{pmatrix} + \lim_{\substack{\zeta \to z \\ |\zeta|>1}} \frac{1}{2\pi i} \oint_{|w|=1} \frac{w^{2n}\bar{\mathbf{N}}'_n(w)\bar{\rho}(w)}{w - \zeta} dw \qquad (5.2.112c)$$

$$\mu'_n(z) = \begin{pmatrix} \mathbf{I}_N \\ \mathbf{0} \end{pmatrix} - \lim_{\substack{\zeta \to z \\ |\zeta|>1}} \frac{1}{2\pi i} \oint_{|w|=1} \frac{w^{-2n}\mathbf{N}'_n(w)\rho(w)}{w - \zeta} dw. \qquad (5.2.112d)$$

To recover $\mathbf{Q}_n$ and $\mathbf{R}_n$, we compute the power series expansions of the eigenfunctions around $z = 0$. By comparing the expansions (5.2.106c) and (5.2.106d) with the expansions about $z = 0$ of the right-hand sides of (5.2.112a) and (5.2.112b), or, equivalently, of (5.2.111),

$$\bar{\mathbf{m}}_n(z) = \mathbf{I}_{N+M} - \frac{z}{2\pi i} \oint_{|w|=1} w^{-2}\bar{\mathbf{m}}_n(w)\mathbf{V}_n(w)dw + O(z^2),$$

we obtain

$$\mathbf{Q}_n = \frac{1}{2\pi i} \oint_{|w|=1} w^{2n}\bar{\mathbf{N}}'^{(\text{up})}_{n+1}(w)\bar{\rho}(w)dw$$

$$\mathbf{R}_n = \frac{1}{2\pi i} \oint_{|w|=1} \left(w^{-2(n+1)}\bar{\mu}'^{(\text{dn})}_n(w)\rho(w) - w^{-2}\bar{\mathbf{N}}'^{(\text{dn})}_n(w)\bar{\rho}(w)\rho(w) \right) dw,$$

where $\bar{\mathbf{N}}'_n(z)$ and $\bar{\mu}'_n(z)$ are determined by the linear system (5.2.112a)–(5.2.112b). Hence, the formulation of the inverse problem without poles is complete.

The method of solution requires an extra step if $\mu'_n(z)$ and $\bar{\mu}'_n(z)$ have poles. Let us denote such poles as $\{z_j\}_{j=1,\ldots,J}$ and $\{\bar{z}_j\}_{j=1,\ldots,\bar{J}}$, respectively. We then proceed as follows. First, we apply the inside projection operator (5.2.110a) to both sides of (5.2.107a) and the outside projection operator (5.2.110b) to both sides of (5.2.107b) to obtain

$$\bar{\mathbf{N}}'_n(z) = \begin{pmatrix} \mathbf{I}_N \\ \mathbf{0} \end{pmatrix} + \sum_{j=1}^{J} z_j^{-2n} \left[\frac{1}{z - z_j} \mathbf{N}'_n(z_j) + \frac{1}{z + z_j} \mathbf{N}'_n(-z_j) \right] \mathbf{C}_j$$

$$- \lim_{\substack{\zeta \to z \\ |\zeta| < 1}} \frac{1}{2\pi i} \oint_{|w|=1} \frac{w^{-2n}}{w - \zeta} \mathbf{N}'_n(w)\rho(w)dw \qquad (5.2.113a)$$

$$\mathbf{N}'_n(z) = \begin{pmatrix} \mathbf{0} \\ \mathbf{I}_M \end{pmatrix} + \sum_{j=1}^{\bar{J}} \bar{z}_j^{2n} \left[\frac{1}{z - \bar{z}_j} \bar{\mathbf{N}}'_n(\bar{z}_j) + \frac{1}{z + \bar{z}_j} \bar{\mathbf{N}}'_n(-\bar{z}_j) \right] \bar{\mathbf{C}}_j$$

$$+ \lim_{\substack{\zeta \to z \\ |\zeta| > 1}} \frac{1}{2\pi i} \oint_{|w|=1} \frac{w^{2n}}{w - \zeta} \bar{\mathbf{N}}'_n(w)\bar{\rho}(w)dw, \qquad (5.2.113b)$$

where $\mathbf{N}'_n(z_j)$ is $\mathbf{N}'_n(z)$ evaluated at the eigenvalue z_j, $\mathbf{N}'_n(-z_j)$ is $\mathbf{N}'_n(z)$ evaluated at the complementary eigenvalue $-z_j$, and similarly for $\bar{\mathbf{N}}'_n(\bar{z}_j)$ and $\bar{\mathbf{N}}'_n(-\bar{z}_j)$. In expressions (5.2.113a)–(5.2.113b), we have taken into account Symmetry 5.1 by including poles in $\pm$ pairs with the appropriate norming constants.

Equations (5.2.113a)–(5.2.113b) constitute a system of linear integral equations on $|z| = 1$. This system depends on the additional matrices $\{\mathbf{N}'_n(z_j), \mathbf{N}'_n(-z_j)\}_{j=1}^{J}$ and $\{\bar{\mathbf{N}}'_n(\bar{z}_j), \bar{\mathbf{N}}'_n(-\bar{z}_j)\}_{j=1}^{\bar{J}}$. We obtain expressions for these matrices by evaluating (5.2.113a) at the points $\pm \bar{z}_j$ and (5.2.113b) at the points $\pm z_j$. This results in a linear algebraic–integral system composed of (5.2.113a)–(5.2.113b) and

$$\bar{\mathbf{N}}'_n(\bar{z}_j) = \begin{pmatrix} \mathbf{I}_N \\ \mathbf{0} \end{pmatrix} + \sum_{k=1}^{J} z_k^{-2n} \left[\frac{1}{\bar{z}_j - z_k} \mathbf{N}'_n(z_k) + \frac{1}{\bar{z}_j + z_k} \mathbf{N}'_n(-z_k) \right] \mathbf{C}_k$$

$$- \frac{1}{2\pi i} \oint_{|w|=1} \frac{w^{-2n}}{w - \bar{z}_j} \mathbf{N}'_n(w)\rho(w)dw \qquad (5.2.113c)$$

$$\bar{\mathbf{N}}'_n(-\bar{z}_j) = \begin{pmatrix} \mathbf{I}_N \\ \mathbf{0} \end{pmatrix} - \sum_{k=1}^{J} z_k^{-2n} \left[\frac{1}{\bar{z}_j + z_k} \mathbf{N}'_n(z_k) + \frac{1}{\bar{z}_j - z_k} \mathbf{N}'_n(-z_k) \right] \mathbf{C}_k$$

$$- \frac{1}{2\pi i} \oint_{|w|=1} \frac{w^{-2n}}{w + \bar{z}_j} \mathbf{N}'_n(w)\rho(w)dw \qquad (5.2.113d)$$

$$\mathbf{N}'_n(z_j) = \begin{pmatrix} \mathbf{0} \\ \mathbf{I}_M \end{pmatrix} + \sum_{k=1}^{\bar{J}} \bar{z}_k^{2n} \left[\frac{1}{z_j - \bar{z}_k} \bar{\mathbf{N}}'_n(\bar{z}_k) + \frac{1}{z_j + \bar{z}_k} \bar{\mathbf{N}}'_n(-\bar{z}_k) \right] \bar{\mathbf{C}}_k$$

$$+ \frac{1}{2\pi i} \oint_{|w|=1} \frac{w^{2n}}{w - z_j} \bar{\mathbf{N}}'_n(w)\bar{\rho}(w)dw \qquad (5.2.113e)$$

$$\mathbf{N}'_n(-z_j) = \begin{pmatrix} \mathbf{0} \\ \mathbf{I}_M \end{pmatrix} - \sum_{k=1}^{\bar{J}} \bar{z}_k^{2n} \left[\frac{1}{z_j + \bar{z}_k} \bar{\mathbf{N}}'_n(\bar{z}_k) + \frac{1}{z_j - \bar{z}_k} \bar{\mathbf{N}}'_n(-\bar{z}_k) \right] \bar{\mathbf{C}}_k$$

$$+ \frac{1}{2\pi i} \oint_{|w|=1} \frac{w^{2n}}{w + z_j} \bar{\mathbf{N}}'_n(w) \bar{\rho}(w) dw, \qquad (5.2.113f)$$

where (5.2.113c)–(5.2.113d) hold for each eigenvalue $\{\bar{z}_j : |\bar{z}_j| < 1, \text{Re}\ \bar{z}_j \geq 0\}_{j=1}^{\bar{J}}$ and (5.2.113e)–(5.2.113f) hold for each eigenvalue $\{z_j : |z_j| > 1, \text{Re}\ z_j \geq 0\}_{j=1}^{J}$.

Note that in the system of equations (5.2.113a)–(5.2.113f), the matrix scattering data (i.e., $\rho(z)$, $\bar{\rho}(z)$, $\mathbf{C}_j$, and $\bar{\mathbf{C}}_j$) always multiply on the right. This implies that the rows of the matrix equations are uncoupled, that is, the first row of (5.2.113a) depends only on the first row of (5.2.113c)–(5.2.113d). In particular, we consider the equations for the first N rows (i.e., the matrices $\bar{\mathbf{N}}'^{(\text{up})}_n(z)$, $\mathbf{N}'^{(\text{up})}_n(z)$, $\bar{\mathbf{N}}'^{(\text{up})}_n(\pm\bar{z}_j)$, and $\mathbf{N}'^{(\text{up})}_n(\pm z_j)$) separately from the equations for the last M rows (i.e., the matrices $\bar{\mathbf{N}}'^{(\text{dn})}_n(z)$, $\mathbf{N}'^{(\text{dn})}_n(z)$, $\bar{\mathbf{N}}'^{(\text{dn})}_n(\pm\bar{z}_j)$, and $\bar{\mathbf{N}}'^{(\text{dn})}_n(\pm z_j)$).

The equations for the first N rows are consistent with the symmetry reductions

$$\bar{\mathbf{N}}'^{(\text{up})}_n(-z) = \bar{\mathbf{N}}'^{(\text{up})}_n(z) \qquad \mathbf{N}'^{(\text{up})}_n(-z) = -\mathbf{N}'^{(\text{up})}_n(z), \qquad (5.2.114a)$$

and the equations for the last M rows are consistent with the symmetry reductions

$$\bar{\mathbf{N}}'^{(\text{dn})}_n(-z) = -\bar{\mathbf{N}}'^{(\text{dn})}_n(z) \qquad \mathbf{N}'^{(\text{dn})}_n(-z) = \mathbf{N}'^{(\text{dn})}_n(z), \qquad (5.2.114b)$$

and these symmetries are consistent with the z-expansions we derived in the direct problem (cf. (5.2.52)–(5.2.54)). However, here the symmetry in the eigenfunctions is a consequence of the symmetry of the scattering data.

By taking into account the symmetry reductions (5.2.114a)–(5.2.114b), we can split the system (5.2.113a)–(5.2.113f) into two smaller algebraic–integral systems. Under the symmetry reduction (5.2.114a), the first N rows of (5.2.113a)–(5.2.113f) become the system

$$\bar{\mathbf{N}}'^{(\text{up})}_n(z) = \mathbf{I}_N + 2 \sum_{j=1}^{J} \frac{z_j^{-2n+1}}{z^2 - z_j^2} \mathbf{N}'^{(\text{up})}_n(z_j) \mathbf{C}_j$$

$$- \lim_{\substack{\zeta \to z \\ |\zeta| < 1}} \frac{1}{2\pi i} \oint_{|w|=1} \frac{w^{-2n}}{w - \zeta} \mathbf{N}'^{(\text{up})}_n(w) \rho(w) dw \qquad (5.2.115a)$$

$$\mathbf{N}'^{(\text{up})}_n(z) = 2 \sum_{j=1}^{\bar{J}} \frac{\bar{z}_j^{2n} z}{z^2 - \bar{z}_j^2} \bar{\mathbf{N}}'^{(\text{up})}_n(\bar{z}_j) \bar{\mathbf{C}}_j$$

$$+ \lim_{\substack{\zeta \to z \\ |\zeta| > 1}} \frac{1}{2\pi i} \oint_{|w|=1} \frac{w^{2n}}{w - \zeta} \bar{\mathbf{N}}'^{(\text{up})}_n(w) \bar{\rho}(w) dw \qquad (5.2.115b)$$

$$\bar{\mathbf{N}}_n'^{(\mathrm{up})}(\bar{z}_j) = \mathbf{I}_N + 2\sum_{k=1}^{J} \frac{z_k^{-2n+1}}{\bar{z}_j^2 - z_k^2}\mathbf{N}_n'^{(\mathrm{up})}(z_k)\mathbf{C}_k$$

$$- \frac{1}{2\pi i}\oint_{|w|=1}\frac{w^{-2n}}{w-\bar{z}_j}\mathbf{N}_n'^{(\mathrm{up})}(w)\rho(w)dw \qquad (5.2.115\mathrm{c})$$

$$\mathbf{N}_n'^{(\mathrm{up})}(z_j) = 2\sum_{k=1}^{\bar{J}} \frac{\bar{z}_k^{2n}z_j}{z_j^2 - \bar{z}_k^2}\bar{\mathbf{N}}_n'^{(\mathrm{up})}(\bar{z}_k)\bar{\mathbf{C}}_k$$

$$+ \frac{1}{2\pi i}\oint_{|w|=1}\frac{w^{2n}}{w-z_j}\bar{\mathbf{N}}_n'^{(\mathrm{up})}(w)\bar{\rho}(w)dw, \qquad (5.2.115\mathrm{d})$$

where (i) equation (5.2.115c) holds for each eigenvalue $\{\bar{z}_j : |\bar{z}_j| < 1,$ $\mathrm{Re}\,\bar{z}_j \geq 0\}_{j=1}^{\bar{J}}$, and (ii) equation (5.2.115d) holds for each eigenvalue $\{z_j : |z_j| > 1,\ \mathrm{Re}\,z_j \geq 0\}_{j=1}^{J}$. The matrices $\bar{\mathbf{N}}_n'^{(\mathrm{up})}(-\bar{z}_j)$ and $\mathbf{N}_n'^{(\mathrm{up})}(-z_j)$ are determined from, respectively, $\bar{\mathbf{N}}_n'^{(\mathrm{up})}(\bar{z}_j)$ and $\mathbf{N}_n'^{(\mathrm{up})}(z_j)$, according to the symmetry (5.2.114a). Hence, while the system (5.2.113a)–(5.2.113f) comprises equations corresponding to $2(J + \bar{J})$ eigenvalues, the system (5.2.115a)–(5.2.115d), which takes the symmetry (5.2.114a) into account, comprises equations corresponding to only $J + \bar{J}$ eigenvalues.

Similarly, under the symmetry reduction (5.2.114b), the last M rows of the system (5.2.113a)–(5.2.113f) become the system

$$\bar{\mathbf{N}}_n'^{(\mathrm{dn})}(z) = 2\sum_{j=1}^{J} \frac{z_j^{-2n}z}{z^2 - z_j^2}\mathbf{N}_n'^{(\mathrm{dn})}(z_j)\mathbf{C}_j$$

$$- \lim_{\substack{\zeta \to z \\ |\zeta|<1}} \frac{1}{2\pi i}\oint_{|w|=1}\frac{w^{-2n}}{w-\zeta}\mathbf{N}_n'^{(\mathrm{dn})}(w)\rho(w)dw \qquad (5.2.116\mathrm{a})$$

$$\mathbf{N}_n'^{(\mathrm{dn})}(z) = \mathbf{I}_M + 2\sum_{j=1}^{\bar{J}} \frac{\bar{z}_j^{2n+1}}{z^2 - \bar{z}_j^2}\bar{\mathbf{N}}_n'^{(\mathrm{dn})}(\bar{z}_j)\bar{\mathbf{C}}_j$$

$$+ \lim_{\substack{\zeta \to z \\ |\zeta|>1}} \frac{1}{2\pi i}\oint_{|w|=1}\frac{w^{2n}}{w-\zeta}\bar{\mathbf{N}}_n'^{(\mathrm{dn})}(w)\bar{\rho}(w)dw \qquad (5.2.116\mathrm{b})$$

$$\bar{\mathbf{N}}_n'^{(\mathrm{dn})}(\bar{z}_j) = 2\sum_{k=1}^{J} \frac{z_k^{-2n}\bar{z}_j}{\bar{z}_j^2 - z_k^2}\mathbf{N}_n'^{(\mathrm{dn})}(z_k)\mathbf{C}_k$$

$$- \frac{1}{2\pi i}\oint_{|w|=1}\frac{w^{-2n}}{w-\bar{z}_j}\mathbf{N}_n'^{(\mathrm{dn})}(w)\rho(w)dw \qquad (5.2.116\mathrm{c})$$

$$\mathbf{N}_n^{\prime(\mathrm{dn})}(z_j) = \mathbf{I}_M + 2\sum_{k=1}^{\bar{J}} \frac{\bar{z}_k^{2n+1}}{z_j^2 - \bar{z}_k^2}\bar{\mathbf{N}}_n^{\prime(\mathrm{dn})}(\bar{z}_k)\bar{\mathbf{C}}_k$$

$$+\frac{1}{2\pi i}\oint_{|w|=1}\frac{w^{2n}}{w-z_j}\bar{\mathbf{N}}_n^{\prime(\mathrm{dn})}(w)\bar{\rho}(w)dw, \quad (5.2.116\mathrm{d})$$

where (i) equation (5.2.116c) holds for each eigenvalue $\{\bar{z}_j : |\bar{z}_j| < 1,$
$\mathrm{Re}\,\bar{z}_j \geq 0\}_{j=1}^{\bar{J}}$, and (ii) equation (5.2.115d) holds for each eigenvalue
$\{z_j : |z_j| > 1,\ \mathrm{Re}\,z_j \geq 0\}_{j=1}^{J}$. The matrices $\bar{\mathbf{N}}_n^{\prime(\mathrm{dn})}(-\bar{z}_j)$ and $\mathbf{N}_n^{\prime(\mathrm{dn})}(-z_j)$ are
determined from, respectively, $\bar{\mathbf{N}}_n^{\prime(\mathrm{dn})}(\bar{z}_j)$ and $\mathbf{N}_n^{\prime(\mathrm{dn})}(z_j)$ according to the symmetry (5.2.114b).

Note that the system (5.2.115a)–(5.2.115d) is uncoupled from the system
(5.2.116a)–(5.2.116d), and each is a closed system of algebraic–integral equations. These systems hold for general potentials. However, if the scattering
data satisfy Symmetry 5.2, one can include this symmetry explicitly in the
systems. Specifically, if Symmetry 5.2 holds, then the set of eigenvalues is
given by

$$\{\pm\bar{z}_j,\ \pm i/\bar{z}_j : |\bar{z}_j| < 1,\ \mathrm{Re}\,\bar{z}_j \geq 0\}_{j=1}^{\bar{J}}$$

$$\cup\{\pm z_j,\ \pm i/z_j : |z_j| > 1,\ \mathrm{Re}\,z_j \geq 0\}_{j=1}^{J} \quad (5.2.117\mathrm{a})$$

with the corresponding norming constants

$$\{\bar{\mathbf{C}}_j,\ \tilde{\mathbf{C}}_j\}_{j=1}^{\bar{J}} \cup \{\mathbf{C}_j,\ \hat{\mathbf{C}}_j\}_{j=1}^{J}, \quad (5.2.117\mathrm{b})$$

and therefore we can rewrite the system (5.2.115a)–(5.2.115b) as

$$\bar{\mathbf{N}}_n^{\prime(\mathrm{up})}(z) = \mathbf{I}_N + 2\sum_{j=1}^{J}\frac{z_j^{-2n+1}}{z^2 - z_j^2}\mathbf{N}_n^{\prime(\mathrm{up})}(z_j)\mathbf{C}_j$$

$$+ 2i\sum_{j=1}^{\bar{J}}\frac{(-1)^n\bar{z}_j^{2n-1}}{z^2 + \bar{z}_j^{-2}}\mathbf{N}_n^{\prime(\mathrm{up})}(i/\bar{z}_j)\tilde{\mathbf{C}}_j$$

$$- \lim_{\substack{\zeta\to z\\ |\zeta|<1}}\frac{1}{2\pi i}\oint_{|w|=1}\frac{w^{-2n}}{w-\zeta}\mathbf{N}_n^{\prime(\mathrm{up})}(w)\rho(w)dw \quad (5.2.118\mathrm{a})$$

$$\mathbf{N}_n^{\prime(\mathrm{up})}(z) = 2\sum_{j=1}^{\bar{J}}\frac{\bar{z}_j^{2n}z}{z^2 - \bar{z}_j^2}\bar{\mathbf{N}}_n^{\prime(\mathrm{up})}(\bar{z}_j)\bar{\mathbf{C}}_j + 2\sum_{j=1}^{J}\frac{(-1)^n z_j^{-2n}z}{z^2 + z_j^{-2}}\bar{\mathbf{N}}_n^{\prime(\mathrm{up})}(i/z_j)\hat{\mathbf{C}}_j$$

$$+ \lim_{\substack{\zeta\to z\\ |\zeta|>1}}\frac{1}{2\pi i}\oint_{|w|=1}\frac{w^{2n}}{w-\zeta}\bar{\mathbf{N}}_n^{\prime(\mathrm{up})}(w)\bar{\rho}(w)dw \quad (5.2.118\mathrm{b})$$

$$\bar{\mathbf{N}}_n'^{(\text{up})}(\bar{z}_j) = \mathbf{I}_N + 2 \sum_{k=1}^{J} \frac{z_k^{-2n+1}}{\bar{z}_j^2 - z_k^2} \mathbf{N}_n'^{(\text{up})}(z_k)\mathbf{C}_k$$

$$+ 2i \sum_{k=1}^{\bar{J}} \frac{(-1)^n \bar{z}_k^{2n-1}}{\bar{z}_j^2 + \bar{z}_k^{-2}} \mathbf{N}_n'^{(\text{up})}(i/\bar{z}_k)\tilde{\mathbf{C}}_k$$

$$- \frac{1}{2\pi i} \oint_{|w|=1} \frac{w^{-2n}}{w - \bar{z}_j} \mathbf{N}_n'^{(\text{up})}(w)\rho(w)dw \quad (5.2.118c)$$

$$\bar{\mathbf{N}}_n'^{(\text{up})}(i/z_j) = \mathbf{I}_N - 2 \sum_{k=1}^{J} \frac{z_k^{-2n+1}}{z_j^{-2} + z_k^2} \mathbf{N}_n'^{(\text{up})}(z_k)\mathbf{C}_k$$

$$- 2i \sum_{k=1}^{\bar{J}} \frac{(-1)^n \bar{z}_k^{2n-1}}{z_j^{-2} - \bar{z}_k^{-2}} \mathbf{N}_n'^{(\text{up})}(i/\bar{z}_k)\tilde{\mathbf{C}}_k$$

$$- \frac{1}{2\pi} \oint_{|w|=1} \frac{z_j w^{-2n}}{1 + i z_j w} \mathbf{N}_n'^{(\text{up})}(w)\rho(w)dw \quad (5.2.118d)$$

$$\mathbf{N}_n'^{(\text{up})}(z_j) = 2 \sum_{k=1}^{\bar{J}} \frac{\bar{z}_k^{2n} z_j}{z_j^2 - \bar{z}_k^2} \bar{\mathbf{N}}_n'^{(\text{up})}(\bar{z}_k)\bar{\mathbf{C}}_k + 2 \sum_{k=1}^{J} \frac{(-1)^n z_k^{-2n} z_j}{z_j^2 + z_k^{-2}} \bar{\mathbf{N}}_n'^{(\text{up})}(i/z_k)\hat{\mathbf{C}}_k$$

$$+ \frac{1}{2\pi i} \oint_{|w|=1} \frac{w^{2n}}{w - z_j} \bar{\mathbf{N}}_n'^{(\text{up})}(w)\bar{\rho}(w)dw \quad (5.2.118e)$$

$$\mathbf{N}_n'^{(\text{up})}(i/\bar{z}_j) = -2i \sum_{k=1}^{\bar{J}} \frac{\bar{z}_k^{2n} \bar{z}_j^{-1}}{\bar{z}_j^{-2} + \bar{z}_k^2} \bar{\mathbf{N}}_n'^{(\text{up})}(\bar{z}_k)\bar{\mathbf{C}}_k$$

$$- 2i \sum_{k=1}^{J} \frac{(-1)^n z_k^{-2n} \bar{z}_j^{-1}}{\bar{z}_j^{-2} - z_k^{-2}} \bar{\mathbf{N}}_n'^{(\text{up})}(i/z_k)\hat{\mathbf{C}}_k$$

$$+ \frac{1}{2\pi} \oint_{|w|=1} \frac{\bar{z}_j w^{2n}}{1 + i\bar{z}_j w} \bar{\mathbf{N}}_n'^{(\text{up})}(w)\bar{\rho}(w)dw. \quad (5.2.118f)$$

Here, as before, $\hat{\mathbf{C}}_j$ is the norming constant associated with $\pm i/z_j$ and $\tilde{\mathbf{C}}_j$ is the norming constant associated with $\pm i/\bar{z}_j$. Similarly, under this symmetry, the system (5.2.116a)–(5.2.115d) for the lower components of the eigenfunctions becomes

$$\bar{\mathbf{N}}_n'^{(\text{dn})}(z) = 2 \sum_{j=1}^{J} \frac{z_j^{-2n} z}{z^2 - z_j^2} \mathbf{N}_n'^{(\text{dn})}(z_j)\mathbf{C}_j + 2 \sum_{j=1}^{\bar{J}} \frac{(-1)^n \bar{z}_j^{2n} z}{z^2 + \bar{z}_j^{-2}} \mathbf{N}_n'^{(\text{dn})}(i/\bar{z}_j)\tilde{\mathbf{C}}_j$$

$$- \lim_{\substack{\zeta \to z \\ |\zeta| < 1}} \frac{1}{2\pi i} \oint_{|w|=1} \frac{w^{-2n}}{w - \zeta} \mathbf{N}_n'^{(\text{dn})}(w)\rho(w)dw \quad (5.2.119a)$$

$$\mathbf{N}_n'^{(\mathrm{dn})}(z) = \mathbf{I}_M + 2 \sum_{j=1}^{\bar{J}} \frac{\bar{z}_j^{2n+1}}{z^2 - \bar{z}_j^2} \bar{\mathbf{N}}_n'^{(\mathrm{dn})}(\bar{z}_j) \bar{\mathbf{C}}_j$$

$$+ 2i \sum_{j=1}^{J} \frac{(-1)^n z_j^{-2n-1}}{z^2 + z_j^{-2}} \bar{\mathbf{N}}_n'^{(\mathrm{dn})}(i/z_j) \hat{\mathbf{C}}_j$$

$$+ \lim_{\substack{\zeta \to z \\ |\zeta| > 1}} \frac{1}{2\pi i} \oint_{|w|=1} \frac{w^{2n}}{w - \zeta} \bar{\mathbf{N}}_n'^{(\mathrm{dn})}(w) \bar{\rho}(w) dw$$

$$(5.2.119\mathrm{b})$$

$$\bar{\mathbf{N}}_n'^{(\mathrm{dn})}(\bar{z}_j) = 2 \sum_{k=1}^{J} \frac{z_k^{-2n}\bar{z}_j}{\bar{z}_j^2 - z_k^2} \mathbf{N}_n'^{(\mathrm{dn})}(z_k) \mathbf{C}_k + 2 \sum_{k=1}^{\bar{J}} \frac{(-1)^n \bar{z}_k^{2n}\bar{z}_j}{\bar{z}_j^2 + \bar{z}_k^{-2}} \mathbf{N}_n'^{(\mathrm{dn})}(i/\bar{z}_k) \tilde{\mathbf{C}}_k$$

$$- \frac{1}{2\pi i} \oint_{|w|=1} \frac{w^{-2n}}{w - \bar{z}_j} \mathbf{N}_n'^{(\mathrm{dn})}(w) \rho(w) dw \qquad (5.2.119\mathrm{c})$$

$$\bar{\mathbf{N}}_n'^{(\mathrm{dn})}(i/z_j) = -2i \sum_{k=1}^{J} \frac{z_k^{-2n} z_j^{-1}}{z_j^{-2} + z_k^2} \mathbf{N}_n'^{(\mathrm{dn})}(z_k) \mathbf{C}_k$$

$$- 2i \sum_{k=1}^{\bar{J}} \frac{(-1)^n \bar{z}_k^{2n} z_j^{-1}}{z_j^{-2} - \bar{z}_k^{-2}} \mathbf{N}_n'^{(\mathrm{dn})}(i/\bar{z}_k) \tilde{\mathbf{C}}_k$$

$$- \frac{1}{2\pi} \oint_{|w|=1} \frac{z_j w^{-2n}}{1 + i z_j w} \mathbf{N}_n'^{(\mathrm{dn})}(w) \rho(w) dw$$

$$(5.2.119\mathrm{d})$$

$$\mathbf{N}_n'^{(\mathrm{dn})}(z_j) = \mathbf{I}_M + 2 \sum_{k=1}^{\bar{J}} \frac{\bar{z}_k^{2n+1}}{z_j^2 - \bar{z}_k^2} \bar{\mathbf{N}}_n'^{(\mathrm{dn})}(\bar{z}_k) \bar{\mathbf{C}}_k$$

$$+ 2i \sum_{k=1}^{J} \frac{(-1)^n z_k^{-2n-1}}{z_j^2 + z_k^{-2}} \bar{\mathbf{N}}_n'^{(\mathrm{dn})}(i/z_k) \hat{\mathbf{C}}_k$$

$$+ \frac{1}{2\pi i} \oint_{|w|=1} \frac{w^{2n}}{w - z_j} \bar{\mathbf{N}}_n'^{(\mathrm{dn})}(w) \bar{\rho}(w) dw \qquad (5.2.119\mathrm{e})$$

$$\mathbf{N}_n'^{(\mathrm{dn})}(i/\bar{z}_j) = \mathbf{I}_M - 2 \sum_{k=1}^{\bar{J}} \frac{\bar{z}_k^{2n+1}}{\bar{z}_j^{-2} + \bar{z}_k^2} \bar{\mathbf{N}}_n'^{(\mathrm{dn})}(\bar{z}_k) \bar{\mathbf{C}}_k$$

$$- 2i \sum_{k=1}^{J} \frac{(-1)^n z_k^{-2n-1}}{\bar{z}_j^{-2} - z_k^{-2}} \bar{\mathbf{N}}_n'^{(\mathrm{dn})}(i/z_k) \hat{\mathbf{C}}_k$$

$$+ \frac{1}{2\pi} \oint_{|w|=1} \frac{\bar{z}_j w^{2n}}{1 + i \bar{z}_j w} \bar{\mathbf{N}}_n'^{(\mathrm{dn})}(w) \bar{\rho}(w) dw,$$

$$(5.2.119\mathrm{f})$$

where the eigenvalues are given by (5.2.117a).

The linear algebraic–integral systems (5.2.118a)–(5.2.118f) and (5.2.119a)–(5.2.119f) determine the eigenfunctions $\bar{\mathbf{N}}'_n(z)$ and $\mathbf{N}'_n(z)$. However, these linear systems do not account for Symmetry 5.3, which relates the eigenvalues in the region $|z| > 1$ with the eigenvalues in the region $|z| < 1$. If we require that Symmetry 5.3, associated with $\mathbf{R}_n = \mp \mathbf{Q}_n^H$ holds, then $\bar{J} = J$, $\bar{z}_j = 1/z_j^*$ and $\bar{\mathbf{C}}_j = \pm (z_j^*)^{-2} \mathbf{C}_j^H$ for $j = 1, \ldots, J$ in the system of integral equations (5.2.119a)–(5.2.119f).

Note that we have not proved that the linear systems have a solution in general. Neither have we established conditions on the scattering data that would ensure that these systems are solvable. Rather, we have shown how to obtain a linear algebraic–integral system for the eigenfunctions from the scattering data. In the following, we compute the solution of (5.2.118a)–(5.2.118f) for a reflectionless potential (i.e., $\rho(z) = \bar{\rho}(z) = \mathbf{0}$ on $|z| = 1$) with a single octet of eigenvalues $\{\pm z_1, \pm i/z_1, \pm \bar{z}_1, \pm i/\bar{z}_1 : |z_1| > 1, |\bar{z}_1| < 1\}$. It is outside the scope of this work to rigorously establish conditions under which either systems are guaranteed to have (unique) solutions.

Recovery of the potentials

We show how to recover the potentials $\mathbf{Q}_n$ and $\mathbf{R}_n$ from the eigenfunctions and the scattering data. First, we show how to recover the matrix potential $\mathbf{Q}_n$. Then, we derive two methods for recovering the matrix potential $\mathbf{R}_n$ – one method that is applicable for general $\mathbf{R}_n$ and a simplified method for potentials that satisfy the symmetry $\mathbf{R}_n \mathbf{Q}_n = \mathbf{Q}_n \mathbf{R}_n = \alpha_n \mathbf{I}$ (cf. Symmetry 5.2).

To recover $\mathbf{Q}_n$, we first apply the inside projection operator (5.2.110a) to both sides of the jump condition (5.2.107b). This yields the relations

$$\bar{\mu}_n'^{(\mathrm{up})}(z) = 2 \sum_{j=1}^{\bar{J}} \frac{\bar{z}_j^{2n} z}{z^2 - \bar{z}_j^2} \bar{\mathbf{N}}_n'^{(\mathrm{up})}(\bar{z}_j) \bar{\mathbf{C}}_j + \frac{1}{2\pi i} \oint_{|w|=1} \frac{w^{2n}}{w - z} \bar{\mathbf{N}}_n'^{(\mathrm{up})}(w) \bar{\rho}(w) dw$$

$$(5.2.120\mathrm{a})$$

$$\bar{\mu}_n'^{(\mathrm{dn})}(z) = \mathbf{I}_M + 2 \sum_{j=1}^{\bar{J}} \frac{\bar{z}_j^{2n+1}}{z^2 - \bar{z}_j^2} \bar{\mathbf{N}}_n'^{(\mathrm{dn})}(\bar{z}_j) \bar{\mathbf{C}}_j$$

$$+ \frac{1}{2\pi i} \oint_{|w|=1} \frac{w^{2n}}{w - z} \bar{\mathbf{N}}_n'^{(\mathrm{dn})}(w) \bar{\rho}(w) dw \qquad (5.2.120\mathrm{b})$$

when the symmetries (5.2.114a)–(5.2.114b) are taken into account. Now, by comparing the power series expansion (in z) of the right-hand side of (5.2.120a) and the expansion (5.2.106d) of $\bar{\mu}_n'(z)$, we obtain

$$\mathbf{Q}_{n-1} = -2 \sum_{j=1}^{\bar{J}} \bar{z}_j^{2(n-1)} \bar{\mathbf{N}}_n'^{(\mathrm{up})}(\bar{z}_j) \bar{\mathbf{C}}_j + \frac{1}{2\pi i} \oint_{|w|=1} w^{2(n-1)} \bar{\mathbf{N}}_n'^{(\mathrm{up})}(w) \bar{\rho}(w) dw,$$

$$(5.2.121\mathrm{a})$$

which gives the potential in terms of the eigenfunction $\bar{\mathbf{N}}_n'^{(\mathrm{up})}(z)$ evaluated on $|z| = 1$ and at the eigenvalues $\{\pm\bar{z}_j : |\bar{z}_j| < 1, \ \mathrm{Re}\,\bar{z}_j > 0\}_{j=1}^{\bar{J}}$.

Analogously, we can recover $\mathbf{R}_n$ from the eigenfunctions $\mathbf{N}_n'^{(\mathrm{dn})}(z)$ and $\bar{\mathbf{N}}_n'^{(\mathrm{dn})}(z)$. Indeed, by applying the outside projection operator (5.2.110b) to the jump condition (5.2.107a), we obtain the equation

$$\boldsymbol{\mu}_n'^{(\mathrm{dn})}(z) = 2 \sum_{j=1}^{J} \frac{z_j^{-2n} z}{z^2 - z_j^2} \mathbf{N}_n'^{(\mathrm{dn})}(z_j)\mathbf{C}_j - \frac{1}{2\pi i} \oint_{|w|=1} \frac{w^{-2n}}{w-z} \mathbf{N}_n'^{(\mathrm{dn})}(w)\rho(w)dw.$$

Then, by comparing the Laurent expansion (in z) of the right-hand side of this expression with the expansion (5.2.106b) of $\boldsymbol{\mu}_n'(z)$, we obtain

$$\boldsymbol{\Delta}_n \mathbf{R}_{n-1} = 2 \sum_{j=1}^{J} z_j^{-2n} \mathbf{N}_n'^{(\mathrm{dn})}(z_j)\mathbf{C}_j + \frac{1}{2\pi i} \oint_{|w|=1} w^{-2n} \mathbf{N}_n'^{(\mathrm{dn})}(w)\rho(w)dw.$$

$$(5.2.121\mathrm{b})$$

Similarly, by comparing the power series expansion (in z) of the right-hand side of (5.2.120b) and the expansion (5.2.106d) of $\bar{\boldsymbol{\mu}}_n'(z)$, we obtain

$$\boldsymbol{\Delta}_n = \mathbf{I}_M - 2 \sum_{j=1}^{\bar{J}} \bar{z}_j^{2n-1} \bar{\mathbf{N}}_n'^{(\mathrm{dn})}(\bar{z}_j)\bar{\mathbf{C}}_j + \frac{1}{2\pi i} \oint_{|w|=1} w^{2n-1} \bar{\mathbf{N}}_n'^{(\mathrm{dn})}(w)\bar{\rho}(w)dw.$$

$$(5.2.121\mathrm{c})$$

Hence, with (5.2.121b) and (5.2.121c) we can recover $\mathbf{R}_n$ from the eigenfunctions evaluated on $|z| = 1$ and at their respective eigenvalues $\{\pm z_j : |z_j| > 1, \ \mathrm{Re}\,z_j > 0\}_{j=1}^{J}$, and $\{\pm\bar{z}_j : |\bar{z}_j| < 1, \ \mathrm{Re}\,\bar{z}_j > 0\}_{j=1}^{\bar{J}}$.

If we explicitly include the effect of Symmetry 5.2 on the eigenvalues, then (5.2.121a) becomes

$$\mathbf{Q}_{n-1} = -2 \sum_{j=1}^{\bar{J}} \bar{z}_j^{2(n-1)} \bar{\mathbf{N}}_n'^{(\mathrm{up})}(\bar{z}_j)\bar{\mathbf{C}}_j - 2 \sum_{j=1}^{J} (-1)^{n-1} z_j^{-2(n-1)} \bar{\mathbf{N}}_n'^{(\mathrm{up})}(i/z_j)\hat{\mathbf{C}}_j$$

$$+ \frac{1}{2\pi i} \oint_{|w|=1} w^{2(n-1)} \bar{\mathbf{N}}_n'^{(\mathrm{up})}(w)\bar{\rho}(w)dw, \qquad (5.2.122\mathrm{a})$$

where the eigenvalues are as in (5.2.117a).

Also, there is a simpler procedure for recovering the potential $\mathbf{R}_n$ when the potentials satisfy the symmetry (5.2.79a). Recall that the eigenfunctions $\mathbf{N}_n'^{(\mathrm{dn})}(z)$ and $\bar{\mathbf{N}}_n'^{(\mathrm{dn})}(z)$ satisfy (5.2.116a) and (5.2.116b). Comparing the power series expansion (in z) of the right-hand side of (5.2.116a) with the expansion

(5.2.106c) of $\bar{\mathbf{N}}_n^{\prime(\mathrm{dn})}(z)$ yields

$$\mathbf{\Delta}_n \mathbf{R}_n \mathbf{\Omega}_n^{-1} = 2 \sum_{j=1}^{J} z_j^{-2(n+1)} \mathbf{N}_n^{\prime(\mathrm{dn})}(z_j) \mathbf{C}_j$$

$$+ \frac{1}{2\pi i} \oint_{|w|=1} w^{-2(n+1)} \mathbf{N}_n^{\prime(\mathrm{dn})}(w) \rho(w) dw.$$

In general, this relation is insufficient to recover $\mathbf{R}_n$ because we must also determine $\mathbf{\Delta}_n$ and $\mathbf{\Omega}_n^{-1}$. However, if the potentials satisfy (5.2.79a), then

$$\mathbf{\Omega}_n = \prod_{k=n}^{+\infty} (1 - \alpha_k) \mathbf{I}_N \qquad \mathbf{\Delta}_n = \prod_{k=n}^{+\infty} (1 - \alpha_k) \mathbf{I}_M, \qquad (5.2.122b)$$

and therefore

$$\mathbf{R}_n = 2 \sum_{j=1}^{J} z_j^{-2(n+1)} \mathbf{N}_n^{\prime(\mathrm{dn})}(z_j) \mathbf{C}_j + 2 \sum_{j=1}^{\bar{J}} (-1)^{n+1} \bar{z}_j^{2(n+1)} \mathbf{N}_n^{\prime(\mathrm{dn})}(i/\bar{z}_j) \tilde{\mathbf{C}}_j$$

$$+ \frac{1}{2\pi i} \oint_{|w|=1} w^{-2(n+1)} \mathbf{N}_n^{\prime(\mathrm{dn})}(w) \rho(w) dw. \qquad (5.2.122c)$$

As before, we can take into account Symmetry 5.3 by requiring $\bar{J} = J$, $\bar{z}_j = 1/z_j^*$, and $\bar{\mathbf{C}}_j = \pm (z_j^*)^{-2} \mathbf{C}_j^H$ for any $j = 1, \ldots, J$.

Finally, we remark that the symmetry (5.2.84) for the reflection coefficients can be obtained from the inverse problem formulas (5.2.122a) and (5.2.122c) by reconstructing the potentials in terms of the scattering data. Indeed, let us restrict ourselves for simplicity to the case $N = 2$ and consider "small" potentials $\mathbf{Q}_n$ and $\mathbf{R}_n$ satisfying the symmetry

$$\mathbf{R}_n = (-1)^n \mathbf{P} \mathbf{Q}_n^T \mathbf{P}. \qquad (5.2.123)$$

Moreover, let us assume no discrete eigenvalues. Hence, due to the small norm assumption for the potentials, from (5.2.118a) and (5.2.119a) it follows that $\bar{\mathbf{N}}_n^{\prime(\mathrm{up})}(w) \sim \mathbf{I}_N$ and $\mathbf{N}_n^{\prime(\mathrm{dn})}(w) \sim \mathbf{I}_M$, and substitution into (5.2.122a)–(5.2.122c) yields

$$\mathbf{Q}_n \sim \frac{1}{2\pi i} \oint_{|w|=1} w^{2n} \bar{\rho}(w) dw \qquad (5.2.124a)$$

$$\mathbf{R}_n \sim \frac{1}{2\pi i} \oint_{|w|=1} w^{-2(n+1)} \rho(w) dw. \qquad (5.2.124b)$$

Performing in (5.2.124a) the change of variable $w \to i/w$ and taking into account that the potentials satisfy (5.2.123) immediately yields the symmetry (5.2.84).

Reflectionless potentials

In this subsection, we consider the recovery of potentials from scattering data such that $\rho(z) = \bar{\rho}(z) = \mathbf{0}$ on $|z| = 1$. We refer to such potentials as reflectionless potentials. In particular, we consider scattering data that satisfy Symmetry 5.2 and Symmetry 5.3. We explicitly calculate the potential in the case where there is only one octet of eigenvalues and norming constants. In this case we obtain potentials with the familiar sech profile with a complex modulation. However, we show below that, to obtain this potential, an additional condition on the norming constants is required.

If $\rho(z) = \bar{\rho}(z) = \mathbf{0}$ on $|z| = 1$, then the integrals vanish in the algebraic–integral system (5.2.118a)–(5.2.118f). In this case, this system reduces to the linear algebraic system

$$\bar{\mathbf{N}}_n^{\prime(\text{up})}(\bar{z}_j) = \mathbf{I}_N + 2\sum_{k=1}^{J}\frac{z_k^{-2n+1}}{\bar{z}_j^2 - z_k^2}\mathbf{N}_n^{\prime(\text{up})}(z_k)\mathbf{C}_k$$

$$+ 2i\sum_{k=1}^{J}\frac{(-1)^n \bar{z}_k^{2n-1}}{\bar{z}_j^2 + \bar{z}_k^{-2}}\mathbf{N}_n^{\prime(\text{up})}(i/\bar{z}_k)\tilde{\mathbf{C}}_k \qquad (5.2.125a)$$

$$\bar{\mathbf{N}}_n^{\prime(\text{up})}(i/z_j) = \mathbf{I}_N - 2\sum_{k=1}^{J}\frac{z_k^{-2n+1}}{z_j^{-2} + z_k^2}\mathbf{N}_n^{\prime(\text{up})}(z_k)\mathbf{C}_k$$

$$- 2i\sum_{k=1}^{J}\frac{(-1)^n \bar{z}_k^{2n-1}}{z_j^{-2} - \bar{z}_k^{-2}}\mathbf{N}_n^{\prime(\text{up})}(i/\bar{z}_k)\tilde{\mathbf{C}}_k \qquad (5.2.125b)$$

$$\mathbf{N}_n^{\prime(\text{up})}(z_j) = 2\sum_{k=1}^{J}\frac{\bar{z}_k^{2n}z_j}{z_j^2 - \bar{z}_k^2}\bar{\mathbf{N}}_n^{\prime(\text{up})}(\bar{z}_k)\bar{\mathbf{C}}_k$$

$$+ 2\sum_{k=1}^{J}\frac{(-1)^n z_k^{-2n}z_j}{z_j^2 + z_k^{-2}}\bar{\mathbf{N}}_n^{\prime(\text{up})}(i/z_k)\hat{\mathbf{C}}_k \qquad (5.2.125c)$$

$$\mathbf{N}_n^{\prime(\text{up})}(i/\bar{z}_j) = -2i\sum_{k=1}^{J}\frac{\bar{z}_k^{2n}\bar{z}_j^{-1}}{\bar{z}_j^{-2} + \bar{z}_k^2}\bar{\mathbf{N}}_n^{\prime(\text{up})}(\bar{z}_k)\bar{\mathbf{C}}_k$$

$$- 2i\sum_{k=1}^{J}\frac{(-1)^n z_k^{-2n}\bar{z}_j^{-1}}{\bar{z}_j^{-2} - z_k^{-2}}\bar{\mathbf{N}}_n^{\prime(\text{up})}(i/z_k)\hat{\mathbf{C}}_k. \qquad (5.2.125d)$$

Moreover, the expression (5.2.122a) for the potential $\mathbf{Q}_n$ reduces to

$$\mathbf{Q}_{n-1} = -2\sum_{j=1}^{J}\bar{z}_j^{2(n-1)}\bar{\mathbf{N}}_n^{\prime(\text{up})}(\bar{z}_j)\bar{\mathbf{C}}_j - 2\sum_{j=1}^{J}(-1)^{n-1}z_j^{-2(n-1)}\bar{\mathbf{N}}_n^{\prime(\text{up})}(i/z_j)\hat{\mathbf{C}}_j.$$

$$(5.2.126)$$

Analogously, the system (5.2.119c)–(5.2.119f) reduces to the linear algebraic system

$$\bar{\mathbf{N}}_n'^{(\mathrm{dn})}(\bar{z}_j) = 2 \sum_{k=1}^{J} \frac{z_k^{-2n}\bar{z}_j}{\bar{z}_j^2 - z_k^2} \mathbf{N}_n'^{(\mathrm{dn})}(z_k)\mathbf{C}_k$$

$$+ 2 \sum_{k=1}^{\bar{J}} \frac{(-1)^n \bar{z}_k^{2n}\bar{z}_j}{\bar{z}_j^2 + \bar{z}_k^{-2}} \mathbf{N}_n'^{(\mathrm{dn})}(i/\bar{z}_k)\tilde{\mathbf{C}}_k \qquad (5.2.127\mathrm{a})$$

$$\bar{\mathbf{N}}_n'^{(\mathrm{dn})}(i/z_j) = -2i \sum_{k=1}^{J} \frac{z_k^{-2n}z_j^{-1}}{z_j^{-2} + z_k^2} \mathbf{N}_n'^{(\mathrm{dn})}(z_k)\mathbf{C}_k$$

$$- 2i \sum_{k=1}^{\bar{J}} \frac{(-1)^n \bar{z}_k^{2n}z_j^{-1}}{z_j^{-2} - \bar{z}_k^{-2}} \mathbf{N}_n'^{(\mathrm{dn})}(i/\bar{z}_k)\tilde{\mathbf{C}}_k \qquad (5.2.127\mathrm{b})$$

$$\mathbf{N}_n'^{(\mathrm{dn})}(z_j) = \mathbf{I}_M + 2 \sum_{k=1}^{\bar{J}} \frac{\bar{z}_k^{2n+1}}{z_j^2 - \bar{z}_k^2} \bar{\mathbf{N}}_n'^{(\mathrm{dn})}(\bar{z}_k)\bar{\mathbf{C}}_k$$

$$+ 2i \sum_{k=1}^{J} \frac{(-1)^n z_k^{-2n-1}}{z_j^2 + z_k^{-2}} \bar{\mathbf{N}}_n'^{(\mathrm{dn})}(i/z_k)\hat{\mathbf{C}}_k \qquad (5.2.127\mathrm{c})$$

$$\mathbf{N}_n'^{(\mathrm{dn})}(i/\bar{z}_j) = \mathbf{I}_M - 2 \sum_{k=1}^{\bar{J}} \frac{\bar{z}_k^{2n+1}}{\bar{z}_j^{-2} + \bar{z}_k^2} \bar{\mathbf{N}}_n'^{(\mathrm{dn})}(\bar{z}_k)\bar{\mathbf{C}}_k$$

$$- 2i \sum_{k=1}^{J} \frac{(-1)^n z_k^{-2n-1}}{\bar{z}_j^{-2} - z_k^{-2}} \bar{\mathbf{N}}_n'^{(\mathrm{dn})}(i/z_k)\hat{\mathbf{C}}_k, \qquad (5.2.127\mathrm{d})$$

and (5.2.122c) gives the reconstruction of the potential $\mathbf{R}_n$,

$$\mathbf{R}_n = 2 \sum_{j=1}^{J} z_j^{-2(n+1)}\mathbf{N}_n'^{(\mathrm{dn})}(z_j)\mathbf{C}_j + 2 \sum_{j=1}^{\bar{J}}(-1)^{n+1}\bar{z}_j^{2(n+1)}\mathbf{N}_n'^{(\mathrm{dn})}(i/\bar{z}_j)\tilde{\mathbf{C}}_j.$$

$$(5.2.128)$$

Therefore, in the reflectionless case (i.e. when $\rho(z) = \bar{\rho}(z) = \mathbf{0}$ on $|z| = 1$) and with Symmetry 5.2 taken into account, the potentials are determined by the solutions of the $2(J + \bar{J})$-dimensional linear-algebraic systems (5.2.125a)–(5.2.125d) and (5.2.127a)–(5.2.127d).

In particular, if there is one octet of eigenvalues and norming constants, that is,

$$\left\{ (\pm z_1, \mathbf{C}_1),\ (\pm\bar{z}_1, \bar{\mathbf{C}}_1),\ \left(\pm i/z_1, \hat{\mathbf{C}}_1\right),\ \left(\pm i/\bar{z}_1, \tilde{\mathbf{C}}_1\right) \right\}, \qquad (5.2.129)$$

the system (5.2.125a)–(5.2.125d), after eliminating the matrices $\mathbf{N}_n'^{(\mathrm{up})}(z_1)$ and $\mathbf{N}_n'^{(\mathrm{up})}(i/\bar{z}_1)$, reduces to

$$\bar{\mathbf{N}}_n'^{(\mathrm{up})}(\bar{z}_1)$$

$$= \mathbf{I}_N - 4\bar{\mathbf{N}}_n'^{(\mathrm{up})}(\bar{z}_1)\left[\frac{\bar{z}_1^{2n}z_1^{-2(n-1)}}{(\bar{z}_1^2 - z_1^2)^2}\bar{\mathbf{C}}_1\mathbf{C}_1 - \frac{(-1)^n\bar{z}_1^{4n+2}}{(1+\bar{z}_1^4)^2}\bar{\mathbf{C}}_1\tilde{\mathbf{C}}_1\right]$$

$$- 4\bar{\mathbf{N}}_n'^{(\mathrm{up})}(i/z_1)\left[\frac{(-1)^n z_1^{-4(n-1)}}{(z_1^2 - \bar{z}_1^2)(1+z_1^4)}\hat{\mathbf{C}}_1\mathbf{C}_1 + \frac{z_1^{-2(n-1)}\bar{z}_1^{2(n+1)}}{(1+\bar{z}_1^4)(\bar{z}_1^2 - z_1^2)}\hat{\mathbf{C}}_1\tilde{\mathbf{C}}_1\right]$$

$$\tag{5.2.130a}$$

$$\bar{\mathbf{N}}_n'^{(\mathrm{up})}(i/z_1)$$

$$= \mathbf{I}_N - 4\bar{\mathbf{N}}_n'^{(\mathrm{up})}(\bar{z}_1)\left[\frac{\bar{z}_1^{2n}z_1^{-2(n-2)}}{(z_1^2 - \bar{z}_1^2)(1+z_1^4)}\bar{\mathbf{C}}_1\mathbf{C}_1 - \frac{(-1)^n z_1^2\bar{z}_1^{4n+2}}{(1+\bar{z}_1^4)(z_1^2 - \bar{z}_1^2)}\bar{\mathbf{C}}_1\tilde{\mathbf{C}}_1\right]$$

$$- 4\bar{\mathbf{N}}_n'^{(\mathrm{up})}(i/z_1)\left[\frac{(-1)^n z_1^{-4n+6}}{(1+z_1^4)^2}\hat{\mathbf{C}}_1\mathbf{C}_1 - \frac{z_1^{-2(n-2)}\bar{z}_1^{2(n+1)}}{(z_1^2 - \bar{z}_1^2)^2}\hat{\mathbf{C}}_1\tilde{\mathbf{C}}_1\right].$$

$$\tag{5.2.130b}$$

So far, in deriving the system (5.2.130a)–(5.2.130b), we have accounted for the effect of Symmetry 5.2 on the eigenvalues, but we have not considered the effect of this symmetry on the norming constants. We now use the symmetry in the norming constants to further simplify the solution of this system.

For $N = 2$, let us define

$$\mathbf{C}_1 = \begin{pmatrix} \gamma_1^{(1)} & \delta_1^{(2)} \\ \gamma_1^{(2)} & -\delta_1^{(1)} \end{pmatrix}, \qquad \bar{\mathbf{C}}_1 = \bar{z}_1^2\begin{pmatrix} \bar{\gamma}_1^{(1)} & \bar{\gamma}_1^{(2)} \\ \bar{\delta}_1^{(2)} & -\bar{\delta}_1^{(1)} \end{pmatrix}. \tag{5.2.131a}$$

Then, the symmetries in the norming constants given by Symmetry 5.2, namely, the relations (5.2.92b)–(5.2.92c), imply that

$$\hat{\mathbf{C}}_1 = -z_1^{-2}\begin{pmatrix} \delta_1^{(1)} & \delta_1^{(2)} \\ \gamma_1^{(2)} & -\gamma_1^{(1)} \end{pmatrix}, \qquad \tilde{\mathbf{C}}_1 = \begin{pmatrix} \bar{\delta}_1^{(1)} & \bar{\gamma}_1^{(2)} \\ \bar{\delta}_1^{(2)} & -\bar{\gamma}_1^{(1)} \end{pmatrix}. \tag{5.2.131b}$$

Correspondingly,

$$\hat{\mathbf{C}}_1\mathbf{C}_1 = \mathbf{C}_1\hat{\mathbf{C}}_1 = -z_1^{-2}\left(\gamma_1 \cdot \delta_1\right)\mathbf{I} \tag{5.2.132a}$$

$$\tilde{\mathbf{C}}_1\bar{\mathbf{C}}_1 = \bar{\mathbf{C}}_1\tilde{\mathbf{C}}_1 = \bar{z}_1^2\left(\bar{\gamma}_1 \cdot \bar{\delta}_1\right)\mathbf{I}, \tag{5.2.132b}$$

where we find it convenient to introduce the two-component vectors

$$\boldsymbol{\gamma}_1 = \begin{pmatrix} \gamma_1^{(1)} \\ \gamma_1^{(2)} \end{pmatrix}, \qquad \boldsymbol{\delta}_1 = \begin{pmatrix} \delta_1^{(1)} \\ \delta_1^{(2)} \end{pmatrix}, \qquad \bar{\boldsymbol{\gamma}}_1 = \begin{pmatrix} \bar{\gamma}_1^{(1)} \\ \bar{\gamma}_1^{(2)} \end{pmatrix}, \qquad \bar{\boldsymbol{\delta}}_1 = \begin{pmatrix} \bar{\delta}_1^{(1)} \\ \bar{\delta}_1^{(2)} \end{pmatrix}$$

$$(5.2.133)$$

and · denotes, as usual, the scalar product.

If (as explained later) we impose the additional condition

$$\boldsymbol{\delta}_1 = \bar{\boldsymbol{\delta}}_1 = 0, \tag{5.2.134}$$

then the system (5.2.130a)–(5.2.130b) has the unique solution

$$\bar{\mathbf{N}}_n^{\prime(\mathrm{up})}(\bar{z}_1) = \begin{pmatrix} \frac{1}{1+g_n} & 0 \\ 0 & 1 - \frac{(z_1^2 - \bar{z}_1^2)g_n}{z_1^2(1+\bar{z}_1^4)(1+g_n)} \end{pmatrix} \tag{5.2.135a}$$

$$\bar{\mathbf{N}}_n^{\prime(\mathrm{up})}(i\bar{z}_1^{-1}) = \begin{pmatrix} 1 - \frac{z_1^2(z_1^2 - \bar{z}_1^2)g_n}{(1+z_1^4)(1+g_n)} & 0 \\ 0 & \frac{1}{1+g_n} \end{pmatrix}, \tag{5.2.135b}$$

where

$$g_n = 4\left(\boldsymbol{\gamma}_1 \cdot \bar{\boldsymbol{\gamma}}_1\right)\left(z_1^2 - \bar{z}_1^2\right)^{-2} z_1^{-2(n-1)}\bar{z}_1^{2(n+1)}. \tag{5.2.135c}$$

Substituting these expressions into (5.2.125c)–(5.2.125d) yields

$$\mathbf{N}_n^{\prime(\mathrm{up})}(z_1) = \frac{2}{1+g_n}\begin{pmatrix} \bar{\gamma}_1^{(1)}\frac{\bar{z}_1^{2(n+1)}z_1}{z_1^2 - \bar{z}_1^2} & \bar{\gamma}_1^{(2)}\frac{\bar{z}_1^{2(n+1)}z_1}{z_1^2 - \bar{z}_1^2} \\ (-1)^{n+1}\gamma_1^{(2)}\frac{z_1^{-2n-1}}{z_1^2 + z_1^{-2}} & (-1)^n\gamma_1^{(1)}\frac{z_1^{-2n-1}}{z_1^2 + z_1^{-2}} \end{pmatrix}$$

$$\mathbf{N}_n^{\prime(\mathrm{up})}(i\bar{z}_1^{-1}) = \frac{-2i}{1+g_n}\begin{pmatrix} \bar{\gamma}_1^{(1)}\frac{\bar{z}_1^{2n+1}}{\bar{z}_1^{-2} + \bar{z}_1^2} & \bar{\gamma}_1^{(2)}\frac{\bar{z}_1^{2n+1}}{\bar{z}_1^{-2} - \bar{z}_1^2} \\ (-1)^{n+1}\gamma_1^{(2)}\frac{z_1^{-2(n+1)}\bar{z}_1^{-1}}{\bar{z}_1^{-2} - z_1^{-2}} & (-1)^n\gamma_1^{(1)}\frac{z_1^{-2(n+1)}\bar{z}_1^{-1}}{\bar{z}_1^{-2} - z_1^{-2}} \end{pmatrix},$$

so that finally

$$\bar{\mathbf{N}}_n^{\prime(\mathrm{up})}(z)$$
$$= \begin{pmatrix} 1 + 4\boldsymbol{\gamma}_1 \cdot \bar{\boldsymbol{\gamma}}_1 \frac{\bar{z}_1^{2(n+1)}z_1^{-2(n-1)}}{(z^2 - z_1^2)(z_1^2 - \bar{z}_1^2)(1+g_n)} & 0 \\ 0 & 1 - 4\boldsymbol{\gamma}_1 \cdot \bar{\boldsymbol{\gamma}}_1 \frac{\bar{z}_1^{2(n-1)}z_1^{-2(n+1)}}{(z^2 + \bar{z}_1^{-2})(\bar{z}_1^{-2} - z_1^{-2})(1+g_n)} \end{pmatrix}.$$

$$(5.2.136)$$

By (5.2.122a), taking into account (5.1.6) and (5.2.126), we obtain the following expressions for the potentials:

$$
\begin{pmatrix} Q_n^{(1)} \\ Q_n^{(2)} \end{pmatrix} = -2 \frac{\bar{z}_1^{2(n+1)}}{1 + 4\left(\gamma_1 \cdot \bar{\gamma}_1\right)\left(z_1^2 - \bar{z}_1^2\right)^{-2} z_1^{-2n} \bar{z}_1^{2(n+2)}} \begin{pmatrix} \bar{\gamma}_1^{(1)} \\ \bar{\gamma}_1^{(2)} \end{pmatrix} \tag{5.2.137a}
$$

$$
\begin{pmatrix} R_n^{(1)} \\ R_n^{(2)} \end{pmatrix} = 2 \frac{z_1^{-2(n+1)}}{1 + 4\left(\gamma_1 \cdot \bar{\gamma}_1\right)\left(z_1^2 - \bar{z}_1^2\right)^{-2} z_1^{-2n} \bar{z}_1^{2(n+2)}} \begin{pmatrix} \gamma_1^{(1)} \\ \gamma_1^{(2)} \end{pmatrix} . \tag{5.2.137b}
$$

We now consider the effect of Symmetry 5.3 on the inverse problem. Specifically, we restrict ourselves to the case $\mathbf{R}_n = -\mathbf{Q}_n^H$, since we showed that in the defocusing case no discrete eigenvalues are allowed for decaying potentials.

Recall that if Symmetry 5.3 holds, then $\bar{J} = J$ and $\bar{z}_j = 1/z_j^*$ for any $j = 1, \ldots, J$. Hence, in this case, the eigenvalue/norming constants octet (5.2.129) takes the form

$$
\left\{ (\pm z_1, \mathbf{C}_1), \ (\pm 1/z_1^*, \bar{\mathbf{C}}_1), \ \left(\pm i/z_1, \hat{\mathbf{C}}_1\right), \ (\pm i z_1^*, \tilde{\mathbf{C}}_1) \right\}. \tag{5.2.138}
$$

Moreover, in terms of the norming constants, Symmetry 5.3 implies that $\bar{\gamma}_1 = \gamma_1^*$ in (5.2.131a) and (5.2.131b).

With these symmetries explicitly included, as well as the substitution $z_1 = e^{\alpha_1 + i\beta_1}$, the potentials (5.2.137a)–(5.2.137b) are

$$
\begin{pmatrix} Q_n^{(1)} \\ Q_n^{(2)} \end{pmatrix} = -\frac{\gamma_1^*}{\|\gamma_1\|} e^{2i\beta_1(n+1)} \sinh(2\alpha_1) \operatorname{sech}\left(2\alpha_1(n+1) - d\right) \tag{5.2.139a}
$$

$$
\begin{pmatrix} R_n^{(1)} \\ R_n^{(2)} \end{pmatrix} = \frac{\gamma_1}{\|\gamma_1\|} e^{-2i\beta_1(n+1)} \sinh(2\alpha_1) \operatorname{sech}\left(2\alpha_1(n+1) - d\right), \tag{5.2.139b}
$$

where $d = \log \|\gamma_1\| - \log\left[\sinh(2\alpha_1)\right]$. These potentials are a discrete, vector version of the familiar sech profile with complex modulation.

If, instead of the condition (5.2.134), that is, $\delta_1 = \bar{\delta}_1 = 0$, we choose

$$
\gamma_1 = \bar{\gamma}_1 = 0, \tag{5.2.140}
$$

then, according to (5.1.6) and (5.2.126), the solution of the system (5.2.130a)–(5.2.130b) yields the potentials

$$
\begin{pmatrix} Q_n^{(1)} \\ Q_n^{(2)} \end{pmatrix} = 2 \frac{(-1)^n z_1^{-2(n+1)}}{1 + 4\left(\delta_1 \cdot \bar{\delta}_1\right)\left(z_1^2 - \bar{z}_1^2\right)^{-2} z_1^{-2n} \bar{z}_1^{2(n+2)}} \begin{pmatrix} \delta_1^{(1)} \\ \delta_1^{(2)} \end{pmatrix} \tag{5.2.141a}
$$

$$
\begin{pmatrix} R_n^{(1)} \\ R_n^{(2)} \end{pmatrix} = -2 \frac{(-1)^n \bar{z}_1^{2(n+1)}}{1 + 4\left(\delta_1 \cdot \bar{\delta}_1\right)\left(z_1^2 - \bar{z}_1^2\right)^{-2} z_1^{-2n} \bar{z}_1^{2(n+2)}} \begin{pmatrix} \bar{\delta}_1^{(1)} \\ \bar{\delta}_1^{(2)} \end{pmatrix} . \tag{5.2.141b}
$$

If Symmetry 5.3 holds, then $\bar{\delta}_1 = \delta_1^*$ and, as before, $\bar{z}_1 = 1/z_1^*$, and, with this symmetry and the substitution $z_1 = e^{\alpha_1 + i\beta_1}$, the potentials (5.2.141a)–(5.2.141b) are

$$\begin{pmatrix} Q_n^{(1)} \\ Q_n^{(2)} \end{pmatrix} = -\frac{\delta_1}{\|\delta_1\|} e^{-2i(\beta_1 + \pi/2)(n+1)} \sinh(2\alpha_1) \operatorname{sech}\left(2\alpha_1(n+1) - \hat{d}\right)$$

$$(5.2.142a)$$

$$\begin{pmatrix} R_n^{(1)} \\ R_n^{(2)} \end{pmatrix} = \frac{\delta_1^*}{\|\delta_1\|} e^{2i(\beta_1 + \pi/2)(n+1)} \sinh(2\alpha_1) \operatorname{sech}\left(2\alpha_1(n+1) - \hat{d}\right), \quad (5.2.142b)$$

where $\hat{d} = \log \|\delta_1\| - \log[\sinh(2\alpha_1)]$. Like (5.2.139a)–(5.2.139b), the potentials (5.2.142a)–(5.2.142b) are a discrete version of the familiar sech profile with complex modulation.

Note that, to obtain the above complex-modulated sech potential, we imposed the condition (5.2.134) or (5.2.140) on the norming constants. However, even in the case of a single octet of eigenvalues/norming constants, there exists a more general solution of the linear system (5.2.130a)–(5.2.130b) corresponding to the case when we do not restrict the norming constants to obey the condition $\gamma_1 = \bar{\gamma}_1 = 0$ or $\delta_1 = \bar{\delta}_1 = 0$. The general case $\gamma_1 \neq 0, \bar{\gamma}_1 \neq 0$ and $\delta_1 \neq 0, \bar{\delta}_1 \neq 0$ corresponds to solutions that we refer to as "composite" solitons, as opposed to the "fundamental" solitons that we obtain for $\gamma_1 = \bar{\gamma}_1 = 0$ or $\delta_1 = \bar{\delta}_1 = 0$. Even though a single octet is the minimal number of discrete eigenvalues, the composite soliton essentially differs from the fundamental solitons. In fact, one composite soliton could be regarded as a "two-fundamental soliton solution." In a sense, the composite soliton is composed of two fundamental solitons "glued" together.

The solution of the linear system (5.2.130a)–(5.2.130b) when γ_1 and δ_1 and $\bar{\gamma}_1$ and $\bar{\delta}_1$ are possibly all different from zero but orthogonal to each other, that is, such that $\gamma_1 \cdot \delta_1 = \bar{\gamma}_1 \cdot \bar{\delta}_1 = 0$, is given by

$$\bar{\mathbf{N}}_n'^{(\mathrm{up})}(i/z_1)$$

$$= \frac{1}{1 + \tilde{g}_n}\left[\mathbf{I} + \frac{\tilde{g}_n}{\gamma_1 \cdot \bar{\gamma}_1 + \delta_1 \cdot \bar{\delta}_1} \frac{\bar{z}_1^2 + z_1^{-2}}{z_1^2 + z_1^{-2}}\begin{pmatrix} \gamma_1 \cdot \bar{\gamma}_1 & W\left(\bar{\gamma}_1, \delta_1\right) \\ W\left(\gamma_1, \bar{\delta}_1\right) & \delta_1 \cdot \bar{\delta}_1 \end{pmatrix}\right]$$

$$\bar{\mathbf{N}}_n'^{(\mathrm{up})}(\bar{z}_1)$$

$$= \frac{1}{1 + \tilde{g}_n}\left[\mathbf{I} + \frac{\tilde{g}_n}{\gamma_1 \cdot \bar{\gamma}_1 + \delta_1 \cdot \bar{\delta}_1} \frac{\bar{z}_1^2 + z_1^{-2}}{\bar{z}_1^2 + \bar{z}_1^{-2}}\begin{pmatrix} \delta_1 \cdot \bar{\delta}_1 & -W\left(\bar{\gamma}_1, \delta_1\right) \\ -W\left(\gamma_1, \bar{\delta}_1\right) & \gamma_1 \cdot \bar{\gamma}_1 \end{pmatrix}\right],$$

where

$$\tilde{g}_n = 4\frac{z_1^{-2(n-1)}\bar{z}_1^{2(n+1)}}{\left(z_1^2 - \bar{z}_1^2\right)^2}\left(\gamma_1 \cdot \bar{\gamma}_1 + \delta_1 \cdot \bar{\delta}_1\right) \qquad (5.2.143)$$

and, as usual,

$$W(\gamma_1, \delta_1) = \gamma_1^{(1)} \delta_1^{(2)} - \gamma_1^{(2)} \delta_1^{(1)}. \qquad (5.2.144)$$

Hence (5.2.126) and (5.1.6) yield

$$\begin{pmatrix} Q_n^{(1)} \\ Q_n^{(2)} \end{pmatrix} = -2 \frac{\bar{z}_1^{2(n+1)}}{1 + \tilde{g}_{n+1}} \bar{\gamma}_1 + 2(-1)^n \frac{z_1^{-2(n+1)}}{1 + \tilde{g}_{n+1}} \delta_1 \qquad (5.2.145a)$$

$$\begin{pmatrix} R_n^{(1)} \\ R_n^{(2)} \end{pmatrix} = 2 \frac{z_1^{-2(n+1)}}{1 + \tilde{g}_{n+1}} \gamma_1 - 2(-1)^n \frac{\bar{z}_1^{2(n+1)}}{1 + \tilde{g}_{n+1}} \bar{\delta}_1. \qquad (5.2.145b)$$

Taking into account Symmetry 5.3, according to which $\bar{z}_1 = 1/z_1^*$ and $\bar{\gamma}_1 = \gamma_1^*$, $\bar{\delta}_1 = \delta_1^*$, and substituting $z_1 = e^{\alpha_1 + i\beta_1}$ yields

$$\begin{pmatrix} Q_n^{(1)} \\ Q_n^{(2)} \end{pmatrix} = \frac{\sinh 2\alpha_1}{\left(\|\gamma_1\|^2 + \|\delta_1\|^2 \right)^{1/2}} \operatorname{sech} \left(2(n+1)\alpha_1 - \tilde{d} \right)$$
$$\times \left[e^{i\pi + 2i(n+1)\beta_1} \gamma_1^* + e^{-2i(n+1)\beta_1 - in\pi} \delta_1 \right], \qquad (5.2.146)$$

where

$$\tilde{d} = \log \left(\|\gamma_1\|^2 + \|\delta_1\|^2 \right)^{\frac{1}{2}} - \log (\sinh 2\alpha_1). \qquad (5.2.147)$$

Note that the solution (5.2.146) has two complex carrier modulation terms, unlike the fundamental soliton solutions (5.2.139) and (5.2.142).

Gel'fand–Levitan–Marchenko integral equations

Like in the scalar case, we can also provide a reconstruction for the potentials by means of Gel'fand–Levitan–Marchenko integral equations. Indeed, let us represent the eigenfunctions ψ_n and $\bar{\psi}_n$ in terms of triangular kernels,

$$\psi_n(z) = \sum_{j=n}^{+\infty} z^{-j} \mathbf{K}(n, j) \qquad |z| > 1 \qquad (5.2.148a)$$

$$\bar{\psi}_n(z) = \sum_{j=n}^{+\infty} z^{j} \bar{\mathbf{K}}(n, j) \qquad |z| < 1, \qquad (5.2.148b)$$

where

$$\mathbf{K}(n, j) = \begin{pmatrix} \mathbf{K}^{(\mathrm{up})}(n, j) \\ \mathbf{K}^{(\mathrm{dn})}(n, j) \end{pmatrix}, \qquad \bar{\mathbf{K}}(n, j) = \begin{pmatrix} \bar{\mathbf{K}}^{(\mathrm{up})}(n, j) \\ \bar{\mathbf{K}}^{(\mathrm{dn})}(n, j) \end{pmatrix}, \qquad (5.2.149)$$

and write the equations (5.2.59a)–(5.2.59b) in the form

$$\phi_n(z) \mathbf{a}^{-1}(z) - \bar{\psi}_n(z) = \psi_n(z) \rho(z) \qquad (5.2.150a)$$

$$\bar{\phi}_n(z) \bar{\mathbf{a}}^{-1}(z) - \psi_n(z) = \bar{\psi}_n(z) \bar{\rho}(z). \qquad (5.2.150b)$$

Applying the operator $\frac{1}{2\pi i} \oint_{|z|=1} dz\, z^{-m-1}$ for $m \geq n$ to equation (5.2.150a) and taking into account the asymptotics (5.2.49a)–(5.2.49b), (5.2.52), (5.2.54), and (5.2.66a)–(5.2.66b), as well as the triangular representations (5.2.148a)–(5.2.148b), we obtain

$$\bar{\mathbf{K}}(n, m) + \sum_{j=n}^{+\infty} \mathbf{K}(n, j)\mathbf{F}(m+j) = \begin{pmatrix} \mathbf{I}_N \\ \mathbf{0} \end{pmatrix} \delta_{m,n} \qquad m \geq n, \qquad (5.2.151)$$

where

$$\mathbf{F}(n) = \sum_{j=1}^{J} z_j^{-n-1}\mathbf{C}_j + \frac{1}{2\pi i} \oint_{|z|=1} z^{-n-1}\rho(z)dz. \qquad (5.2.152)$$

Analogously, operating on equation (5.2.150b) with $\frac{1}{2\pi i} \oint_{|z|=1} dz\, z^{m-1}$ for $m \geq n$ yields

$$\mathbf{K}(n, m) + \sum_{j=n}^{+\infty} \bar{\mathbf{K}}(n, j)\bar{\mathbf{F}}(m+j) = \begin{pmatrix} \mathbf{0} \\ \mathbf{I}_M \end{pmatrix} \delta_{m,n} \qquad m \geq n, \qquad (5.2.153)$$

where

$$\bar{\mathbf{F}}(n) = -\sum_{j=1}^{\bar{J}} \bar{z}_j^{n-1}\bar{\mathbf{C}}_j + \frac{1}{2\pi i} \oint_{|z|=1} z^{n-1}\bar{\rho}(z)dz. \qquad (5.2.154)$$

Equations (5.2.151) and (5.2.153) constitute the Gel'fand–Levitan–Marchenko equations. Note that the sum into (5.2.152) (resp. (5.2.154)) is performed over all the discrete eigenvalues that are outside (resp. inside) the unit circle.

Comparing the representations (5.2.148a)–(5.2.148b) for the eigenfunctions with the asymptotics (5.2.52) and (5.2.54) and recalling (5.2.19b) yields the reconstruction of the potentials in terms of the kernels of GLM equations, that is,

$$\mathbf{K}^{(\mathrm{up})}(n, n) = \bar{\mathbf{K}}^{(\mathrm{dn})}(n, n) = \mathbf{0}, \qquad \bar{\mathbf{K}}^{(\mathrm{up})}(n, n) = \mathbf{\Omega}_n^{-1}, \qquad \mathbf{K}^{(\mathrm{dn})}(n, n) = \mathbf{\Delta}_n^{-1}$$

$$(5.2.155)$$

$$\mathbf{Q}_n = -\mathbf{K}^{(\mathrm{up})}(n, n+1)\left(\mathbf{K}^{(\mathrm{dn})}(n, n)\right)^{-1},$$
$$\mathbf{R}_n = -\bar{\mathbf{K}}^{(\mathrm{dn})}(n, n+1)\left(\bar{\mathbf{K}}^{(\mathrm{up})}(n, n)\right)^{-1}. \qquad (5.2.156)$$

It is more convenient to write equations (5.2.151) and (5.2.153) as forced summation equations, which is accomplished if we introduce

$$\kappa(n, n) = \begin{pmatrix} \mathbf{0} \\ \mathbf{I}_M \end{pmatrix} \qquad \bar{\kappa}(n, n) = \begin{pmatrix} \mathbf{I}_N \\ \mathbf{0} \end{pmatrix} \qquad (5.2.157a)$$

and for $m > n$

$$\mathbf{K}(n, m) = \begin{pmatrix} \boldsymbol{\Omega}_n^{-1} & \mathbf{0} \\ \mathbf{0} & \boldsymbol{\Delta}_n^{-1} \end{pmatrix} \boldsymbol{\kappa}(n, m) \qquad \bar{\mathbf{K}}(n, m) = \begin{pmatrix} \boldsymbol{\Omega}_n^{-1} & \mathbf{0} \\ \mathbf{0} & \boldsymbol{\Delta}_n^{-1} \end{pmatrix} \bar{\boldsymbol{\kappa}}(n, m).$$

$$(5.2.157b)$$

Equations (5.2.151) and (5.2.153) then yield for $m > n$

$$\bar{k}(n, m) + \begin{pmatrix} \mathbf{0} \\ \mathbf{I}_M \end{pmatrix} \mathbf{F}(n + m) + \sum_{j=n+1}^{+\infty} \boldsymbol{\kappa}(n, j)\mathbf{F}(m + j) = \mathbf{0} \qquad (5.2.158a)$$

$$\boldsymbol{\kappa}(n, m) + \begin{pmatrix} \mathbf{I}_N \\ \mathbf{0} \end{pmatrix} \bar{\mathbf{F}}(n + m) + \sum_{j=n+1}^{+\infty} \bar{\boldsymbol{\kappa}}(n, j)\bar{\mathbf{F}}(m + j) = \mathbf{0}, \qquad (5.2.158b)$$

and from (5.2.155)–(5.2.156) we obtain

$$\mathbf{Q}_n = -\boldsymbol{\Omega}_n^{-1} \boldsymbol{\kappa}^{(\mathrm{up})}(n, n + 1)\boldsymbol{\Delta}_n \qquad \mathbf{R}_n = -\boldsymbol{\Delta}_n^{-1} \bar{\boldsymbol{\kappa}}^{(\mathrm{dn})}(n, n + 1)\boldsymbol{\Omega}_n.$$

$$(5.2.159)$$

In general, these relations are insufficient to recover $\mathbf{Q}_n$ and $\mathbf{R}_n$ because we must also determine $\boldsymbol{\Delta}_n$ and $\boldsymbol{\Omega}_n$. However, if the potentials satisfy Symmetry 5.2, that is, if (5.2.79a) holds, taking into account (5.2.122b), from (5.2.155) and (5.2.156) it follows that $\boldsymbol{\kappa}$ and $\bar{\boldsymbol{\kappa}}$ are related to the potentials as follows:

$$\mathbf{Q}_n = -\boldsymbol{\kappa}^{(\mathrm{up})}(n, n + 1) \qquad \mathbf{R}_n = -\bar{\boldsymbol{\kappa}}^{(\mathrm{dn})}(n, n + 1). \qquad (5.2.160)$$

Note that if the potentials satisfy Symmetry 5.2, the eigenvalues are paired and the corresponding norming constants satisfy (5.2.78). Therefore, the GLM equations can be simplified as follows:

$$\bar{k}(n, m) + \begin{pmatrix} \mathbf{0} \\ \mathbf{I}_M \end{pmatrix} \mathbf{F}_{\mathrm{R}}(n + m) + \sum_{\substack{j=n+1 \\ j+m=\mathrm{odd}}}^{+\infty} \boldsymbol{\kappa}(n, j)\mathbf{F}_{\mathrm{R}}(m + j) = \mathbf{0} \qquad (5.2.161a)$$

$$\boldsymbol{\kappa}(n, m) + \begin{pmatrix} \mathbf{I}_N \\ \mathbf{0} \end{pmatrix} \bar{\mathbf{F}}_{\mathrm{R}}(n + m) + \sum_{\substack{j=n+1 \\ j+m=\mathrm{odd}}}^{+\infty} \bar{\boldsymbol{\kappa}}(n, j)\bar{\mathbf{F}}_{\mathrm{R}}(m + j) = \mathbf{0}, \qquad (5.2.161b)$$

where

$$\mathbf{F}_{\mathrm{R}}(n) = \begin{cases} 2\sum_{j=1}^{J} z_j^{-n-1}\mathbf{C}_j + \frac{1}{\pi i}\int_{C_{\mathrm{R}}} z^{-n-1}\boldsymbol{\rho}(z)dz & n = \mathrm{odd} \\ \mathbf{0} & n = \mathrm{even} \end{cases}$$

$$(5.2.162a)$$

$$\bar{\mathbf{F}}_R(n) = \begin{cases} -2\sum_{j=1}^{\bar{J}} \bar{z}_j^{n-1}\bar{\mathbf{C}}_j + \frac{1}{\pi i}\int_{C_R} z^{n-1}\bar{\rho}(z)dz & n = \text{odd} \\ 0 & n = \text{even} \end{cases}$$

$$(5.2.162b)$$

and C_R denotes the right half of the unit circle.

Finally, we note that if, in addition, the potentials are such that $\mathbf{R}_n = \mp\mathbf{Q}_n^H$, that is, if Symmetry 5.3 applies, then from (5.2.98a)–(5.2.98b) it also follows that

$$\bar{\mathbf{F}}(n) = \mp\mathbf{F}^H(n). \qquad (5.2.163)$$

Existence and uniqueness of solution

Consider the homogeneous equations corresponding to (5.2.158a)–(5.2.158b),

$$\mathbf{h}_1(m) + \sum_{j=n+1}^{+\infty} \mathbf{h}_2(j)\mathbf{F}(m+j) = \mathbf{0} \qquad (5.2.164a)$$

$$\mathbf{h}_2(m) + \sum_{j=n+1}^{+\infty} \mathbf{h}_1(j)\bar{\mathbf{F}}(m+j) = \mathbf{0}, \qquad (5.2.164b)$$

and suppose $\mathbf{h}(n) = (\mathbf{h}_1(n), \mathbf{h}_2(n))$ is a solution of (5.2.164a)–(5.2.164b) that vanish identically for $m < n$. For any matrix element,

$$h_1^{(i,k)}(m) + \sum_{j=n+1}^{+\infty} \sum_{\ell=1}^{M} h_2^{(i,\ell)}(j)F^{(\ell,k)}(m+j) = 0$$

$$i = 1, \ldots, (N+M), \quad k = 1, \ldots, N \qquad (5.2.165a)$$

$$h_2^{(i,k)}(m) + \sum_{\ell=1}^{N} \sum_{j=n+1}^{+\infty} h_1^{(i,\ell)}(j)\bar{F}^{(\ell,k)}(m+j) = 0$$

$$i = 1, \ldots, (N+M), \quad k = 1, \ldots, M. \qquad (5.2.165b)$$

Multiplying (5.2.165a)–(5.2.165b) by $(h_1^{(i,k)}, h_2^{(i,k)})^*$ and summing over all i, k and over all integers m, one obtains

$$\sum_{m=-\infty}^{\infty} \left\{ \sum_{i=1}^{N+M} \sum_{k=1}^{N} |h_1^{(i,k)}(m)|^2 + \sum_{i=1}^{N+M} \sum_{k=1}^{M} |h_2^{(i,k)}(m)|^2 \right.$$

$$+ \sum_{j=-\infty}^{\infty} \left[\sum_{i=1}^{N+M} \sum_{k=1}^{N} \sum_{\ell=1}^{M} h_2^{(i,\ell)}(m) \left(h_1^{(i,k)}(j) \right)^* F^{(\ell,k)}(m+j) \right.$$

$$\left. \left. + \sum_{i=1}^{N+M} \sum_{k=1}^{M} \sum_{\ell=1}^{N} h_1^{(i,\ell)}(j) \left(h_2^{(i,k)}(m) \right)^* \bar{F}^{(\ell,k)}(m+j) \right] \right\} = 0. \quad (5.2.166)$$

If Symmetry 5.3 holds and $\mathbf{R}_n = -\mathbf{Q}_n^H$, (5.2.163) yields $\bar{\mathbf{F}} = -\mathbf{F}^H$, and then (5.2.166) can be simplified as follows:

$$
0 = \sum_{m=-\infty}^{\infty} \left\{ \sum_{i=1}^{N+M} \sum_{k=1}^{N} |h_1^{(i,k)}(m)|^2 + \sum_{i=1}^{N+M} \sum_{k=1}^{M} |h_2^{(i,k)}(m)|^2 \right.
$$
$$
\left. + 2i \operatorname{Im} \left[\sum_{j=-\infty}^{\infty} \sum_{i=1}^{N+M} \sum_{k=1}^{N} \sum_{\ell=1}^{M} h_2^{(i,\ell)}(m) \left(h_1^{(i,k)}(j) \right)^* F^{(\ell,k)}(m+j) \right] \right\}.
$$

The real and imaginary parts must both vanish, from which it follows that

$$
\mathbf{h}_1(m) \equiv \mathbf{0}, \qquad \mathbf{h}_2(m) \equiv \mathbf{0}.
$$

On the other hand, if $\mathbf{R}_n = \mathbf{Q}_n^H$, then $\bar{\mathbf{F}} = \mathbf{F}^H$, which allows equation (5.2.166) to be written as

$$
0 = \sum_{m=-\infty}^{\infty} \left\{ \sum_{i=1}^{N+M} \sum_{k=1}^{N} |h_1^{(i,k)}(m)|^2 + \sum_{i=1}^{N+M} \sum_{k=1}^{M} |h_2^{(i,k)}(m)|^2 \right.
$$
$$
\left. + 2 \operatorname{Re} \left[\sum_{j=-\infty}^{\infty} \sum_{i=1}^{N+M} \sum_{k=1}^{N} \sum_{\ell=1}^{M} h_2^{(i,\ell)}(m) \left(h_1^{(i,k)}(j) \right)^* F^{(\ell,k)}(m+j) \right] \right\}.
$$
$$(5.2.167)$$

Like in the continuous case, the scattering problem with this reduction is formally self-adjoint and there are no discrete eigenvalues. This implies that

$$
\mathbf{F}(m) = \frac{1}{2\pi i} \oint_{|z|=1} z^{-m-1} \rho(z) dz. \tag{5.2.168}
$$

A function $h(n)$ of a discrete variable n assuming integer values, and its discrete Fourier transform $\hat{h}(z)$,

$$
\hat{h}(z) = \sum_{n=-\infty}^{\infty} h(n) z^n \qquad h(n) = \frac{1}{2\pi i} \oint_{|z|=1} \hat{h}(z) z^{n-1} dz,
$$

satisfy a discrete version of Parseval's identity, namely,

$$
\sum_{n=-\infty}^{\infty} |h(n)|^2 = \frac{1}{2\pi i} \oint_{|z|=1} z^{-1} |\hat{h}(z)|^2 \equiv \frac{1}{2\pi} \int_0^{2\pi} |\hat{h}(e^{i\theta})|^2 d\theta.
$$

Substituting these results into (5.2.167) yields

$$
0 = \int_0^{2\pi} \left\{ \sum_{i=1}^{N+M} \sum_{j=1}^{N} |\hat{h}_1^{(i,j)}(e^{i\theta})|^2 + \sum_{i=1}^{N+M} \sum_{j=1}^{M} |\hat{h}_2^{(i,j)}(e^{i\theta})|^2 \right.
$$
$$
\left. + 2 \operatorname{Re} \int_0^{2\pi} \sum_{i=1}^{N+M} \sum_{j=1}^{N} \sum_{k=1}^{M} \left(\hat{h}_1^{(i,j)}(e^{i\theta}) \right)^* \hat{h}_2^{(i,k)}(e^{-i\theta}) \rho^{(j,k)}(e^{i\theta}) \right\} d\theta.
$$
$$(5.2.169)$$

Under the reduction $\mathbf{R}_n = \mathbf{Q}_n^H$, from (5.2.101) it follows that

$$\det\left(\mathbf{I}_N - \rho^H(w)\rho(w)\right) > 0 \qquad \text{for } |w| = 1$$

(recall that we assume the potentials are "small" enough so that $1 - \alpha_j > 0$ for any $j \in \mathbb{Z}$). Assuming that this condition is sufficient to ensure that $|\rho^{(j,k)}(w)| < 1$ for any $j = 1, \ldots, M$ and $k = 1, \ldots N$, we have

$$\left| 2\,\mathrm{Re}\left[\hat{h}_1^{(i,\ell)}(e^{-i\theta}) \left(\hat{h}_2^{(i,k)}(e^{i\theta})\right)^* \rho^{(k,\ell)}(e^{i\theta}) \right] \right|$$
$$< |\hat{h}_1^{(i,\ell)}(e^{-i\theta})|^2 + |\hat{h}_2^{(i,k)}(e^{i\theta})|^2.$$

Hence $\hat{\mathbf{h}}_1(n) \equiv \hat{\mathbf{h}}_2(n) \equiv 0$. and $\mathbf{h}_1(y) \equiv \mathbf{h}_2(y) \equiv 0$.

We conclude that, when $\mathbf{R}_n = \mp\mathbf{Q}_n^H$, the integral equations (5.2.158a), admit no homogenous solutions besides the trivial one. The complete study of the inverse scattering problem is outside the scope of this work. However, we can say that when the kernel $\mathbf{F}(m)$ is compact, as it would be when the class of potentials is suitably restricted, then the Gel'fand–Levitan–Marchenko equations are Fredholm. In this case, the Fredholm alternative applies, and the previous result implies that the solution of the GLM equation exists and is unique.

5.2.3 Time evolution

The operator (5.1.2b) determines the evolution of the eigenfunctions. From this we deduce the time evolution of the scattering data. Since we have assumed that $\mathbf{Q}_n, \mathbf{R}_n \to 0$ as $n \to \pm\infty$, then the time-dependence (5.1.2b) is asymptotically of the form

$$\partial_\tau \mathbf{v}_n = \begin{pmatrix} -i\mu\mathbf{I}_N - i\mathbf{A} & 0 \\ 0 & i\mu\mathbf{I}_M + i\mathbf{B} \end{pmatrix} \mathbf{v}_n \qquad \text{as } n \to \pm\infty, \qquad (5.2.170)$$

where

$$\mu = \tfrac{1}{2}\left(z - z^{-1}\right)^2.$$

The constant matrices $\mathbf{A}$ and $\mathbf{B}$ can be absorbed by a gauge transformation (5.1.3), but, as we pointed out in Section 5.1, the symmetries in the potentials impose restrictions on the choice of such gauge matrices. Therefore, we determine the time-dependence of the scattering data in the gauge $\mathbf{A} = \mathbf{I}_N, \mathbf{B} = \mathbf{I}_M$. In this case, the system (5.2.170) has solutions that are linear combinations of the solutions

$$\mathbf{v}_n^+ = e^{-\frac{i}{2}(z^2 + z^{-2})\tau} \begin{pmatrix} \mathbf{I}_N \\ 0 \end{pmatrix}, \qquad \mathbf{v}_n^- = e^{\frac{i}{2}(z^2 + z^{-2})\tau} \begin{pmatrix} 0 \\ \mathbf{I}_M \end{pmatrix}.$$

However, such solutions are not compatible with the fixed boundary conditions of the eigenfunctions (5.2.19a)–(5.2.19b), and therefore we define the time-dependent functions

$$\mathcal{M}_n(z, \tau) = e^{-\frac{i}{2}(z^2 + z^{-2})\tau} \mathbf{M}_n(z, \tau), \qquad \bar{\mathcal{M}}_n(z, \tau) = e^{\frac{i}{2}(z^2 + z^{-2})\tau} \bar{\mathbf{M}}_n(z, \tau)$$

$$\mathcal{N}_n(z, \tau) = e^{\frac{i}{2}(z^2 + z^{-2})\tau} \mathbf{N}_n(z, \tau), \qquad \bar{\mathcal{N}}_n(z, \tau) = e^{\frac{i}{2}(z^2 + z^{-2})\tau} \bar{\mathbf{N}}_n(z, \tau)$$

to be solutions of the time-dependence equation (5.1.2b). These τ-dependent functions satisfy the relations

$$\mathcal{M}_n(z, \tau) = z^{-2n} e^{-i(z^2 + z^{-2})\tau} \mathcal{N}_n(z, \tau) \mathbf{b}(z, \tau) + \bar{\mathcal{N}}_n(z, \tau) \mathbf{a}(z, \tau) \qquad (5.2.171a)$$

$$\bar{\mathcal{M}}_n(z, \tau) = z^{2n} e^{i(z^2 + z^{-2})\tau} \bar{\mathcal{N}}_n(z, \tau) \bar{\mathbf{b}}(z, \tau) + \mathcal{N}_n(z, \tau) \bar{\mathbf{a}}(z, \tau), \qquad (5.2.171b)$$

which are obtained from equations (5.2.61a)–(5.2.61b). Like in the scalar case, this yields, for the time evolution of the scattering data,

$$\mathbf{b}(z, \tau) = e^{i(z^2 + z^{-2})\tau} \mathbf{b}(z, 0) \qquad \mathbf{a}(z, \tau) = \mathbf{a}(z, 0) \qquad (5.2.172a)$$

$$\bar{\mathbf{a}}(z, \tau) = \bar{\mathbf{a}}(z, 0) \qquad \bar{\mathbf{b}}(z, \tau) = e^{-i(z^2 + z^{-2})\tau} \bar{\mathbf{b}}(z, 0). \qquad (5.2.172b)$$

The evolution of the reflection coefficients is thus given by

$$\rho(z, \tau) = e^{i(z^2 + z^{-2})\tau} \rho(z, 0) \qquad (5.2.173a)$$

$$\bar{\rho}(z, \tau) = e^{-i(z^2 + z^{-2})\tau} \bar{\rho}(z, 0). \qquad (5.2.173b)$$

From (5.2.172a)–(5.2.172b) it is clear that, for the IDMNLS, the eigenvalues (i.e., the zeros of det $\mathbf{a}(z)$ and det $\bar{\mathbf{a}}(z)$) are also constant as the solution evolves. Not only the number of eigenvalues, but also their locations, are fixed. Thus, the eigenvalues are time-independent discrete states of the evolution.

The evolution of the norming constants follows from the definitions (5.2.75), (5.2.77), and the relations (5.2.172a)–(5.2.172b), that is,

$$\mathbf{C}_j(\tau) = e^{i\mu_j\tau} \mathbf{C}_j(0), \qquad \bar{\mathbf{C}}_j(\tau) = e^{-i\bar{\mu}_j\tau} \bar{\mathbf{C}}_j(0), \qquad (5.2.174)$$

where

$$\mu_j = z_j^2 + z_j^{-2} \qquad \bar{\mu}_j = \bar{z}_j^2 + \bar{z}_j^{-2}. \qquad (5.2.175)$$

Note that the expressions (5.2.172a)–(5.2.172b) for the evolution of the scattering data are valid only for a time-dependence (5.1.2b) with gauge $\mathbf{A} = \mathbf{I}_N$, $\mathbf{B} = \mathbf{I}_M$. To obtain the solutions of (5.1.1a)–(5.1.1b) with gauge $\mathbf{A}, \mathbf{B} = \mathbf{0}$, one first determines the evolution of the potentials $\hat{\mathbf{Q}}_n$, $\hat{\mathbf{R}}_n$ with gauges $\mathbf{A} = \mathbf{B} = \mathbf{I}$ such that

$$\hat{\mathbf{Q}}_n(\tau_0) = \mathbf{Q}_n(\tau_0), \qquad \hat{\mathbf{R}}_n(\tau_0) = \mathbf{R}_n(\tau_0)$$

and then uses the transformations (5.1.14a)–(5.1.14b), that is,

$$\mathbf{Q}_n(\tau) = e^{-i(\tau-\tau_0)\mathbf{A}} \hat{\mathbf{Q}}_n(\tau) e^{-i(\tau-\tau_0)\mathbf{B}} = e^{-2i(\tau-\tau_0)} \hat{\mathbf{Q}}_n(\tau) \quad (5.2.176a)$$

$$\mathbf{R}_n(\tau) = e^{i(\tau-\tau_0)\mathbf{B}} \hat{\mathbf{R}}_n(\tau) e^{i(\tau-\tau_0)\mathbf{A}} = e^{2i(\tau-\tau_0)} \hat{\mathbf{R}}_n(\tau). \quad (5.2.176b)$$

5.3 Soliton solutions

5.3.1 One-soliton solutions

The one-soliton solution of the two-component system (5.1.1a)–(5.1.1b) with $\mathbf{A} = \mathbf{0}, \mathbf{B} = \mathbf{0}$, the symmetries (5.2.79a), and $\mathbf{R}_n = -\mathbf{Q}_n^H$ is the reflection-less potential associated with a single octet of eigenvalues/norming constants (5.2.129) and the condition $\bar{z}_1 = 1/z_1^*$. In (5.2.146) we computed the explicit expression of these potentials for the special choice of the norming constants corresponding to $\boldsymbol{\gamma}_1 \cdot \boldsymbol{\delta}_1 = 0$. The evolution of the norming constants is determined by (5.2.174) plus the symmetry relation $\bar{\mathbf{C}}_1 = (z_1^*)^{-2} \mathbf{C}_1^H$. Therefore, by taking into account (5.2.176a), we get from (5.2.146)

$$\begin{pmatrix} Q_n^{(1)}(\tau) \\ Q_n^{(2)}(\tau) \end{pmatrix} = \frac{\sinh 2\alpha_1}{\left(\|\boldsymbol{\gamma}_1(0)\|^2 + \|\boldsymbol{\delta}_1(0)\|^2 \right)^{1/2}} \operatorname{sech}\left[2(n+1)\alpha_1 - 2v\tau - d(0) \right]$$

$$\times \left[e^{i\pi + 2i(n+1)\beta_1 - 2i\omega_-\tau} \boldsymbol{\gamma}_1^*(0) + e^{-2i(n+1)\beta_1 - in\pi + 2i\omega_+\tau} \boldsymbol{\delta}_1(0) \right],$$

$$(5.3.177)$$

where

$$z_1 = e^{\alpha_1 + i\beta_1} \quad (5.3.178a)$$

$$v = -\sinh(2\alpha_1)\sin(2\beta_1), \qquad \omega_\pm = \cosh(2\alpha_1)\cos(2\beta_1) \pm 1 \quad (5.3.178b)$$

$$d(0) = \log\left(\|\boldsymbol{\gamma}_1(0)\|^2 + \|\boldsymbol{\delta}_1(0)\|^2 \right)^{1/2} - \log \sinh 2\alpha_1. \quad (5.3.178c)$$

Note that, as discussed earlier, this is a special type of "composite" soliton in which the two terms have different carrier frequencies but the peaks and (now, with the time-dependence added) the velocities are coincident. By letting either $\boldsymbol{\gamma}_1 = \mathbf{0}$ or $\boldsymbol{\delta}_1 = \mathbf{0}$ in (5.3.177), one obtains the solutions we referred to as "fundamental" solitons.

Note that, by substituting

$$\tau = h^{-2}t, \qquad Q_n^{(j)} = h q_n^{(j)} = h q^{(j)}(nh), \quad j = 1, 2 \quad (5.3.179a)$$

and

$$z_1 = e^{-ik_1 h}, \qquad \boldsymbol{\gamma}_1(0) = ih\tilde{\boldsymbol{\gamma}}_1(0), \qquad \boldsymbol{\delta}_1(0) = -ih^2\tilde{\boldsymbol{\delta}}_1(0) \quad (5.3.179b)$$

into (5.3.177) we obtain the expression of the one-soliton solution for the vector equation (5.1.10):

$$
\begin{pmatrix} q_n^{(1)}(t) \\ q_n^{(2)}(t) \end{pmatrix} = \frac{1}{\left(\|\tilde{\gamma}_1(0)\|^2 + h^2 \|\tilde{\delta}_1(0)\|^2 \right)^{1/2}}
$$
$$
\times \frac{\sinh 2\eta h}{h} \operatorname{sech} \left(2(n+1)h\xi - 2\tilde{v}t - d(0) \right)
$$
$$
\times \left[i\tilde{\gamma}_1^*(0) e^{-2i(n+1)h\xi - 2i\tilde{\omega}_- t} + i h \tilde{\delta}_1(0) e^{2i(n+1)h\xi - in\pi + 2i\tilde{\omega}_+ t} \right],
$$
$$(5.3.180)$$

where

$$
k_1 = \xi + i\eta, \qquad \eta = \frac{\alpha_1}{h} = \frac{\log |z_1|}{h}, \qquad \xi = -\frac{\beta_1}{h} = -\frac{\arg z_1}{h}, \qquad (5.3.181a)
$$

$$
\tilde{w}_\pm = \frac{\cosh(2\eta h)\cos(2\xi h) \pm 1}{h^2}, \qquad \tilde{v} = \frac{\sinh(2\eta h)\sin(2\xi h)}{h^2}, \qquad (5.3.181b)
$$

$$
d(0) = \log \left(\|\tilde{\gamma}_1(0)\|^2 + h^2 \|\tilde{\delta}_1(0)\|^2 \right)^{\frac{1}{2}} - \log \frac{\sinh(2\eta h)}{h}. \qquad (5.3.181c)
$$

In the limit $h \to 0$, $nh \to x$ (5.3.180) goes to the one-soliton solution (4.3.84) of VNLS.

5.3.2 Transmission coefficients for the pure one-soliton potential

We can use the representations (5.2.65e) and (5.2.65g) to reconstruct the transmission coefficients $\mathbf{c}(z)$ and $\bar{\mathbf{c}}(z)$. We are interested in the reflectionless case with $J = 1$ when Symmetry 5.3 holds for the case $\mathbf{Q}_n = -\mathbf{R}_n^H$. For definiteness, we consider the case when $\delta_1 = \bar{\delta}_1 = \mathbf{0}$. Taking into account the symmetry (5.2.79b), equation (5.2.126) gives

$$
\mathbf{R}_n = -\mathbf{Q}_n^H = 2z_1^{-2(n+1)} \mathbf{C}_1 \left(\bar{\mathbf{N}}_{n+1}^{\prime(\mathrm{up})}(\bar{z}_1) \right)^H + 2(-1)^n \bar{z}_1^{2n} \hat{\mathbf{C}}_1^H \left(\bar{\mathbf{N}}_{n+1}^{\prime(\mathrm{up})}(i/z_1) \right)^H,
$$

where we also used the condition

$$
\bar{z}_1 = 1/z_1^*,
$$

following from Symmetry 5.3. Moreover, equation (5.2.118b) with $J = \bar{J} = 1$ and $\rho(w) = \mathbf{0}$ for $|w| = 1$ gives

$$
\mathbf{N}_n^{\prime(\mathrm{up})}(z) = \mathbf{N}_n^{(\mathrm{up})}(z) = 2 \frac{\bar{z}_1^{2n} z}{z^2 - \bar{z}_1^2} \bar{\mathbf{N}}_n^{\prime(\mathrm{up})}(\bar{z}_1) \bar{\mathbf{C}}_1 + 2(-1)^n \frac{z_1^{-2n} z}{z^2 + z_1^{-2}} \bar{\mathbf{N}}_n^{\prime(\mathrm{up})}(i/z_1) \hat{\mathbf{C}}_1,
$$

so that substituting into (5.2.65e) yields

$$\mathbf{c}(z) = \mathbf{I} - \sum_{k=-\infty}^{+\infty} z \left[2 z_1^{-2(k+1)} \mathbf{C}_1 \left(\bar{\mathbf{N}}_{k+1}^{\prime(\mathrm{up})}(\bar{z}_1) \right)^H \right.$$

$$\left. + 2(-1)^k z_1^{2k} \hat{\mathbf{C}}_1^H \left(\bar{\mathbf{N}}_{k+1}^{\prime(\mathrm{up})}(i/z_1) \right)^H \right]$$

$$\times \left[2 \frac{\bar{z}_1^{2k} z}{z^2 - \bar{z}_1^2} \bar{\mathbf{N}}_k^{\prime(\mathrm{up})}(\bar{z}_1) \bar{\mathbf{C}}_1 + 2(-1)^k \frac{z_1^{-2k} z}{z^2 + z_1^{-2}} \bar{\mathbf{N}}_k^{\prime(\mathrm{up})}(i/z_1) \hat{\mathbf{C}}_1 \right].$$

Writing the norming constants as in (5.2.131a)–(5.2.131b) with the constraint $\delta_1 = \bar{\delta}_1 = 0$, taking into account that $\bar{\boldsymbol{\gamma}}_1 = \boldsymbol{\gamma}_1^*$, and using the corresponding explicit expressions (5.2.135a)–(5.2.135b) for the eigenfunctions, we obtain

$$\mathbf{c}(z) = \mathbf{I} - 4 \sum_{k=-\infty}^{+\infty} \frac{z_1^{-2(k+1)} \bar{z}_1^{2(k+1)}}{(1 + g_k)(1 + g_{k+1})}$$

$$\times \left[\frac{z^2}{z^2 - \bar{z}_1^2} \boldsymbol{\gamma}_1 \boldsymbol{\gamma}_1^H + \frac{z^2}{z^2 + z_1^{-2}} \mathrm{cof} \left(\boldsymbol{\gamma}_1 \boldsymbol{\gamma}_1^H \right) \right], \qquad (5.3.182)$$

where cof denotes the cofactor matrix, g_n is given by (5.2.135c), and we have also used the fact that when Symmetry 5.3 holds, $g_k^* = g_k$. Then, using the identity

$$\frac{1}{(1 + g_k)(1 + g_{k+1})} = \frac{\bar{z}_1^2 - z_1^2}{4 \left\| \boldsymbol{\gamma}_1 \right\|^2 \bar{z}_1^{2(k+1)} z_1^{-2k}} \left[\frac{1}{1 + g_k} - \frac{1}{1 + g_{k+1}} \right], \qquad (5.3.183)$$

we can write

$$\mathbf{c}(z) = \mathbf{I} - \sum_{k=-\infty}^{+\infty} \left[\frac{1}{1 + g_k} - \frac{1}{1 + g_{k+1}} \right]$$

$$\times \left\{ z^2 z_1^{-2} \frac{\bar{z}_1^2 - z_1^2}{z^2 - \bar{z}_1^2} \frac{1}{\|\boldsymbol{\gamma}_1\|^2} \boldsymbol{\gamma}_1 \boldsymbol{\gamma}_1^H + z^2 z_1^{-2} \frac{\bar{z}_1^2 - z_1^2}{z^2 + z_1^{-2}} \frac{1}{\|\boldsymbol{\gamma}_1\|^2} \mathrm{cof} \left(\boldsymbol{\gamma}_1 \boldsymbol{\gamma}_1^H \right) \right\}.$$

The series in the right-hand side is convergent since

$$\lim_{k \to +\infty} \frac{1}{1 + g_k} = 1, \qquad \lim_{k \to -\infty} \frac{1}{1 + g_k} = 0, \qquad (5.3.184)$$

and therefore we can explicitly compute the sum

$$\sum_{k=-\infty}^{+\infty} \left[\frac{1}{1 + g_k} - \frac{1}{1 + g_{k+1}} \right] = \lim_{j \to +\infty} \left[\frac{1}{1 + g_{-j}} - \frac{1}{1 + g_{j+1}} \right] = -1,$$

so that

$$\mathbf{c}(z) = \mathbf{I} + z^2 z_1^{-2} \frac{\bar{z}_1^2 - z_1^2}{z^2 - \bar{z}_1^2} \frac{1}{||\gamma_1||^2} \gamma_1 \gamma_1^H + z^2 z_1^{-2} \frac{\bar{z}_1^2 - z_1^2}{z^2 + z_1^{-2}} \frac{1}{||\gamma_1||^2} \mathrm{cof}\left(\gamma_1 \gamma_1^H\right).$$
(5.3.185)

Using the identity

$$\mathrm{cof}\left(\frac{\gamma \gamma^H}{||\gamma||^2}\right) = \mathbf{I} - \frac{\gamma \gamma^H}{||\gamma||^2},$$
(5.3.186)

one can further simplify the expression of $\mathbf{c}(z)$ to

$$\mathbf{c}(z) = \frac{\bar{z}_1^2 + z^{-2}}{z_1^2 + z^{-2}}\left[\mathbf{I} + \frac{(\bar{z}_1^2 - z_1^2)(\bar{z}_1^2 + z_1^{-2})}{(z^2 - \bar{z}_1^2)(\bar{z}_1^2 + z^{-2})} \frac{1}{||\gamma_1||^2} \gamma_1 \gamma_1^H\right].$$
(5.3.187)

Note that, as expected, $\mathbf{c}(z)$ is analytic outside the unit circle, and when $z \to \infty$

$$\mathbf{c}(z) \sim \bar{z}_1^2 z_1^{-2} \mathbf{I};$$
(5.3.188)

and comparing (5.3.188) with the asymptotic expansion of $\mathbf{c}(z)$ given by (5.2.66b), we get the relation

$$1 + \sum_{k=-\infty}^{+\infty} \alpha_k \prod_{j=k}^{+\infty} \left(1 - \alpha_j\right)^{-1} = \bar{z}_1^2 z_1^{-2}.$$
(5.3.189)

Moreover, one can check that

$$\det \mathbf{c}(z) = \bar{z}_1^4 z_1^{-4} \frac{\left(z^2 - z_1^2\right)\left(z^2 + \bar{z}_1^{-2}\right)}{\left(z^2 - \bar{z}_1^2\right)\left(z^2 + z_1^{-2}\right)},$$

which means that indeed $\det \mathbf{c}\,(z)$ has zeros at the proper points. Then, from the symmetry relation (5.2.96a), that is,

$$\bar{\mathbf{a}}(z) = \prod_{j=-\infty}^{+\infty} (1 - \alpha_j)(\mathbf{c}(1/z^*))^H,$$

we get from (5.3.187) the expression for $\bar{\mathbf{a}}(z)$,

$$\bar{\mathbf{a}}(z) = \prod_{j=-\infty}^{+\infty} (1 - \alpha_j) \frac{z_1^{-2} + z^2}{\bar{z}_1^{-2} + z^2}$$

$$\times \left[\mathbf{I} + z_1^2 \bar{z}_1^{-2} \frac{(z_1^2 - \bar{z}_1^2)(z_1^{-2} + \bar{z}_1^2)}{(z^2 - z_1^2)(z_1^2 + z^{-2})} \frac{1}{||\gamma_1||^2} \gamma_1 \gamma_1^H\right].$$
(5.3.190)

From (5.2.96a) it also follows that

$$\det \bar{\mathbf{a}}(1/z^*) = \prod_{j=-\infty}^{+\infty} (1 - \alpha_j)^2 \left[\det \mathbf{c}(z)\right]^*,$$

and evaluating this relation for $|z| \to \infty$ and taking into account that $\bar{\mathbf{a}}(1/z^*) \to \mathbf{I}$ and $\mathbf{c}(z) \to \bar{z}_1^2 z_1^{-2}\mathbf{I}$ yields

$$\prod_{j=-\infty}^{+\infty} (1-\alpha_j)^2 = z_1^4 \bar{z}_1^{-4}. \tag{5.3.191}$$

Substituting it into (5.3.190), we finally get

$$\bar{\mathbf{a}}(z) = z_1^2 \bar{z}_1^{-2} \frac{\bar{z}_1^{-2} + z^2}{\bar{z}_1^{-2} + z^2} \left[\mathbf{I} + z_1^2 \bar{z}_1^{-2} \frac{(z_1^2 - \bar{z}_1^2)(z_1^{-2} + \bar{z}_1^2)}{(z^2 - z_1^2)(z_1^2 + z^{-2})} \frac{1}{\|\gamma_1\|^2} \gamma_1 \gamma_1^H \right]. \tag{5.3.192}$$

To reconstruct $\bar{\mathbf{c}}(z)$ from (5.2.65g) and, consequently, $\mathbf{a}(z)$ by means of (5.2.96b), we need to solve the system (5.2.119a)–(5.2.119f) with $J = \bar{J} = 1$ and $\rho(w) = \bar{\rho}(w) = \mathbf{0}$ for $w = 1$ to determine $\bar{\mathbf{N}}_n'^{(\mathrm{dn})}(z)$. A few calculations yield

$$\bar{\mathbf{N}}_n'^{(\mathrm{dn})}(z) = \frac{2z}{1 + \bar{z}_1^2 z_1^{-2} g_n} \begin{pmatrix} \frac{z_1^{-2n}}{z^2 - z_1^2} \gamma_1^{(1)} & (-1)^n \frac{\bar{z}_1^{2n}}{z^2 + \bar{z}_1^{-2}} \bar{\gamma}_1^{(2)} \\ \frac{z_1^{-2n}}{z^2 - z_1^2} \gamma_1^{(2)} & (-1)^{n+1} \frac{\bar{z}_1^{2n}}{z^2 + \bar{z}_1^{-2}} \bar{\gamma}_1^{(1)} \end{pmatrix}. \tag{5.3.193}$$

Evaluating (5.2.136) as $z \to 0$ and comparing it with the asymptotic expansion (5.2.106c), we get

$$\Omega_n^{-1} = \Delta_n^{-1} = \frac{1 + \bar{z}_1^2 z_1^{-2} g_n}{1 + g_n} \mathbf{I}, \tag{5.3.194}$$

where the first equality follows from (5.2.122b). Note that, when $n \to -\infty$, (5.3.194) yields

$$\mathbf{I} \prod_{j=-\infty}^{+\infty} (1-\alpha_j)^{-1} = \lim_{n\to-\infty} \Delta_n^{-1} = \bar{z}_1^2 z_1^{-2}\mathbf{I},$$

which is in agreement with (5.3.191). From (5.2.106c) and (5.3.193)–(5.3.194) it follows that

$$\bar{\mathbf{N}}_n^{(\mathrm{dn})}(z) = \Delta_n^{-1}\bar{\mathbf{N}}_n'^{(\mathrm{dn})}(z) = \frac{2z}{1 + g_n} \begin{pmatrix} \frac{z_1^{-2n}}{z^2 - z_1^2} \gamma_1^{(1)} & (-1)^n \frac{\bar{z}_1^{2n}}{z^2 + \bar{z}_1^{-2}} \bar{\gamma}_1^{(2)} \\ \frac{z_1^{-2n}}{z^2 - z_1^2} \gamma_1^{(2)} & (-1)^{n+1} \frac{\bar{z}_1^{2n}}{z^2 + \bar{z}_1^{-2}} \bar{\gamma}_1^{(1)} \end{pmatrix},$$

and the potential matrix $\mathbf{Q}_n$ is given by (5.2.137a)–(5.2.137b) taking into account (5.1.6), that is,

$$\mathbf{Q}_n = -\frac{2}{1 + g_{n+1}} \begin{pmatrix} \bar{z}_1^{2(n+1)} \bar{\gamma}_1^{(1)} & \bar{z}_1^{2(n+1)} \bar{\gamma}_1^{(2)} \\ (-1)^{n+1} z_1^{-2(n+1)} \gamma_1^{(2)} & (-1)^{n+1} z_1^{-2(n+1)} \gamma_1^{(1)} \end{pmatrix}.$$

Substituting these expressions into (5.2.65g) yields

$$\bar{\mathbf{c}}(z) = \mathbf{I} + 4 \sum_{k=-\infty}^{+\infty} \frac{1}{(1+g_k)(1+g_{k+1})} \begin{pmatrix} \frac{z_1^{-2k} \bar{z}_1^{2(k+1)}}{z^2 - z_1^2} \|\gamma_1\|^2 & 0 \\ 0 & -\frac{z_1^{-2(k+1)} \bar{z}_1^{2k}}{z^2 + \bar{z}_1^{-2}} \|\gamma_1\|^2 \end{pmatrix},$$

and using (5.3.183)–(5.3.184) once more we get

$$\bar{\mathbf{c}}(z) = \begin{pmatrix} \frac{z^2 - \bar{z}_1^2}{z^2 - z_1^2} & 0 \\ 0 & \frac{z^2 + z_1^{-2}}{z^2 + \bar{z}_1^{-2}} \end{pmatrix} \tag{5.3.195}$$

with

$$\lim_{z \to 0} \bar{\mathbf{c}}(z) = \bar{z}_1^2 z_1^{-2} \mathbf{I} \qquad \det \bar{\mathbf{c}}(z) = \frac{(z^2 - \bar{z}_1^2)(z^2 + z_1^{-2})}{(z^2 - z_1^2)(z^2 + \bar{z}_1^{-2})}.$$

Finally, using the symmetry (5.2.96b), (5.3.195) provides the expression for $\mathbf{a}(z)$,

$$\mathbf{a}(z) = \begin{pmatrix} \frac{z^2 - z_1^2}{z^2 - \bar{z}_1^2} & 0 \\ 0 & \frac{z^2 + \bar{z}_1^{-2}}{z^2 + z_1^{-2}} \end{pmatrix}. \tag{5.3.196}$$

Note that, as expected, the transmission coefficients are all time-independent. Moreover, $\mathbf{a}$ and $\bar{\mathbf{c}}$ are diagonal and do not depend on the norming constants, while $\bar{\mathbf{a}}$ and $\mathbf{c}$ do.

Finally, we remark that the roles of $\mathbf{a}$ and $\bar{\mathbf{a}}$ are interchanged if one takes the norming constant of the fundamental soliton to be of the form

$$\mathbf{C}_1 = \begin{pmatrix} \gamma_1^{(1)} & \gamma_1^{(2)} \\ 0 & 0 \end{pmatrix}. \tag{5.3.197}$$

5.3.3 Vector–soliton interactions

The problem of a multisoliton collision can be investigated by looking at the asymptotic states as $\tau \to \pm\infty$. The solutions of integrable discrete VNLS (IDVNLS) are more complex than the solutions of the scalar IDNLS because, in the vector equations, there are more degrees of freedom. In particular, vector solitons are characterized, in part, by a polarization. In direct analogy with the VNLS, the system (1.3.9a)–(1.3.9b) reduces to the scalar IDNLS under the reduction

$$\mathbf{q}_n = q_n \mathbf{p}, \tag{5.3.198}$$

where $\mathbf{p}$ is an N-component vector with $\|\mathbf{p}\| = 1$, which is referred to as the polarization of the reduction solution (5.3.198). Moreover, if $\mathbf{q}_n$ satisfies the symmetric system, then so does $\mathbf{U}\mathbf{q}_n$ where $\mathbf{U}$ is any unitary matrix; hence the polarization of a particular reduction solution depends on the choice of the basis.

For a generic multisoliton solution (in which the solitons travel with different velocities) we can define independently a polarization for each soliton in the long-time limit where the solitons are well separated. Then, when $\tau \to \pm\infty$,

$$\mathbf{q}_n^\pm \sim \sum_{j=1}^J \mathbf{q}_n^{(j),\pm}, \qquad \mathbf{q}_n^{(j),\pm} = q_n^{(j),\pm}\mathbf{p}_j^\pm, \tag{5.3.199}$$

where $q_n^{(j),\pm}$ is a one-soliton solution of the IDNLS. We then identify $\mathbf{p}_j^\pm$ as the polarization of the j-th soliton before $(-)$ and after $(+)$ the soliton interaction. For the matrix system (5.1.1a)–(5.1.1b) with Symmetry 5.3, we will write

$$\mathbf{Q}_n^\pm \sim \sum_{j=1}^J \mathbf{Q}_n^{(j),\pm}. \tag{5.3.200}$$

We can investigate the problem of a multisoliton collision proceeding in a similar way as for the continuous VNLS equation (cf. Chapter 4). We consider the solution of integrable discrete matrix NLS (IDMNLS) (5.1.1a)–(5.1.1b) under Symmetry 5.3 corresponding to the scattering data $\{z_j : |z_j| > 1, \mathbf{C}_j\}_{j=1}^J$, and for simplicity we restrict ourselves to the case of fundamental solitons when all the norming constants $\mathbf{C}_j$ have the form

$$\mathbf{C}_j(0) = \begin{pmatrix} \gamma_j^{(1)}(0) & 0 \\ \gamma_j^{(2)}(0) & 0 \end{pmatrix} \qquad j = 1, \ldots, J. \tag{5.3.201}$$

Let us fix the values of the soliton parameters for $\tau \to -\infty$, that is, for each eigenvalue z_j we assign a matrix $\mathbf{S}_j^-$ that completely determines $\mathbf{Q}_n^{(j),-}$. The matrices $\mathbf{S}_j^-$ are the "effective" norming constants for the pure one-soliton potential $\mathbf{Q}_n^{(j),-}$. For $\tau \to +\infty$ we denote the corresponding matrices by $\mathbf{S}_j^+$. For definiteness, let us assume that the discrete eigenvalues are such that the velocities of the individual solitons satisfy the condition $v_1 < v_2 < \cdots < v_J$ (cf. (5.3.177)–(5.3.178b)). Then, as $\tau \to -\infty$, the solitons are distributed along the n-axis in the order corresponding to $v_J, v_{J-1}, \ldots, v_1$; the order of the soliton sequence is reversed as $\tau \to +\infty$. To determine the result of the interaction between solitons, that is, to calculate $\mathbf{S}_j^+$ given $\mathbf{S}_l^-$ for $l = 1, \ldots, J$, we trace the passage of the eigenfunctions through the asymptotic states. We denote the "soliton coordinates" (i.e., the center of the solitons) at the instant of time τ by $n_j(\tau)$ ($|\tau|$ is assumed large enough so that one can talk about individual solitons). If $\tau \to -\infty$, then $n_J \ll n_{J-1} \ll \cdots \ll n_1$. The matrix-valued function $\phi_n(z_j)$ has the form

$$\phi_n(z_j) \sim z_j^n \begin{pmatrix} \mathbf{I}_N \\ \mathbf{0} \end{pmatrix} \qquad n \ll n_J.$$

After passing through the J-th soliton, it will be of the form

$$\phi_n(z_j) \sim z_j^n \begin{pmatrix} \mathbf{I}_N \\ \mathbf{0} \end{pmatrix} \mathbf{a}_J(z_j, \mathbf{S}_J^-),$$

where $\mathbf{a}_J(z, \mathbf{S}_J^-)$ is the transmission coefficient relative to the J-th soliton, and the additional argument is to take into account the dependence on the relative norming constant. Repeating the argument in tracing the passage of the eigenfunctions through the solitons $J-1, J-2, \ldots, j+1$, we find that

$$\phi_n(z_j) \sim z_j^n \begin{pmatrix} \mathbf{I}_N \\ \mathbf{0} \end{pmatrix} \mathbf{a}_{j+1}(z_j, \mathbf{S}_{j+1}^-)\mathbf{a}_{j+2}(z_j, \mathbf{S}_{j+2}^-)\ldots \mathbf{a}_J(z_j, \mathbf{S}_J^-)$$

$$\equiv z_j^n \begin{pmatrix} \mathbf{I}_N \\ \mathbf{0} \end{pmatrix} \prod_{\substack{l=j+1 \\ \text{right}}}^{J} \mathbf{a}_l(z_j, \mathbf{S}_l^-) \qquad n_{j+1} \ll n \ll n_j,$$

where the notation "right" indicates that the product is performed such that the matrix with index ℓ occurs to the right of the matrix with index $\ell-1$. Note that, due to our choice of the norming constants, according to (5.3.196) the transmission coefficients $\mathbf{a}_\ell(z, \mathbf{S}_\ell^-)$ are diagonal and independent of $\mathbf{S}_\ell^-$. Since the j-th soliton corresponds to a bound state, passing through the j-th soliton yields

$$z_j^n \begin{pmatrix} \mathbf{I}_N \\ \mathbf{0} \end{pmatrix} \rightarrow z_j^{-n} \begin{pmatrix} \mathbf{0} \\ \mathbf{I}_M \end{pmatrix} \mathbf{S}_j^-,$$

and therefore

$$\phi_n(z_j) \sim z_j^{-n} \begin{pmatrix} \mathbf{0} \\ \mathbf{I}_M \end{pmatrix} \mathbf{S}_j^- \prod_{\substack{l=j+1 \\ \text{right}}}^{J} \mathbf{a}_l(z_j, \mathbf{S}_l^-) \qquad n_j \ll n \ll n_{j-1}. \quad (5.3.202)$$

On the other hand, starting from $n \gg n_1$ and proceeding in a similar way, we find for the eigenfunction ψ_n the following asymptotic behavior:

$$\psi_n(z_j) \sim z_j^{-n} \begin{pmatrix} \mathbf{0} \\ \mathbf{I}_M \end{pmatrix} \mathbf{c}_{j-1}(z_j, \mathbf{S}_{j-1}^-)\mathbf{c}_{j-2}(z_j, \mathbf{S}_{j-2}^-)\ldots \mathbf{c}_1(z_j, \mathbf{S}_1^-) \quad (5.3.203)$$

$$\equiv z_j^{-n} \begin{pmatrix} \mathbf{0} \\ \mathbf{I}_M \end{pmatrix} \prod_{\substack{l=1 \\ \text{left}}}^{j-1} \mathbf{c}_l(z_j, \mathbf{S}_l^-) \qquad n_j \ll n \ll n_{j-1}, \quad (5.3.204)$$

where we have used (5.2.62a), and the notation "left" indicates that the product is performed such that the matrix with index ℓ occurs to the left of the matrix with index $\ell-1$.

Assuming that at an eigenvalue $\phi_n(z_j) = \psi_n(z_j)\mathbf{B}_j$ and comparing (5.3.202) and (5.3.204), we obtain

$$\mathbf{B}_j \sim \prod_{\substack{l=1 \\ \text{right}}}^{j-1} \left(\mathbf{c}_l(z_j, \mathbf{S}_l^-)\right)^{-1} \mathbf{S}_j^- \prod_{\substack{l=j+1 \\ \text{right}}}^{J} \mathbf{a}_l(z_j, \mathbf{S}_l^-) \qquad \tau \to -\infty. \qquad (5.3.205)$$

Proceeding in a similar fashion as $\tau \to +\infty$ and taking into account that the order of solitons is reversed, we obtain

$$\mathbf{B}_j \sim \prod_{\substack{l=j+1 \\ \text{left}}}^{J} \left(\mathbf{c}_l(z_j, \mathbf{S}_l^+)\right)^{-1} \mathbf{S}_j^+ \prod_{\substack{l=1 \\ \text{left}}}^{j-1} \mathbf{a}_l(z_j, \mathbf{S}_l^+) \qquad \tau \to +\infty, \qquad (5.3.206)$$

and the comparison between the two representations (5.3.205)–(5.3.206) for $\mathbf{B}_j$ yields for $j = J$

$$\mathbf{S}_J^+ = \prod_{\substack{l=1 \\ \text{right}}}^{J-1} \left(\mathbf{c}_l(z_J, \mathbf{S}_l^-)\right)^{-1} \mathbf{S}_J^- \prod_{\substack{l=1 \\ \text{right}}}^{J-1} (\mathbf{a}_l(z_J, \mathbf{S}_l^+))^{-1} \qquad (5.3.207a)$$

and for $j = 1, \ldots, J-1$

$$\mathbf{S}_j^+ = \prod_{\substack{l=j+1 \\ \text{right}}}^{J} \mathbf{c}_l(z_j, \mathbf{S}_l^+) \prod_{\substack{l=1 \\ \text{right}}}^{j-1} \left(\mathbf{c}_l(z_j, \mathbf{S}_l^-)\right)^{-1} \mathbf{S}_j^- \prod_{\substack{l=j+1 \\ \text{right}}}^{J} \mathbf{a}_l(z_j, \mathbf{S}_l^-) \prod_{\substack{l=1 \\ \text{right}}}^{j-1} \left(\mathbf{a}_l(z_j, \mathbf{S}_l^+)\right)^{-1}.$$
$$(5.3.207b)$$

Since, in the case of fundamental solitons that we are considering here, the $\mathbf{a}_l$'s actually do not depend on $\mathbf{S}_l$, the formulas (5.3.207a) and (5.3.207b) determine $\mathbf{S}_j^+$ in terms of $\mathbf{S}_j^-$. Indeed, given $\mathbf{S}_j^-$ for $j = 1, \ldots, J$, from (5.3.207a) we can determine $\mathbf{S}_J^+$. In (5.3.207b) only the knowledge of $\mathbf{S}_l^+$ for $l > j$ is required to determine $\mathbf{S}_j^+$. Hence we can iteratively solve the system (5.3.207a)–(5.3.207b) for $\mathbf{S}_j^+$, $j = 1, \ldots J$.

Two-soliton interaction

Let us consider in detail the interaction of two solitons. In this case, (5.3.207a)–(5.3.207b) give

$$\mathbf{S}_2^+ = (\mathbf{c}_1(z_2, \mathbf{S}_1^-))^{-1} \mathbf{S}_2^- (\mathbf{a}_1(z_2))^{-1} \qquad (5.3.208a)$$

$$\mathbf{S}_1^+ = \mathbf{c}_2(z_1, \mathbf{S}_2^+) \mathbf{S}_1^- \mathbf{a}_2(z_1), \qquad (5.3.208b)$$

where we have taken into account that, due to (5.3.196), the matrices $\mathbf{a}_j(z)$ are independent of $\mathbf{S}_j$. Note that the formulas (5.3.208a)–(5.3.208b) are not symmetric with respect to the exchange of the subscripts 1 and 2. However, such a notation expresses the invariance of the system under the substitution $\tau \rightarrow -\tau$, $\mathbf{Q}_n \rightarrow \mathbf{Q}_n^H$ (here a "fast" soliton becomes a "slow" soliton, and vice versa). Taking into account the explicit expressions of $\mathbf{a}_j(z)$ and $\mathbf{c}_j(z, \mathbf{S}_j)$ for the pure soliton case, one can solve (5.3.208a) for $\mathbf{S}_2^+$ and then substitute it into the right-hand side of (5.3.208b) to get $\mathbf{S}_1^+$. We observe that, in the case of "fundamental solitons" we are considering here, the elements in the first column of matrix $\mathbf{S}_j$ are proportional to the vectors $\boldsymbol{\gamma}_j$, which, in turn, are related to the polarization of the j-th vector soliton (cf. (5.3.177) with $\delta_1 = 0$). Note also that, if $\mathbf{S}_j^-$ and $\mathbf{S}_k^-$ are of the form (5.3.201), since the $\mathbf{a}$'s are diagonal, $\mathbf{S}_j^+$ and $\mathbf{S}_k^+$ also have the same form, which is essential for the "iteration." Therefore, if we denote by $\mathbf{s}_j^{\pm}$ the vectors composing the first column of the matrices $\mathbf{S}_j^{\pm}$, the unit polarization vectors before and after the interaction are written, respectively, as

$$\mathbf{p}_j^{\pm} = \frac{\left(\mathbf{S}_j^{\pm}\right)^*}{\left\|\mathbf{S}_j^{\pm}\right\|}. \tag{5.3.209}$$

For convenience, let us rewrite the formula (5.3.187) for the transmission coefficient of a pure one-soliton solution corresponding to a discrete eigenvalue z_j and polarization vector $\mathbf{p}_j$ as

$$\mathbf{c}_j(z, \mathbf{p}_j) = \frac{\bar{z}_j^2 + z^{-2}}{z_j^2 + z^{-2}} \left[\mathbf{I} + \frac{(\bar{z}_j^2 - z_j^2)(\bar{z}_j^2 + z_j^{-2})}{(z^2 - \bar{z}_j^2)(\bar{z}_j^2 + z^{-2})} \mathbf{p}_j^* \mathbf{p}_j^T \right]. \tag{5.3.210}$$

Note that the $\mathbf{c}$'s depend on the $\mathbf{p}$'s (or, equivalently, the $\mathbf{S}$'s) but their determinant does not and precisely

$$\det \mathbf{c}_j(z, \mathbf{p}_j) = \bar{z}_j^4 z_j^{-4} \frac{\left(z^2 - z_j^2\right)\left(z^2 + \bar{z}_j^{-2}\right)}{\left(z^2 - \bar{z}_j^2\right)\left(z^2 + z_j^{-2}\right)}; \tag{5.3.211}$$

therefore

$$(\mathbf{c}_j(z, \mathbf{p}_j))^{-1} = \frac{1}{\det \mathbf{c}_j} \operatorname{cof} \mathbf{c}_j(z, \mathbf{p}_j),$$

that is,

$$(\mathbf{c}_j(z, \mathbf{p}_j))^{-1} = \frac{z_j^2 + z^{-2}}{\bar{z}_j^2 + z^{-2}} \left[\mathbf{I} - \frac{(\bar{z}_j^2 - z_j^2)(z_j^2 + \bar{z}_j^{-2})}{(z^2 - z_j^2)(z_j^2 + z^{-2})} \mathbf{p}_j^* \mathbf{p}_j^T \right]. \tag{5.3.212}$$

Then, taking into account the explicit expression of $\mathbf{a}_1(z_2)$ and $\mathbf{a}_2(z_1)$ provided by (5.3.196), from (5.3.208a)–(5.3.208b) it follows that

$$\mathbf{S}_2^+ = \frac{z_2^2 - \bar{z}_1^2}{z_2^2 - z_1^2} (\mathbf{c}_1(z_2, \mathbf{p}_1^-))^{-1} \mathbf{S}_2^- \tag{5.3.213a}$$

$$\mathbf{S}_1^+ = \frac{z_1^2 - z_2^2}{\bar{z}_1^2 - \bar{z}_2^2} \mathbf{c}_2(z_1, \mathbf{p}_2^+) \, \mathbf{S}_1^-. \qquad (5.3.213b)$$

Introduce

$$\chi^2 = \frac{\|\mathbf{S}_2^+\|^2}{\|\mathbf{S}_2^-\|^2} = \left| \frac{z_2^2 - \bar{z}_1^2}{\bar{z}_2^2 - z_1^2} \right|^2 (\mathbf{p}_2^-)^T \left(\mathbf{c}_1(z_2, \mathbf{p}_1^-) \right)^{-H} (\mathbf{c}_1(z_2, \mathbf{p}_1^-))^{-1} (\mathbf{p}_2^-)^*.$$
$$(5.3.214)$$

Substituting (5.3.212) into (5.3.214), one can explicitly express χ^2 in terms of the parameters of the colliding solitons

$$\chi^2 = \left| \frac{(z_1^2 - \bar{z}_2^2)(\bar{z}_1^2 + \bar{z}_2^{-2})}{(\bar{z}_1^2 - \bar{z}_2^2)(z_1^2 + z_2^{-2})} \right|^2 \tilde{\chi}^2 \qquad (5.3.215a)$$

$$\tilde{\chi}^2 = 1 + \frac{(z_1^2 - \bar{z}_1^2)(z_2^2 - \bar{z}_2^2)(\bar{z}_1^2 + z_1^{-2})(\bar{z}_2^{-2} + z_2^2)}{(z_2^2 - z_1^2)(\bar{z}_1^2 - \bar{z}_2^2)(z_2^2 + z_1^{-2})(\bar{z}_1^2 + \bar{z}_2^{-2})} \left| \mathbf{p}_1^{-*} \cdot \mathbf{p}_2^- \right|^2. \qquad (5.3.215b)$$

Moreover, from (5.3.209), (5.3.213a), and (5.3.214) it follows that

$$\mathbf{p}_2^+ = \frac{1}{\chi} z_1^{-2} \bar{z}_1^2 \frac{z_1^2 - z_2^2}{\bar{z}_1^2 - \bar{z}_2^2} \left(\mathbf{c}_1(z_2, \mathbf{p}_1^-)^{-1} \right)^* \mathbf{p}_2^-,$$

so that, again substituting (5.3.212), we finally get

$$\mathbf{p}_2^+ = \frac{1}{\chi} \frac{(z_1^2 - \bar{z}_2^2)(\bar{z}_1^2 + \bar{z}_2^{-2})}{(\bar{z}_1^2 - \bar{z}_2^2)(z_1^2 + z_2^{-2})} \left(\mathbf{p}_2^- + \frac{(z_1^2 - \bar{z}_1^2)(\bar{z}_1^2 + z_1^{-2})}{(\bar{z}_1^2 - \bar{z}_2^2)(\bar{z}_1^2 + \bar{z}_2^{-2})} (\mathbf{p}_1^{-*} \cdot \mathbf{p}_2^-) \mathbf{p}_1^- \right).$$
$$(5.3.216)$$

Analogously, from (5.3.213b) and (5.3.209) it follows that

$$\mathbf{p}_1^+ = \frac{(\mathbf{s}_1^+)^*}{\|\mathbf{s}_1^+\|} = z_2^2 \bar{z}_2^{-2} \frac{(\bar{z}_2^2 - \bar{z}_1^2)}{(z_2^2 - z_1^2)} \frac{\|\mathbf{s}_1^-\|}{\|\mathbf{s}_1^+\|} \mathbf{c}_2^*(z_1, \mathbf{p}_2^+) \mathbf{p}_1^-,$$

and since

$$\frac{\|\mathbf{s}_1^+\|^2}{\|\mathbf{s}_1^-\|^2} = \left| \frac{z_1^2 - z_2^2}{\bar{z}_1^2 - \bar{z}_2^2} \right|^2 (\mathbf{p}_1^-)^T \left(\mathbf{c}_2(z_1, \mathbf{p}_2^+) \right)^H \mathbf{c}_2(z_1, \mathbf{p}_2^+) (\mathbf{p}_1^-)^* = \frac{1}{\chi^2},$$

one obtains

$$\mathbf{p}_1^+ = \chi \, z_2^2 \bar{z}_2^{-2} \frac{(\bar{z}_2^2 - \bar{z}_1^2)(\bar{z}_1^2 + z_2^{-2})}{(z_2^2 - z_1^2)(\bar{z}_1^2 + \bar{z}_2^{-2})} \left(\mathbf{p}_1^- + \frac{(\bar{z}_2^2 - z_2^2)(z_2^2 + \bar{z}_2^{-2})}{(z_2^2 - z_1^2)(z_2^2 + \bar{z}_1^{-2})} (\mathbf{p}_2^{+*} \cdot \mathbf{p}_1^-) \mathbf{p}_2^+ \right).$$
$$(5.3.217)$$

From (5.3.216) it then follows that

$$\mathbf{p}_2^{+*} \cdot \mathbf{p}_1^- = \frac{1}{\chi} z_1^2 \bar{z}_1^{-2} \left(\frac{\bar{z}_1^2 - z_2^2}{\bar{z}_2^2 - z_2^2} \right)^2 (\mathbf{p}_1^- \cdot \mathbf{p}_2^{-*}), \qquad (5.3.218)$$

and substituting (5.3.218) and (5.3.217) into (5.3.217) we get the explicit expression of $\mathbf{p}_1^+$ in terms of the incoming polarizations $\mathbf{p}_1^-$, $\mathbf{p}_2^-$, namely,

$$\mathbf{p}_1^+ = \frac{1}{\chi} \frac{(z_1^2 - \bar{z}_2^2)(z_2^2 + z_1^{-2})}{(z_1^2 - z_2^2)(\bar{z}_2^2 + z_1^{-2})} \left(\mathbf{p}_1^- + \frac{(\bar{z}_2^2 - z_2^2)(z_2^2 + \bar{z}_2^{-2})}{(z_2^2 - z_1^2)(z_2^2 + z_1^{-2})} (\mathbf{p}_2^{-*} \cdot \mathbf{p}_1^-) \mathbf{p}_2^- \right).$$

$$(5.3.219)$$

Without loss of generality, one can parametrize the incoming polarization vectors of the solitons as

$$\mathbf{p}_1^- = \begin{pmatrix} 1 \\ 0 \end{pmatrix}, \qquad \mathbf{p}_2^- = \begin{pmatrix} \rho e^{i\varphi} \\ \sqrt{1-\rho^2} \end{pmatrix} \qquad (5.3.220)$$

with $0 \le \rho \le 1$ and $0 \le \varphi \le 2\pi$. Hence, from (5.3.215a)–(5.3.215b) and (5.3.216) it follows that the polarization vectors after the interaction are such that

$$\left| p_1^{+\,(1)} \right|^2 = \frac{1}{\chi^2} \left| \frac{(z_1^2 - \bar{z}_2^2)(z_2^2 + z_1^{-2})}{(z_1^2 - z_2^2)(\bar{z}_2^2 + z_1^{-2})} \right|^2 \left| 1 + \frac{(\bar{z}_2^2 - z_2^2)(z_2^2 + \bar{z}_2^{-2})}{(z_2^2 - z_1^2)(z_2^2 + z_1^{-2})} \rho^2 \right|^2$$

$$(5.3.221a)$$

$$\left| p_2^{+\,(1)} \right|^2 = \frac{1}{\chi^2} \bar{z}_1^4 z_1^{-4} \left| \frac{z_1^2 - \bar{z}_2^2}{\bar{z}_1^2 - \bar{z}_2^2} \right|^4 \rho^2 \qquad (5.3.221b)$$

with

$$\chi^2 = \left| \frac{(z_1^2 - \bar{z}_2^2)(\bar{z}_1^2 + \bar{z}_2^{-2})}{(\bar{z}_1^2 - \bar{z}_2^2)(z_1^2 + \bar{z}_2^{-2})} \right|^2 \left[1 + \frac{(z_1^2 - \bar{z}_1^2)(z_2^2 - \bar{z}_2^2)(\bar{z}_1^2 + z_1^{-2})(\bar{z}_2^{-2} + z_2^2)}{(z_2^2 - z_1^2)(\bar{z}_1^2 - \bar{z}_2^2)(z_2^2 + z_1^{-2})(\bar{z}_1^2 + \bar{z}_2^{-2})} \rho^2 \right].$$

$$(5.3.221c)$$

Note that, like in the continuous case, the magnitudes of the soliton polarizations are invariant only in the case when their initial polarizations are either parallel or orthogonal. In the continuous limit, that is, for $z_j = e^{-ik_j h}$, $\bar{z}_j \equiv 1/z_j^* = e^{-ik_j^* h}$ with $h \to 0$, the formulas (5.3.216) and (5.3.219) coincide with (4.3.107a)–(4.3.107b). In Figure 5.2 we plot both $|p_1^{+\,(1)}|^2$ and $|p_2^{+\,(1)}|^2$ as functions of ρ^2 for some specific choice of the eigenvalues. The continuous curve represents the continuous limit.

Note that, as in all of the previous cases, the "center" of each soliton shifts during the collision process. However, in constrast to the scalar soliton interaction, but like the continuous vector soliton interaction, these shifts depend on both z_1, z_2 and on the epolarization vectors. Indeed, if we denote by $d_1^\pm$, $d_2^\pm$ the centers of the two solitons before/after the interaction, by (5.3.177) and

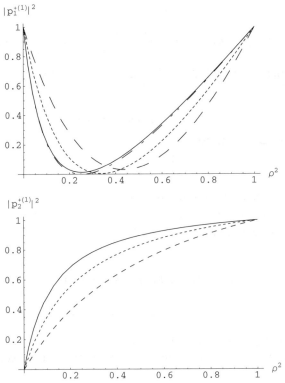

Figure 5.2: Intensity shift induced by two-soliton interaction of two-component IDMNLS. The soliton parameters are $z_j = \exp(a_j + ib_j)h$ with $a_1 = 1$, $a_2 = 2$, $b_1 = -b_2 = 0.1$ and $h = 1$ (dashed), $h = 0.5$ (dotted), $h = 0.1$ (dot-dashed). The solid curves represent the continuous limit, i.e., $h \to 0$. The initial polarizations are given by (5.3.220).

(5.3.178) with $\boldsymbol{\delta}_1 = \mathbf{0}$ it follows that

$$e^{d_1^\pm} = \frac{\left\|\mathbf{S}_1^\pm\right\|}{2\alpha_1} \qquad e^{d_2^\pm} = \frac{\left\|\mathbf{S}_2^\pm\right\|}{2\alpha_2},$$

where $z_j = e^{\alpha j + i\beta j}$. Hence

$$e^{d_2^+ - d_2^-} = \frac{\left\|\mathbf{S}_2^+\right\|}{\left\|\mathbf{S}_2^-\right\|} \equiv \chi \qquad e^{d_1^+ - d_1^-} = \frac{\left\|\mathbf{S}_1^+\right\|}{\left\|\mathbf{S}_1^-\right\|} \equiv \frac{1}{\chi}.$$

As a consequence, the asymptotic separation between the soliton centers, $n_{1,2}^\pm$, also varies due to the collision. Precisely,

$$n_{1,2}^- = \frac{d_2^-}{2\alpha_2} - \frac{d_1^-}{2\alpha_1} \qquad n_{1,2}^+ = \frac{d_2^+}{2\alpha_2} - \frac{d_1^+}{2\alpha_1};$$

hence

$$\Delta n = n_{1,2}^+ - n_{1,2}^- = \frac{d_2^+ - d_2^-}{2\alpha_2} - \frac{d_1^+ - d_1^-}{2\alpha_1} = \frac{\alpha_1 + \alpha_2}{2\alpha_1 \alpha_2} \log \chi$$

with χ given by (5.3.215), since $d_2^+ - d_2^- = -(d_1^+ - d_1^-) = \log \chi$.

Multisoliton interaction

Take $J = 3$ solitons and assume

$$v_1 < v_2 < v_3, \tag{5.3.222}$$

so that the solitons are distributed along the n-axis according to $3 - 2 - 1$ as $\tau \to -\infty$ and $1 - 2 - 3$ as $\tau \to \infty$. According to equations (5.3.208a)–(5.3.208b), for a pairwise interaction between soliton j and soliton k with $v_j > v_k$,

$$\mathbf{S}_j^+ = \left(\mathbf{c}_k(z_j, \mathbf{S}_k^-)\right)^{-1} \mathbf{S}_j^- \left(\mathbf{a}_k(z_j)\right)^{-1} \tag{5.3.223a}$$

$$\mathbf{S}_k^+ = \mathbf{c}_j(z_k, \mathbf{S}_j^+) \mathbf{S}_k^- \mathbf{a}_j(z_k). \tag{5.3.223b}$$

First note that the expression for the **a**'s and the **c**'s have been derived under the assumption of "fundamental solitons," that is, all the norming constants have the form

$$\mathbf{C}_j(0) = \begin{pmatrix} \gamma_j^{(1)}(0) & 0 \\ \gamma_j^{(2)}(0) & 0 \end{pmatrix} \qquad j = 1, \dots, J. \tag{5.3.224}$$

Note also that, if $\mathbf{S}_j^-$ and $\mathbf{S}_k^-$ have this form, since the **a**'s are diagonal, $\mathbf{S}_j^+$ and $\mathbf{S}_k^+$ also have the same form, which is essential for the "iteration" and to give meaning to $\mathbf{c}_j(z_k, \mathbf{S}_j^+)$.

The formulas for a three-soliton interaction, under the assumption (5.3.222) for the velocities, are obtained from (5.3.207a)–(5.3.207b):

$$\mathbf{S}_3^+ = \left(\mathbf{c}_1(z_3, \mathbf{S}_1^-)\right)^{-1} \left(\mathbf{c}_2(z_3, \mathbf{S}_2^-)\right)^{-1} \mathbf{S}_3^- \left(\mathbf{a}_1(z_3)\right)^{-1} \left(\mathbf{a}_2(z_3)\right)^{-1} \tag{5.3.225a}$$

$$\mathbf{S}_2^+ = \mathbf{c}_3(z_2, \mathbf{S}_3^+) \left(\mathbf{c}_1(z_2, \mathbf{S}_1^-)\right)^{-1} \mathbf{S}_2^- \mathbf{a}_3(z_2) \left(\mathbf{a}_1(z_2)\right)^{-1} \tag{5.3.225b}$$

$$\mathbf{S}_1^+ = \mathbf{c}_2(z_1, \mathbf{S}_2^+) \mathbf{c}_3(z_1, \mathbf{S}_3^+) \mathbf{S}_1^- \mathbf{a}_2(z_1) \mathbf{a}_3(z_1). \tag{5.3.225c}$$

Note that the **a**'s are diagonal, and hence they all commute with each other, while two **c**'s do not commute with each other, unless the corresponding polarization vectors are either parallel or orthogonal.

According to equations (5.3.223a)–(5.3.223b), (5.3.225a)–(5.3.225c) can be looked at in terms of compositions of pairwise interactions, namely: (5.3.225a) corresponds to 3 interacting first with 2 and then with 1 (eventually 2 will interact with 1, but this is not relevant as far as S_3^+ is concerned); (5.3.225b) corresponds to 2 interacting first with 1 and then with 3 (but since in c_3 we have S_3^+, this means, as it should be from a physical point of view, that before $2 \leftrightarrow 3$, the interaction $3 \leftrightarrow 1$ takes place); finally, (5.3.225c) corresponds to 3 interacting with 2 (because again in c_3 we have S_3^+), then 1 interacting with 3, and finally 1 interacting with 2.

Let us denote by $S_{\{j,\ell\}}$ the polarization of soliton j after the interaction with soliton ℓ. Note that, according to (5.3.223a)–(5.3.223b), $S_{\{j,\ell\}} \neq S_{\{\ell,j\}}$. Adopting this notation (and dropping the superscript $^-$ for the polarization of the incoming solitons, that is, taking $S_j^- \to S_j$), we can write (5.3.225a) for the polarization of soliton 3 as

$$S_3^+ = (c_1(z_3, S_1))^{-1} S_{\{3,2\}} (a_1(z_3))^{-1} \equiv S_{\{\{3,2\},1\}}. \tag{5.3.226}$$

That is, the net interaction corresponds to the composition of pairwise interactions $3 \leftrightarrow 2$, $3 \leftrightarrow 1$ (eventually, $2 \leftrightarrow 1$ will follow). On the other hand, depending on the initial positions and relative velocities of the solitons, one could have first 2 interacting with 1, giving an $S_{\{2,1\}}$ for soliton 2 and an $S_{\{1,2\}}$ for soliton 1; then 3 interacting with 1, giving $S_{\{3,\{1,2\}\}}$ and $S_{\{\{1,2\},3\}}$; and, finally, 3 interacting with 2, thus giving, for the outcoming polarization of soliton 3,

$$S_3^+ = S_{\{\{3,\{1,2\}\},\{2,1\}\}}.$$

However, we show that the final result is independent of the order of interaction, that is,

$$S_{\{\{3,2\},1\}} = S_{\{\{3,\{1,2\}\},\{2,1\}\}}. \tag{5.3.227}$$

Because

$$S_{\{3,\{1,2\}\}} = \left(c_1\left(z_3, S_{\{1,2\}}\right)\right)^{-1} S_3 (a_1(z_3))^{-1} \tag{5.3.228}$$

and

$$S_{\{\{3,\{1,2\}\}\{2,1\}\}} = \left(c_2\left(z_3, S_{\{2,1\}}\right)\right)^{-1} S_{\{3,\{1,2\}\}} (a_2(z_3))^{-1}, \tag{5.3.229}$$

(5.3.227) corresponds to

$$(c_1(z_3, S_1))^{-1} (c_2(z_3, S_2))^{-1} S_3 (a_1(z_3))^{-1} (a_2(z_3))^{-1}$$
$$= \left(c_2(z_3, S_{\{2,1\}})\right)^{-1} \left(c_1(z_3, S_{\{1,2\}})\right)^{-1} S_3 (a_1(z_3))^{-1} (a_2(z_3))^{-1},$$

that is,

$$c_2(z_3, p_2)c_1(z_3, p_1) = c_1\left(z_3, p_{\{1,2\}}\right) c_2\left(z_3, p_{\{2,1\}}\right). \tag{5.3.230}$$

Taking into account (5.3.210), (5.3.230) amounts to showing that

$$\alpha_2 \mathbf{p}_2^* \mathbf{p}_2^T + \alpha_1 \mathbf{p}_1^* \mathbf{p}_1^T + \alpha_1 \alpha_2 \left(\mathbf{p}_2 \cdot \mathbf{p}_1^* \right) \mathbf{p}_2^* \mathbf{p}_1^T$$
$$= \alpha_2 \mathbf{p}_{\{2,1\}}^* \mathbf{p}_{\{2,1\}}^T + \alpha_1 \mathbf{p}_{\{1,2\}}^* \mathbf{p}_{\{1,2\}}^T + \alpha_1 \alpha_2 \left(\mathbf{p}_{\{1,2\}} \cdot \mathbf{p}_{\{2,1\}}^* \right) \mathbf{p}_{\{1,2\}}^* \mathbf{p}_{\{2,1\}}^T,$$
$$\text{(5.3.231)}$$

where

$$\alpha_1 = \frac{(\bar{z}_1^2 - z_1^2)(\bar{z}_1^2 + z_1^{-2})}{(z_3^2 - \bar{z}_1^2)(\bar{z}_1^2 + z_3^{-2})}, \qquad \alpha_2 = \frac{(\bar{z}_2^2 - z_2^2)(\bar{z}_2^2 + z_2^{-2})}{(z_3^2 - \bar{z}_2^2)(\bar{z}_2^2 + z_3^{-2})}.$$

Without loss of generality we can take

$$\mathbf{p}_1 = \begin{pmatrix} 1 \\ 0 \end{pmatrix}, \qquad \mathbf{p}_2 = \begin{pmatrix} \rho e^{i\varphi} \\ \sqrt{1 - \rho^2} \end{pmatrix}. \qquad \text{(5.3.232)}$$

Then, according to (5.3.216) and (5.3.219),

$$\mathbf{p}_{\{2,1\}} = \frac{1}{\chi} \epsilon_2 \begin{pmatrix} \delta_2 \rho e^{i\varphi} \\ \sqrt{1 - \rho^2} \end{pmatrix} \qquad \text{(5.3.233)}$$

$$\mathbf{p}_{\{1,2\}} = \frac{1}{\chi} \epsilon_1 \begin{pmatrix} 1 + \delta_1 \rho^2 \\ \delta_1 \rho \sqrt{1 - \rho^2} e^{-i\varphi} \end{pmatrix}, \qquad \text{(5.3.234)}$$

where

$$\epsilon_1 = \frac{(z_1^2 - \bar{z}_2^2)(z_2^2 + z_1^{-2})}{(z_1^2 - z_2^2)(\bar{z}_2^2 + z_1^{-2})}, \qquad \epsilon_2 = \frac{(z_1^2 - \bar{z}_2^2)(\bar{z}_1^2 + \bar{z}_2^{-2})}{(\bar{z}_1^2 - \bar{z}_2^2)(z_1^2 + \bar{z}_2^{-2})}$$

$$\delta_1 = \frac{(\bar{z}_2^2 - z_2^2)(z_2^2 + \bar{z}_2^{-2})}{(z_2^2 - z_1^2)(z_2^2 + z_1^{-2})}, \qquad \delta_2 = \frac{(z_1^2 - \bar{z}_2^2)(\bar{z}_2^2 + z_1^{-2})}{(\bar{z}_1^2 - \bar{z}_2^2)(\bar{z}_2^2 + \bar{z}_1^{-2})}$$

and

$$\chi^2 = \left| \frac{(z_1^2 - \bar{z}_2^2)(\bar{z}_1^2 + \bar{z}_2^{-2})}{(\bar{z}_1^2 - \bar{z}_2^2)(z_1^2 + \bar{z}_2^{-2})} \right|^2 \tilde{\chi}^2 \qquad \text{(5.3.235)}$$

$$\tilde{\chi}^2 = 1 + \frac{(z_1^2 - \bar{z}_1^2)(z_2^2 - \bar{z}_2^2)(\bar{z}_1^{-2} + z_1^{-2})(\bar{z}_2^{-2} + z_2^2)}{(z_2^2 - z_1^2)(\bar{z}_1^2 - \bar{z}_2^2)(z_2^2 + z_1^{-2})(\bar{z}_1^2 + \bar{z}_2^{-2})} \rho^2. \qquad \text{(5.3.236)}$$

Using these relations and the property

$$|\epsilon_1|^2 = |\epsilon_2|^2 = \chi^2 \tilde{\chi}^{-2},$$

one can show that

$$\mathbf{p}_{\{1,2\}}^* \mathbf{p}_{\{1,2\}}^T$$
$$= \frac{1}{\tilde{\chi}^2} \begin{pmatrix} |1 + \delta_1 \rho^2|^2 & \delta_1 \left(1 + \delta_1^* \rho^2 \right) \rho \sqrt{1 - \rho^2} e^{-i\varphi} \\ \delta_1^* \left(1 + \delta_1 \rho^2 \right) \rho \sqrt{1 - \rho^2} e^{-i\varphi} & |\delta_1|^2 \rho^2 \left(1 - \rho^2 \right) \end{pmatrix}$$

$$\mathbf{p}^*_{\{2,1\}}\mathbf{p}^T_{\{2,1\}} = \frac{1}{\tilde\chi^2}\begin{pmatrix} \rho^2 & \delta_2^*\rho\sqrt{1-\rho^2}e^{-i\varphi} \\ \delta_2\rho\sqrt{1-\rho^2}e^{i\varphi} & 1-\rho^2 \end{pmatrix}$$

$$\mathbf{p}_{\{1,2\}}\cdot\mathbf{p}^*_{\{2,1\}} = \frac{1}{\chi^2}\epsilon_1\epsilon_2^*\left[\delta_2^* + \delta_1 + \delta_1\left(\delta_2^* - 1\right)\rho^2\right]\rho e^{-i\varphi}$$

$$= \left(\epsilon_1^*\right)^{-1}\epsilon_2^*\left(\delta_2^* + \delta_1\right)\rho e^{-i\varphi}$$

$$\mathbf{p}^*_{\{1,2\}}\mathbf{p}^T_{\{2,1\}} = \frac{1}{\chi^2}\epsilon_1^*\epsilon_2\begin{pmatrix} \delta_2(1+\delta_1^*\rho^2) & (1+\delta_1^*\rho^2)\sqrt{1-\rho^2} \\ \delta_1^*\delta_2\rho^2\sqrt{1-\rho^2}e^{2i\varphi} & \delta_1^*\rho(1-\rho^2)e^{i\varphi} \end{pmatrix};$$

and by substituting these identities into (5.3.231) one can check that indeed it is satisfied. According to (5.3.230), this means that, as far as soliton 3 is concerned, the polarization shift is obtained via pairwise interactions and the result is independent of the order in which such interactions occur. In the same way one can check that the result also holds for solitons 2 and 1.

The shift of the center of soliton 3 due to the interaction with the other solitons is given by

$$\chi_3 = e^{2(d_3^+ - d_3^-)} = \frac{\|\mathbf{S}_3^+\|}{\|\mathbf{S}_3^-\|}.$$

One can verify that this shift of the center of soliton 3 is the result of two consecutive pairwise interactions. Indeed, using the expressions (5.3.216) and (5.3.219) for the pairwise collisions, one can verify that

$$\chi_3 = \chi_{\{3,2\}}\chi_{\{\{3,2\},1\}},$$

where, as before, we are using the notation $\{i, j\}$ to denote the interaction between the i-th and j-th solitons, and $\{\{i, j\}, l\}$ to denote the interaction between the i-th soliton as it emerges from the collision with the j-th and the l-th soliton. Therefore $\chi_{\{i,j\}}$ is given by (5.3.215) with $1 \to i, 2 \to j$ with $i < j$. The same also can be shown for the shifts of centers of the other solitons.

As a consequence of these results, it follows by induction on equations (5.3.207a)–(5.3.207b) (cf. the discussion at the end of Section 4.3.4) that a J-soliton collision is equivalent to the composition of $J(J-1)/2$ pairwise interactions taking place in an arbitrary order.

5.4 Conserved quantities

In this section we calculate polynomial (in $\mathbf{Q}_n$ and $\mathbf{R}_n$) conserved quantities of the system (5.1.1a)–(5.1.1b). The method given here applies to the general matrix system (5.1.1a)–(5.1.1b), including the case $M \neq N$.

We proved that, in the gauge $\mathbf{A}, \mathbf{B} = \mathbf{I}$, the scattering coefficient $\mathbf{a}(z)$ is time-independent. Since $\mathbf{a}(z)$ is analytic for $|z| > 1$, it admits a Laurent expansion whose coefficients are constants of the motion, as well. From the representation (5.2.65a) for $\mathbf{a}(z)$, it follows that the quantities

$$\boldsymbol{\Gamma}_j = \sum_{n=-\infty}^{+\infty} \mathbf{Q}_n \mathbf{M}_n^{(\mathrm{dn}),-2j+1}, \qquad (5.4.237)$$

where $\mathbf{M}_n^{(\mathrm{dn}),-2j+1}$ are the coefficients of the Laurent expansion of $\mathbf{M}_n^{(\mathrm{dn})}(z)$, are conserved for any integer $j \geq 1$. Hence, to find conserved quantities, we need only to express these coefficients in terms of the potentials, which are given by (5.2.44)–(5.2.45). For instance, the first two coefficients are

$$\mathbf{M}_n^{(\mathrm{dn}),-1} = \mathbf{R}_{n-1}, \qquad \mathbf{M}_n^{(\mathrm{dn}),-3} = \mathbf{R}_{n-2} + \mathbf{R}_{n-1} \sum_{k=-\infty}^{n-2} \mathbf{Q}_k \mathbf{R}_{k-1},$$

and therefore the first two (matrix) constants of the motion are given by

$$\boldsymbol{\Gamma}_1 = \sum_{n=-\infty}^{+\infty} \mathbf{Q}_n \mathbf{R}_{n-1}, \qquad \boldsymbol{\Gamma}_2 = \sum_{n=-\infty}^{+\infty} \mathbf{Q}_n \mathbf{R}_{n-2} + \sum_{n=-\infty}^{+\infty} \mathbf{Q}_n \mathbf{R}_{n-1} \sum_{k=-\infty}^{n-2} \mathbf{Q}_k \mathbf{R}_{k-1}.$$
$$(5.4.238)$$

The scattering coefficient $\bar{\mathbf{a}}(z)$ is also a constant of the motion in the gauge $\mathbf{A}, \mathbf{B} = \mathbf{I}$, and proceeding exactly as before one can obtain a second set of conserved quantities given by

$$\bar{\boldsymbol{\Gamma}}_j = \sum_{n=-\infty}^{+\infty} \mathbf{R}_n \bar{\mathbf{M}}_n^{(\mathrm{up}),2j-1}$$

for any $j \geq 1$. Hence, from (5.2.47)–(5.2.48) it follows that

$$\bar{\boldsymbol{\Gamma}}_1 = \sum_{n=-\infty}^{+\infty} \mathbf{R}_n \mathbf{Q}_{n-1}, \qquad \bar{\boldsymbol{\Gamma}}_2 = \sum_{n=-\infty}^{+\infty} \mathbf{R}_n \mathbf{Q}_{n-2} + \sum_{n=-\infty}^{+\infty} \mathbf{R}_n \mathbf{Q}_{n-1} \sum_{k=-\infty}^{n-2} \mathbf{R}_k \mathbf{Q}_{k-1}.$$
$$(5.4.239)$$

Note also that, taking into account the τ-dependence of the scattering matrix defined by (5.2.60), the determinant of the scattering matrix is a constant of the motion in the gauge $\mathbf{A}, \mathbf{B} = \mathbf{I}$, that is,

$$\det \begin{pmatrix} \mathbf{a}(z, \tau) & \bar{\mathbf{b}}(z, \tau) \\ \mathbf{b}(z, \tau) & \bar{\mathbf{a}}(z, \tau) \end{pmatrix} = \det \begin{pmatrix} \mathbf{a}(z, 0) & \bar{\mathbf{b}}(z, 0) \\ \mathbf{b}(z, 0) & \bar{\mathbf{a}}(z, 0) \end{pmatrix}.$$

To obtain a conserved quantity in terms of the potentials, we observe that from

(5.2.57)–(5.2.59a) it follows that

$$\prod_{j=-\infty}^{+\infty} \det\left(\mathbf{I}_M - \mathbf{R}_j \mathbf{Q}_j\right) = \det\begin{pmatrix} \mathbf{a} & \bar{\mathbf{b}} \\ \mathbf{b} & \bar{\mathbf{a}} \end{pmatrix}, \tag{5.4.240}$$

and therefore (5.4.240) is a constant of the motion when $\mathbf{A}, \mathbf{B} = \mathbf{I}$.

In the general gauge $\mathbf{A}, \mathbf{B} \neq \mathbf{I}$, the method illustrated so far generates conserved quantities in terms of the transformed potentials $\hat{\mathbf{Q}}_n$, and $\hat{\mathbf{R}}_n$. The gauge transformation (cf. Section 5.1)

$$\mathbf{Q}_n(\tau) = e^{i\tau\mathbf{A}}\hat{\mathbf{Q}}_n(\tau) e^{i\tau\mathbf{B}} \tag{5.4.241a}$$

$$\mathbf{R}_n(\tau) = e^{-i\tau\mathbf{B}}\hat{\mathbf{R}}_n(\tau) e^{-i\tau\mathbf{A}} \tag{5.4.241b}$$

allows us to express such constants of the motions in terms of the original potentials $\mathbf{Q}_n$ and $\mathbf{R}_n$. In particular, the system (5.1.9a)–(5.1.9b) is a reduction of the system (5.1.1a)–(5.1.1b) with $\mathbf{A} = \mathbf{B} = \mathbf{0}$. For instance, under the reduction (5.1.4)–(5.1.6), the trace of the first of equations (5.4.238) is

$$\sum_{n=-\infty}^{+\infty} \sum_{j=1}^{2} \left[Q_{n-1}^{(j)} R_n^{(j)} + Q_n^{(j)} R_{n-1}^{(j)} \right]$$

or, with the rescaling $t = h^2\tau$ and $\mathbf{q}_n = h^{-1}\left(Q_n^{(1)}, Q_n^{(2)}\right)^T$, $\mathbf{r}_n = h^{-1}\left(R_n^{(1)}, R_n^{(2)}\right)^T$,

$$\sum_{n=-\infty}^{+\infty} \sum_{j=1}^{N} \left[\mathbf{q}_{n-1} \cdot \mathbf{r}_n + \mathbf{q}_n \cdot \mathbf{r}_{n-1} \right], \tag{5.4.242}$$

which is therefore a constant of the motion. Similarly, by taking the trace of the second of (5.4.238), one obtains the conserved quantity

$$\sum_{n=-\infty}^{+\infty} \left\{ \mathbf{q}_{n-2} \cdot \mathbf{r}_n + \mathbf{q}_n \cdot \mathbf{r}_{n-2} - \frac{h^2}{2}\left[\left(\mathbf{q}_{n-1} \cdot \mathbf{r}_n - \mathbf{q}_n \cdot \mathbf{r}_{n-1}\right)^2 \right.\right.$$
$$\left.\left. + 2\left(\mathbf{q}_n \cdot \mathbf{r}_n\right)\left(\mathbf{q}_{n-1} \cdot \mathbf{r}_{n-1}\right)\right]^2 \right\}, \tag{5.4.243}$$

and so on.

Finally, since $\det(\mathbf{I}_M - \hat{\mathbf{R}}_n\hat{\mathbf{Q}}_n) = \det(\mathbf{I}_M - \mathbf{R}_n\mathbf{Q}_n)$, it follows that it is a conserved quantity regardless of the gauge, and under the reduction (5.1.4)–(5.1.6) this conserved quantity can be written as

$$\Gamma = \prod_{j=-\infty}^{+\infty} \left(1 - h^2 \mathbf{q}_n \cdot \mathbf{r}_n\right). \tag{5.4.244}$$

Appendix A
Summation by parts formula

Lemma A.1 *For any sequences $\{a_j\}_{j=-\infty}^{+\infty}$ and $\{b_j\}_{j=-\infty}^{+\infty}$ such that $\sum_{-\infty}^{+\infty} a_j < \infty$ and $\sum_{-\infty}^{+\infty} b_j < \infty$,*

$$\sum_{k=-\infty}^{n} \left\{ a_k \sum_{j=-\infty}^{k-1} b_j \right\} = \left(\sum_{k=-\infty}^{n} a_k \right) \left(\sum_{k=-\infty}^{n} b_k \right) - \sum_{k=-\infty}^{n} \left\{ b_k \sum_{j=-\infty}^{k} a_j \right\}. \qquad \text{(A.1.1)}$$

Proof

$$\sum_{k=-\infty}^{n} \left\{ a_k \sum_{j=-\infty}^{k-1} b_j \right\} = \sum_{k=-\infty}^{n} \left\{ \left(\sum_{j=-\infty}^{k} (a_j - a_{j-1}) \right) \sum_{j=-\infty}^{k-1} b_j \right\}$$

$$= \sum_{k=-\infty}^{n} \left\{ \left(\sum_{j=-\infty}^{k} a_j \right) \left(\sum_{j=-\infty}^{k} b_j \right) - \left(\sum_{j=-\infty}^{k-1} a_j \right) \left(\sum_{j=-\infty}^{k-1} b_j \right) \right\}$$

$$- \sum_{k=-\infty}^{n} \left\{ b_k \sum_{j=-\infty}^{k} a_j \right\}$$

$$= \left(\sum_{k=-\infty}^{n} a_k \right) \left(\sum_{k=-\infty}^{n} b_k \right) - \sum_{k=-\infty}^{n} \left\{ b_k \sum_{j=-\infty}^{k} a_j \right\}$$

The formula for a finite lower bound and an infinite upper bound can be obtained by essentially the same approach.

If $a_k = b_k$, Lemma A1 yields the identity

$$\sum_{k=-\infty}^{n} \left\{ b_k \sum_{j=-\infty}^{k-1} b_j \right\} = \frac{1}{2} \left(\sum_{k=-\infty}^{n} b_k \right)^2 - \frac{1}{2} \sum_{k=-\infty}^{n} b_k^2;$$

therefore

$$\sum_{k=-\infty}^{n} \left\{ b_k \sum_{j=-\infty}^{k-1} b_j \right\} \le \frac{1}{2} \left(\sum_{k=-\infty}^{n} b_k \right)^2. \qquad \text{(A.1.2)}$$

Now, we use the summation by parts formula (A.1.1) to prove a generalization of (A.1.2).

Lemma A.2 *For any real positive sequence* $\{b_j\}_{j=-\infty}^{+\infty}$ *such that the series* $\sum_{-\infty}^{+\infty} b_j$ *is convergent and for any* $m \in \mathbb{N}_0$,

$$\sum_{k=-\infty}^{n} b_k \left(\sum_{j=-\infty}^{k-1} b_j \right)^m \leq \frac{1}{m+1} \left(\sum_{j=-\infty}^{n} b_j \right)^{m+1}. \qquad (A.1.3)$$

Proof We show by induction that

$$\sum_{k=-\infty}^{n} b_k \left(\sum_{j=-\infty}^{k-1} b_j \right)^m = \frac{1}{m+1} \left(\sum_{j=-\infty}^{n} b_j \right)^{m+1} - B_n^{(m)},$$

where

$$B_n^{(m)} \geq 0, \qquad B_{n+1}^{(m)} \geq B_n^{(m)}.$$

The result is trivially true for $m = 0$, with $B_n^{(0)} = 0$. Assuming it holds for $m - 1$, we have

$$\sum_{k=-\infty}^{n} b_k \left(\sum_{j=-\infty}^{k-1} b_j \right)^m$$

$$= \sum_{k=-\infty}^{n} \left[b_k \left(\sum_{j=-\infty}^{k-1} b_j \right)^{m-1} \right] \sum_{j=-\infty}^{k-1} b_j$$

$$= \left\{ \sum_{k=-\infty}^{n} \left[b_k \left(\sum_{j=-\infty}^{k-1} b_j \right)^{m-1} \right] \right\} \left\{ \sum_{j=-\infty}^{k-1} b_j \right\} - \sum_{k=-\infty}^{n} \left\{ b_k \sum_{j=-\infty}^{k-1} b_j \left[b_j \left(\sum_{l=-\infty}^{j-1} b_l \right)^{m-1} \right] \right\}$$

$$- \sum_{k=-\infty}^{n} \left[b_k^2 \left(\sum_{j=-\infty}^{k-1} b_j \right)^{m-1} \right],$$

and by using the inductive hypothesis we get that the result also holds for m with

$$B_n^{(m)} = \frac{m}{m+1} \sum_{k=-\infty}^{n} \left[\left(B_n^{(m-1)} - B_{k-1}^{(m-1)} \right) b_k + b_k^2 \left(\sum_{j=-\infty}^{k-1} b_j \right)^{m-1} \right].$$

Note that, for $b_j \geq 0$, the estimate

$$\sum_{k=-\infty}^{n} b_k \left(\sum_{j=-\infty}^{k-1} b_j \right)^m \leq \left(\sum_{j=-\infty}^{n} b_j \right)^{m+1}$$

is elementary. However, with such an estimate we would only be able to establish that the Neumann series for the Jost functions converge for $\|Q\|_1$, $\|R\|_1 < 1$. It is the additional factor $(m + 1)^{-1}$ in (A.1.3) that extends the convergence of these series to potentials $Q, R \in \ell^1$.

Appendix B

Transmission of the Jost function through a localized potential

Suppose the asymptotic form of the Jost function $\tilde{\mathbf{M}}(x, k)$ for a VNLS is given by

$$\tilde{\mathbf{M}}(x, k) \sim \begin{pmatrix} \mathbf{I}_N \\ \mathbf{0} \end{pmatrix} \mathbf{a}_L(k) \qquad x \ll -L \tag{B.1.1}$$

and we want to find the "transmitted" function through a barrier of finite extension $2L$. $\tilde{\mathbf{M}}(x, k)$ is solution of the integral equation (4.2.13a),

$$\tilde{\mathbf{M}}(x, k) = \begin{pmatrix} \mathbf{I}_N \\ \mathbf{0} \end{pmatrix} \mathbf{a}_L(k) + \int_{-\infty}^{+\infty} \mathbf{G}_+(x - \xi, k) \mathbf{Q}(\xi) \tilde{\mathbf{M}}(\xi, k) d\xi, \tag{B.1.2}$$

where $\mathbf{G}_+(x, k)$ is given by (4.2.12), that is,

$$\tilde{\mathbf{M}}(x, k) = \begin{pmatrix} \mathbf{I}_N \\ \mathbf{0} \end{pmatrix} \mathbf{a}_L(k) + \int_{-\infty}^{x} \begin{pmatrix} \mathbf{q}(\xi) \tilde{\mathbf{M}}^{(dn)}(\xi, k) \\ e^{2ik(x-\xi)} \mathbf{r}(\xi) \tilde{\mathbf{M}}^{(up)}(\xi, k) \end{pmatrix} d\xi. \tag{B.1.3}$$

Introduce $\hat{\mathbf{M}}(x, k)$ such that

$$\tilde{\mathbf{M}}(x, k) = \hat{\mathbf{M}}(x, k) \mathbf{a}_L(k). \tag{B.1.4}$$

$\hat{\mathbf{M}}$ solves the integral equation

$$\hat{\mathbf{M}}(x, k) = \begin{pmatrix} \mathbf{I}_N \\ \mathbf{0} \end{pmatrix} + \int_{-\infty}^{x} \begin{pmatrix} \mathbf{q}(\xi) \hat{\mathbf{M}}^{(dn)}(\xi, k) \\ e^{2ik(x-\xi)} \mathbf{r}(\xi) \hat{\mathbf{M}}^{(dup)}(\xi, k) \end{pmatrix} d\xi, \tag{B.1.5}$$

and, according to (4.2.26a), the transmission coefficient through the barrier is given by

$$\mathbf{a}(k) = \mathbf{I}_N + \int_{-\infty}^{+\infty} \mathbf{q}(\xi) \hat{\mathbf{M}}^{(dn)}(\xi, k) d\xi.$$

Therefore

$$\hat{\mathbf{M}}^{(up)}(x, k) \mathbf{a}(k) \qquad x \gg L,$$

and from (B.1.4) it follows that

$$\tilde{\mathbf{M}}^{(up)}(x, k) \sim \mathbf{a}(k) \mathbf{a}_L(k).$$

Moreover, the integral equation (B.1.5) yields

$$\hat{\mathbf{M}}^{(\text{dn})}(x, k) \to \mathbf{0} \qquad x \to +\infty \tag{B.1.6}$$

for k in the upper k-plane, and we conclude that the transmitted Jost function has the form:

$$\tilde{\mathbf{M}}(x, k) \sim \begin{pmatrix} \mathbf{I}_N \\ \mathbf{0} \end{pmatrix} \mathbf{a}(k)\mathbf{a}_L(k) \qquad x \gg -L. \tag{B.1.7}$$

A similar result can be obtained for the integrable discrete matrix NLS equation.

Appendix C

Scattering theory for the discrete Schrödinger equation

C.1 Introduction

In a certain sense, the stationary solutions of the IDNLS equation (1.3.8) can be related to the solutions of a discretized version of the Schrödinger equation. Indeed, let us consider the differential-difference equation

$$i\frac{dq_n}{dt} + \frac{(q_{n+1} + q_{n-1} - 2q_n)}{h^2} + |q_n|^2 (q_{n+1} + q_{n-1}) = 0 \qquad (C.1.1)$$

and let us look for stationary solutions of the form

$$q_n = w_n e^{i\left(\lambda - \frac{2}{h^2}\right)t} \qquad (C.1.2)$$

for some $\lambda \in \mathbb{R}$ and where w_n does not depend on t. Then w_n satisfies the equation

$$(w_{n+1} + w_{n-1})\gamma_n = \tilde{\lambda} w_n \qquad (C.1.3)$$

with

$$\tilde{\lambda} = h^2 \lambda \qquad (C.1.4)$$

$$\gamma_n = 1 + h^2 |q_n|^2 . \qquad (C.1.5)$$

In (C.1.3) we consider the term $|q_n|^2$ in (C.1.5) to be a known (i.e., given) potential, as opposed to solving a differential-difference equation. Note also that, if $q_n \to 0$ as $n \to \pm\infty$, then $\gamma_n \to 1$ as $n \to \pm\infty$.

Let us introduce a new function φ_n defined by

$$w_n = c_n \varphi_n \qquad (C.1.6)$$

with $c_n \to 1$ for $n \to -\infty$. Then φ_n satisfies

$$\gamma_n \frac{c_{n+1}}{c_n} \varphi_{n+1} + \gamma_n \frac{c_{n-1}}{c_n} \varphi_{n-1} = \tilde{\lambda} \varphi_n, \qquad (C.1.7)$$

and requiring $\frac{c_n}{c_{n-1}} = \gamma_n$, or equivalently,

$$c_n = \prod_{j=-\infty}^{n} \gamma_j, \qquad (C.1.8)$$

yields the discrete Schrödinger equation

$$\alpha_n \varphi_{n+1} + \varphi_{n-1} = \tilde{\lambda} \varphi_n \qquad (C.1.9)$$

208

with potential

$$\alpha_n = \gamma_n \gamma_{n+1}. \tag{C.1.10}$$

From the point of view of the IST, (C.1.9) can be associated with a nonlinear equation called the "nonlinear ladder network" (for the derivation cf. [121] and [98] and equations (C.1.23)–(C.1.24) below). A generalization of (C.1.9) is to add a term $\beta_n \varphi_n$ to the left-hand side, that is, to consider

$$\alpha_n \varphi_{n+1} + \beta_n \varphi_n + \varphi_{n-1} = \tilde{\lambda} \varphi_n, \tag{C.1.11}$$

or, in terms of the function w_n defined in (C.1.6),

$$(w_{n+1} + w_{n-1}) \gamma_n + \beta_n w_n = \tilde{\lambda} w_n. \tag{C.1.12}$$

Indeed, Flaschka [78, 79] studied the IST associated with this modification and showed that it is related to the Toda lattice equation [165], namely,

$$\frac{d^2 Q_n}{dt^2} = e^{-(Q_n - Q_{n-1})} - e^{-(Q_{n+1} - Q_n)}. \tag{C.1.13}$$

See also the derivation and equations (C.1.22) below.

Equation (C.1.11) can be written in an equivalent form by means of the following transformation:

$$\varphi_n = S_n v_n, \tag{C.1.14}$$

where

$$\frac{S_{n+1}}{S_n} \alpha_n = a_n, \qquad \frac{S_{n-1}}{S_n} = a_{n-1}, \qquad a_n \xrightarrow[|n| \to \infty]{} 1. \tag{C.1.15}$$

Solving this recursion relation yields

$$S_n = \prod_{k=-\infty}^{n-1} \frac{a_k}{\alpha_k}, \qquad \alpha_n = a_n^2, \tag{C.1.16}$$

and by means of the transformation (C.1.14) the scattering problem (C.1.11) is cast into the form

$$a_n v_{n+1} + a_{n-1} v_{n-1} + b_n v_n = \tilde{\lambda} v_n, \tag{C.1.17}$$

where β_n has been renamed b_n in order to meet the traditional notation. We note that (C.1.17) reduces to the continuous time-independent Schrödinger equation. Indeed, letting $a_n = e^{-h^2(u_{n+1} + u_n)/2}$, $b_n = 0$, $v_n = e^{h^2 u_n/2} \psi_n$, $\tilde{\lambda} = 2 + h^2 \hat{\lambda}$, we see that, as $h \to 0$, (C.1.17) becomes

$$\psi_{xx} - u(x) \psi = \hat{\lambda} \psi.$$

This equation is the time-independent Schrödinger equation and, of course, is well known in the study of direct and inverse scattering theory (cf. [73], [63]).

To derive the Toda lattice equation (C.1.13) from (C.1.17), we consider [11] the associated time evolution equation

$$\frac{d}{dt} v_n = A_n v_{n+1} + B_n v_n. \tag{C.1.18}$$

Expanding A_n and B_n as

$$A_n = A_n^{(0)} + \tilde{\lambda} A_n^{(1)}, \qquad B_n = B_n^{(0)} + \tilde{\lambda} B_n^{(1)}, \tag{C.1.19}$$

the compatibility condition between (C.1.17) and (C.1.18) yields the following evolution equations:

$$\frac{d}{dt}a_n = \frac{1}{2}A^{(0)}_\infty a_n(b_{n+1} - b_n) + \frac{1}{2}A^{(1)}_\infty a_n\big(a^2_{n+1} - a^2_{n-1} + b^2_{n+1} - b^2_n\big) \qquad \text{(C.1.20a)}$$

$$\frac{d}{dt}b_n = A^{(0)}_\infty(a^2_n - a^2_{n-1}) + A^{(1)}_\infty \big[a^2_n(b_{n+1} + b_n) - a^2_{n-1}(b_n + b_{n-1})\big], \qquad \text{(C.1.20b)}$$

where

$$A_n = A^{(1)}_\infty a_n \tilde{\lambda} + A^{(0)}_\infty a_n + A^{(1)}_\infty a_n b_n, \qquad \text{(C.1.21a)}$$

$$B_n = B^{(1)}_\infty \tilde{\lambda} + B^{(0)}_\infty + A^{(1)}_\infty(4 - a^2_n) + \sum_{k=-\infty}^{n} \partial_t \log a_{k-1}, \qquad \text{(C.1.21b)}$$

and $A^{(j)}_\infty$, $B^{(j)}_\infty$, $j = 0, 1$ are constants. The Toda lattice is arrived at by taking $A^{(1)}_\infty = 0$, $A^{(0)}_\infty = 1$ and arbitrary $B^{(j)}_\infty$ $j = 1, 2$, thus obtaining from (C.1.20a)–(C.1.20b)

$$\frac{d}{dt}a_n = \frac{1}{2}a_n(b_{n+1} - b_n) \qquad \text{(C.1.22a)}$$

$$\frac{d}{dt}b_n = a^2_n - a^2_{n-1} \qquad \text{(C.1.22b)}$$

and relating Q_n in the Toda lattice (C.1.13) to a_n, b_n via

$$a_n = e^{-(Q_n - Q_{n-1})/2} \qquad \text{(C.1.22c)}$$

$$b_n = -\frac{d}{dt}Q_{n-1} \qquad \text{(C.1.22d)}$$

(cf. on the Toda and relativistic Toda lattice, for instance, the papers by Suris [152–155] and Ruijsenaars [145]).

Another interesting lattice equation, the so-called nonlinear ladder network, is obtained from (C.1.20a)–(C.1.20b) by taking $b_n = 0$, $A^{(0)}_\infty = 0$, $A^{(1)}_\infty = 1$,

$$\frac{d}{dt}a_n = \frac{1}{2}a_n(a^2_{n+1} - a^2_{n-1}),$$

which, choosing

$$a_n = e^{-u_n/2}, \qquad \text{(C.1.23)}$$

gives [121], [98]

$$\frac{d}{dt}u_n = e^{-u_{n-1}} - e^{-u_{n+1}}. \qquad \text{(C.1.24)}$$

C.2 Direct scattering problem

Let us write equation (C.1.11) in the form

$$\alpha_n\varphi_{n+1} + \varphi_{n-1} + \beta_n\varphi_n = \big(z + z^{-1}\big)\varphi_n. \qquad \text{(C.2.25)}$$

In the following we assume that $\alpha_n \neq 0$ for any $n \in \mathbb{Z}$. If $\alpha_n \to 1$ and $\beta_n \to 0$ as $n \to \pm\infty$, this equation is asymptotic to

$$\varphi_{n+1} + \varphi_{n-1} = \big(z + z^{-1}\big)\varphi_n;$$

therefore it is natural to introduce the solutions satisfying the boundary conditions

$$\phi_n \sim z^n, \qquad \bar{\phi}_n \sim z^{-n} \qquad n \to -\infty \qquad \text{(C.2.26a)}$$

$$\psi_n \sim z^{-n}, \qquad \bar{\psi}_n \sim z^n \qquad n \to +\infty. \qquad \text{(C.2.26b)}$$

It is also convenient to define the functions

$$M_n(z) = z^{-n}\phi_n(z), \qquad \bar{M}_n(z) = z^{-n}\bar{\phi}_n(z) \qquad \text{(C.2.27a)}$$

$$N_n(z) = z^{-n}\psi_n(z), \qquad \bar{N}_n(z) = z^{-n}\bar{\psi}_n(z) \qquad \text{(C.2.27b)}$$

satisfying the following difference equation:

$$z\,\alpha_n v_{n+1} + z^{-1}v_{n-1} + \beta_n v_n = \left(z + z^{-1}\right)v_n \qquad \text{(C.2.28)}$$

with boundary conditions

$$M_n \sim 1, \qquad \bar{M}_n \sim z^{-2n} \qquad n \to -\infty \qquad \text{(C.2.29a)}$$

$$N_n \sim z^{-2n}, \qquad \bar{N}_n \sim 1 \qquad n \to +\infty. \qquad \text{(C.2.29b)}$$

Now we construct summation equations for these solutions using the method of the Green's functions. As usual, we represent the solutions of (C.2.28) in the form

$$v_n = w + \sum_{k=-\infty}^{+\infty} G_{n-k}(z)\left\{(\alpha_k - 1)v_{k+1} + z^{-1}\beta_k v_k\right\}, \qquad \text{(C.2.30)}$$

where $G_n(z)$ satisfies the difference equation

$$zG_{n+1} + z^{-1}G_{n-1} - (z + z^{-1})G_n = -z\delta_{n,0}. \qquad \text{(C.2.31)}$$

Let us represent G_n and $\delta_{n,0}$ as Fourier integrals,

$$G_n = \frac{1}{2\pi i}\oint_{|p|=1} p^{n-1}\hat{g}(p)dp \qquad \text{(C.2.32)}$$

$$\delta_{n,0} = \frac{1}{2\pi i}\oint_{|p|=1} p^{n-1}dp. \qquad \text{(C.2.33)}$$

From equation (C.2.31) it then follows that

$$\hat{g}(p) = -\frac{z^2 p}{z^2 p^2 - (z^2 + 1)p + 1}.$$

$\hat{g}$ has poles for $p = 1$ and $p = z^{-2}$, and the value of the integral in (C.2.32) and, consequently, of the Green's function G_n, depends on the location of the poles of $\hat{g}$. When $|z| = 1$, both poles are on the contour of integration, and therefore we consider two different deformations of this contour, C^{out}, enclosing both 1 and z^{-2} as well as $p = 0$, and C^{in}, enclosing $p = 0$ but neither 1 nor z^{-2}. Correspondingly, we have the two Green's functions

$$G_n(z) = \theta(n)\frac{z^2}{1 - z^2}\left(1 - z^{-2n}\right) = \theta(n - 1)\frac{z^2}{1 - z^2}\left(1 - z^{-2n}\right) \qquad \text{(C.2.34)}$$

$$\bar{G}_n(z) = -\theta(-n - 1)\frac{z^2}{1 - z^2}\left(1 - z^{-2n}\right) \qquad \text{(C.2.35)}$$

solving the difference equation (C.2.31). As before, $\theta(n)$ denotes the discrete version of the Heaviside function, that is,

$$\theta(n) = \sum_{k=-\infty}^{n} \delta_{k,0} = \begin{cases} 1 & n \geq 0 \\ 0 & n < 0. \end{cases}$$

Note that $G_n(z)$ is analytic outside the unit circle, $\bar{G}_n(z)$ is analytic inside, and

$$G_n(z) = -\theta(n-1) + O(z^{-2}) \qquad |z| \to \infty$$
$$\bar{G}_n(z) = -\theta(-n-1) z^2 + O(z^4) \qquad z \to 0.$$

Taking into account the asymptotics (C.2.29a)–(C.2.29b), the summation equations for M_n and $\bar{N}_n$ can be written as

$$M_n(z) = 1 + \frac{z^2}{1-z^2} \sum_{\ell=-\infty}^{n-1} \left(1 - z^{-2(n-\ell)}\right) \left\{ (\alpha_\ell - 1) M_{\ell+1}(z) + z^{-1} \beta_\ell M_\ell(z) \right\}$$

$$\text{(C.2.36a)}$$

$$\bar{N}_n(z) = 1 - \frac{z^2}{1-z^2} \sum_{\ell=n+1}^{+\infty} \left(1 - z^{-2(n-\ell)}\right) \left\{ (\alpha_\ell - 1) \bar{N}_{\ell+1}(z) + z^{-1} \beta_\ell \bar{N}_\ell(z) \right\}.$$

$$\text{(C.2.36b)}$$

Note that when, β_n is identically zero, both $M_n(z)$ and $\bar{N}_n(z)$ are even functions of z. We will prove that if the potentials $(\alpha_n - 1)$ and β_n decay sufficiently rapidly as $|n| \to \infty$, the summation equations (C.2.36a)–(C.2.36b) are well defined and admit a unique solution. Moreover, $M_n(z)$ is analytic for $|z| > 1$ and continuous for $|z| = 1$, while $\bar{N}_n(z)$ is analytic for $|z| < 1$ and continuous for $|z| = 1$. Under this assumption, M_n has a convergent Laurent series expansion in the annulus centered on $z = 0$,

$$M_n(z) = M_n^{(0)} + z^{-1} M_n^{(-1)} + z^{-2} M_n^{(-2)} + \cdots, \qquad \text{(C.2.37)}$$

and substituting this expansion into the summation equation (C.2.36a) and matching the corresponding powers of z^{-1} yields

$$M_n^{(0)} = 1 - \sum_{\ell=-\infty}^{n-1} (\alpha_\ell - 1) M_{\ell+1}^{(0)}$$

or, equivalently,

$$M_n^{(0)} = \alpha_n M_{n+1}^{(0)}.$$

Taking into account that $M_n \to 1$ as $n \to -\infty$, we finally get

$$M_n^{(0)} = \prod_{\ell=-\infty}^{n-1} \alpha_\ell^{-1}. \qquad \text{(C.2.38)}$$

From the other side, since $\bar{N}_n(z)$ is analytic inside the unit circle, it admits a Taylor series expansion about $z = 0$,

$$\bar{N}_n(z) = \bar{N}_n^{(0)} + z \bar{N}_n^{(1)} + z^2 \bar{N}_n^{(2)} + \cdots, \qquad \text{(C.2.39)}$$

and substituting this expansion into the summation equation (C.2.36b) and matching the corresponding powers of z yields

$$\bar{N}_n^{(0)} = 1. \qquad \text{(C.2.40)}$$

C.2.1 *Existence and analyticity of the Jost functions*

Lemma C.1 *Assuming the potentials α_n, β_n satisfy the conditions*
$\sum_{-\infty}^{+\infty} (1 + |n|) |1 - \alpha_n^{-1}| < \infty$, $\sum_{-\infty}^{+\infty} (1 + |n|) |\beta_n/\alpha_{n-1}| < \infty$, *the function $M_n(z)$ defined by the summation equation (C.2.36a) is analytic for $|z| > 1$ and continuous for $|z| \geq 1$. Similarly, $\bar{N}_n(z)$ defined by (C.2.36b) is analytic for $|z| < 1$ and continuous for $|z| \leq 1$. Moreover, the solutions of these discrete integral equations are unique in the space of functions continuous in z and bounded in n.*

Proof It is convenient to introduce

$$\hat{M}_n(z) = \alpha_{n-1} M_n(z) \tag{C.2.41}$$

satisfying the summation equation

$$
\hat{M}_n(z) = 1 + \frac{z^2}{1 - z^2} \sum_{\ell=-\infty}^{n-1} \left\{ \left(1 - \alpha_{\ell-1}^{-1}\right) \left(1 - z^{-2(n-\ell+1)}\right) \right.
$$
$$
\left. + z^{-1} \frac{\beta_\ell}{\alpha_{\ell-1}} \left(1 - z^{-2(n-\ell)}\right) \right\} \hat{M}_\ell(z) \tag{C.2.42}
$$

and look for solutions in the form of a Neumann series,

$$\hat{M}_n(z) = \sum_{j=0}^{+\infty} \hat{M}_n^{(j)}(z), \tag{C.2.43}$$

where

$$\hat{M}_n^{(0)} = 1$$

$$
\hat{M}_n^{(j+1)}(z) = \frac{z^2}{1 - z^2} \sum_{\ell=-\infty}^{n-1} \left\{ \left(1 - \alpha_{\ell-1}^{-1}\right) \left(1 - z^{-2(n-\ell+1)}\right) \right.
$$
$$
\left. + z^{-1} \frac{\beta_\ell}{\alpha_{\ell-1} \left(1 - z^{-2(n-\ell)}\right)} \right\} \hat{M}_\ell^{(j)}(z).
$$

For any z such that $|z| > 1$ and for any positive integer k,

$$\left| \frac{z^2}{1 - z^2} (1 - z^{-2k}) \right| \leq 2 \frac{|z|^2}{|z|^2 - 1},$$

and using this bound and the summation by parts formula (A.1.3), one can show by induction on j that

$$\left| \hat{M}_n^{(j)}(z) \right| \leq \frac{1}{j!} \left[2 \frac{|z|^2}{|z|^2 - 1} \right]^j \left\{ \sum_{\ell=-\infty}^{n-1} \left[\left|1 - \alpha_{\ell-1}^{-1}\right| + \left| \frac{\beta_\ell}{\alpha_{\ell-1}} \right| \right] \right\}^j. \tag{C.2.44}$$

Hence, provided $z \neq \pm 1$ and the potentials α_n, β_n are such that

$$C = \sum_{n=-\infty}^{+\infty} \left|1 - \alpha_n^{-1}\right| + \sum_{n=-\infty}^{+\infty} \left| \frac{\beta_n}{\alpha_{n-1}} \right| < \infty, \tag{C.2.45}$$

the Neumann series (C.2.43) is majorized by a uniformly convergent series and is itself uniformly convergent with the bound

$$\left| \hat{M}_n(z) \right| \leq e^{2 \frac{|z|^2}{|z|^2 - 1} C}. \tag{C.2.46}$$

This yields existence and analyticity of the Jost function $M_n(z)$ for $|z| \geq 1$, $z \neq \pm 1$.

From the other side, the kernel of the summation equation (C.2.42),

$$H_n(z) = \frac{z^2}{1 - z^2} \left(1 - z^{-2n}\right) \qquad n \geq 1, \tag{C.2.47}$$

satisfies

$$|H_n(z)| \leq \sum_{k=0}^{n-1} |z|^{-2k} \leq n \tag{C.2.48}$$

for any z outside the unit circle, that is, $|z| \geq 1$. Using this bound, one can show by induction that for any $j \geq 0$

$$\left|\hat{M}_n^{(j)}(z)\right| \leq \frac{1}{j!} \left\{ \sum_{\ell=-\infty}^{n-1} \left[(n - \ell + 1)\left|1 - \alpha_{\ell-1}^{-1}\right| + (n - \ell)\left|\frac{\beta_\ell}{\alpha_{\ell-1}}\right|\right] \right\}^j. \tag{C.2.49}$$

Indeed, from (C.2.44) it follows that

$$\left|\hat{M}_n^{(j+1)}(z)\right| \leq \sum_{\ell=-\infty}^{n-1} \left[\left|1 - \alpha_{\ell-1}^{-1}\right| |H_{n-\ell+1}(z)| + \left|\frac{\beta_\ell}{\alpha_{\ell-1}}\right| |H_{n-\ell}(z)|\right] \left|\hat{M}_\ell^{(j)}(z)\right|,$$

and using the bound (C.2.48) and the inductive hypothesis yields

$$\left|\hat{M}_n^{(j+1)}(z)\right| \leq \sum_{\ell=-\infty}^{n-1} \left\{(n - \ell + 1)\left|1 - \alpha_{\ell-1}^{-1}\right| + (n - \ell)\left|\frac{\beta_\ell}{\alpha_{\ell-1}}\right|\right\}$$

$$\times \frac{1}{j!} \left\{\sum_{k=-\infty}^{\ell-1} \left[(\ell - k + 1)\left|1 - \alpha_{k-1}^{-1}\right| + (\ell - k)\left|\frac{\beta_\ell}{\alpha_{\ell-1}}\right|\right]\right\}^j$$

$$\leq \sum_{\ell=-\infty}^{n-1} \left\{(n - \ell + 1)\left|1 - \alpha_{\ell-1}^{-1}\right| + (n - \ell)\left|\frac{\beta_\ell}{\alpha_{\ell-1}}\right|\right\}$$

$$\times \frac{1}{j!} \left\{\sum_{k=-\infty}^{\ell-1} \left[(n - k + 1)\left|1 - \alpha_{k-1}^{-1}\right| + (n - k)\left|\frac{\beta_\ell}{\alpha_{\ell-1}}\right|\right]\right\}^j.$$

The summation by parts formula (A.1.3) completes the induction. We conclude that

$$\left|\hat{M}_n(z)\right| \leq e^{P_n}, \tag{C.2.50}$$

where

$$P_n = \sum_{\ell=-\infty}^{n-2} \left\{(n - \ell)\left|1 - \alpha_\ell^{-1}\right| + (n - \ell - 1)\left|\frac{\beta_{\ell+1}}{\alpha_\ell}\right|\right\}$$

$$\leq \sum_{\ell=-\infty}^{+\infty} |n - \ell| \left\{\left|1 - \alpha_\ell^{-1}\right| + \left|\frac{\beta_\ell}{\alpha_{\ell-1}}\right|\right\}. \tag{C.2.51}$$

However, P_n diverges as $n \to \pm\infty$, and therefore the uniform convergence of the Neumann series is ensured for any finite n provided the potentials satisfy the condition

$$\sum_{n=-\infty}^{+\infty} (1 + |n|)\left|1 - \alpha_n^{-1}\right| < +\infty, \qquad \sum_{n=-\infty}^{+\infty} (1 + |n|)\left|\frac{\beta_n}{\alpha_{n-1}}\right| < +\infty. \tag{C.2.52}$$

In a similar way one can prove the result for the eigenfunction $\bar{N}_n(z)$. Finally, uniqueness of the solutions of the integral equations can be proved as in Lemma 2.1.

C.2.2 Scattering data

Let us define the Wronskian of two functions u_n and v_n as

$$W(u_n, v_n) = u_{n+1}v_n - u_n v_{n+1}. \tag{C.2.53}$$

If φ_n and $\tilde{\varphi}_n$ are any two solutions of the scattering problem (C.2.25), their Wronskian satisfies the following recursion relation:

$$W(\varphi_{n+1}, \tilde{\varphi}_{n+1}) = \frac{1}{\alpha_{n+1}} W(\varphi_n, \tilde{\varphi}_n).$$

Hence, for any nonnegative integer j we have

$$W\left(\phi_n(z), \bar{\phi}_n(z)\right) = \left[\prod_{s=0}^{j} \alpha_{n-s}^{-1}\right] W\left(\phi_{n-j-1}(z), \bar{\phi}_{n-j-1}(z)\right),$$

and in the limit $j \to +\infty$, taking into account the asymptotics (C.2.26a), we get

$$W\left(\phi_n(z), \bar{\phi}_n(z)\right) = \left(z - z^{-1}\right) \prod_{s=-\infty}^{n} \alpha_s^{-1}. \tag{C.2.54}$$

Analogously, from the asymptotics (C.2.26b) it follows that

$$W(\bar{\psi}_n(z), \psi_n(z)) = \left(z - z^{-1}\right) \prod_{s=n+1}^{+\infty} \alpha_s. \tag{C.2.55}$$

The relations (C.2.54)–(C.2.55) prove that ϕ_n and $\bar{\phi}_n$ are linearly independent, as are ψ_n and $\bar{\psi}_n$. Since the scattering problem (C.2.25) is a linear second-order difference equation, the last two can be written as a linear combination of the former ones (or vice versa), that is,

$$\phi_n(z) = b(z)\psi_n(z) + a(z)\bar{\psi}_n(z) \tag{C.2.56a}$$
$$\bar{\phi}_n(z) = \bar{a}(z)\psi_n(z) + \bar{b}(z)\bar{\psi}_n(z) \tag{C.2.56b}$$

for any z such that all four eigenfunctions exist. In particular, these relations hold for $|z| = 1$ and define the scattering coefficients a, $\bar{a}$, b and $\bar{b}$. In terms of the Jost functions, equations (C.2.56a)–(C.2.56b) are written as

$$M_n(z) = \bar{N}_n(z)a(z) + N_n(z)b(z) \tag{C.2.57a}$$
$$\bar{M}_n(z) = N_n(z)\bar{a}(z) + \bar{N}_n(z)\bar{b}(z) \tag{C.2.57b}$$

or also

$$\mu_n(z) = \bar{N}_n(z) + N_n(z)\rho(z) \tag{C.2.58a}$$
$$\bar{\mu}_n(z) = N_n(z) + \bar{N}_n(z)\bar{\rho}(z), \tag{C.2.58b}$$

where

$$\mu_n(z) = \frac{M_n(z)}{a(z)}, \qquad \bar{\mu}_n(z) = \frac{\bar{M}_n(z)}{\bar{a}(z)}, \tag{C.2.59}$$

and the reflection coefficients have been introduced as

$$\rho(z) = \frac{b(z)}{a(z)} \qquad \bar{\rho}(z) = \frac{\bar{b}(z)}{\bar{a}(z)}. \tag{C.2.60}$$

The scattering coefficients can be related to the Wronskian of the Jost functions. Indeed, from (C.2.56a)–(C.2.56b) it follows that

$$W(\phi_n(z), \psi_n(z)) = a(z)W(\bar{\psi}_n(z), \psi_n(z)),$$

and using (C.2.55) we get

$$a(z) = \frac{z}{z^2 - 1} \left[\prod_{s=n+1}^{+\infty} \alpha_s^{-1} \right] W(\phi_n(z), \psi_n(z)). \tag{C.2.61}$$

Analogously, one can show that

$$\bar{a}(z) = \frac{z}{1 - z^2} \left[\prod_{s=n+1}^{+\infty} \alpha_s^{-1} \right] W(\bar{\phi}_n(z), \bar{\psi}_n(z)). \tag{C.2.62}$$

Comparing (C.2.54) and (C.2.55) yields

$$W(\phi_n(z), \bar{\phi}_n(z)) = \left[\prod_{s=-\infty}^{+\infty} \alpha_s^{-1} \right] W(\bar{\psi}_n(z), \psi_n(z)),$$

and using (C.2.56a)–(C.2.56b) gives the following characterization relation for the scattering coefficients:

$$a(z)\bar{a}(z) - b(z)\bar{b}(z) = \prod_{s=-\infty}^{+\infty} \alpha_s^{-1}. \tag{C.2.63}$$

C.2.3 Symmetries

The scattering problem (C.2.25) is symmetric for the exchange of z to z^{-1}, and so are the asymptotic conditions (C.2.26a)–(C.2.26b); therefore

$$\bar{\phi}_n(z) = \phi_n(1/z) \qquad \bar{\psi}_n(z) = \psi_n(1/z) \tag{C.2.64}$$

or, analogously

$$\bar{M}_n(z) = z^{-2n} M_n(1/z), \qquad N_n(z) = z^{-2n} \bar{N}_n(1/z). \tag{C.2.65}$$

In its turn, the symmetry between the eigenfunctions induces, due to (C.2.56a)–(C.2.56b), the following symmetry in the scattering data:

$$\bar{a}(z) = a(1/z), \qquad \bar{b}(z) = b(1/z), \tag{C.2.66a}$$

and consequently

$$\bar{\rho}(z) = \rho(1/z). \tag{C.2.66b}$$

In addition, when the potentials α_n, β_n are real, if $\varphi_n(z)$ solves the scattering problem (C.2.25), so does $\varphi_n^*(1/z^*)$; therefore, taking into account the asymptotic conditions (C.2.26a)–(C.2.26b), we get the following symmetry relations:

$$\bar{\phi}_n(z) = \phi_n^*(1/z^*), \qquad \bar{\psi}_n(z) = \psi_n^*(1/z^*). \tag{C.2.67}$$

or, analogously,

$$\bar{M}_n(z) = z^{-2n} M_n^* \left(1/z^*\right), \qquad N_n(z) = z^{-2n} \bar{N}_n^* \left(1/z^*\right). \tag{C.2.68}$$

As a consequence, for real potentials, the following symmetries between the scattering data hold:

$$\bar{b}(z) = b^* \left(1/z^*\right), \qquad \bar{a}(z) = a^* \left(1/z^*\right). \tag{C.2.69}$$

We can now reconstruct the scattering data by means of integral representations. Recalling the summation equations (C.2.36a)–(C.2.36b), the quantity $M_n(z) - \bar{N}_n(z)a(z)$ can written as

$$M_n(z) - \bar{N}_n(z)a(z) = 1 - a(z) + \sum_{k=-\infty}^{+\infty} \bar{G}_{n-k}(z) \left\{ (\alpha_k - 1) \left[M_{k+1}(z) - \bar{N}_{k+1}(z)a(z) \right] \right.$$

$$\left. + z^{-1}\beta_k \left[M_k(z) - \bar{N}_k(z)a(z) \right] \right\} + \frac{z^2}{1-z^2} \sum_{k=-\infty}^{+\infty} \left(1 - z^{-2(n-k)}\right)$$

$$\times \left\{ (\alpha_k - 1) M_{k+1}(z) + z^{-1}\beta_k M_k(z) \right\},$$

where we used the identity

$$G_n(z) - \bar{G}_n(z) = \frac{z^2}{1-z^2} \left(1 - z^{-2n}\right).$$

From the other side, we also have $M_n(z) - \bar{N}_n(z)a(z) = N_n(z)b(z)$, and therefore

$$N_n(z)b(z) = 1 - a(z)$$

$$+ \frac{z^2}{1-z^2} \sum_{k=-\infty}^{+\infty} \left(1 - z^{-2(n-k)}\right) \left\{ (\alpha_k - 1) M_{k+1}(z) + z^{-1}\beta_k M_k(z) \right\}$$

$$+ \sum_{k=-\infty}^{+\infty} \bar{G}_{n-k}(z) \left\{ (\alpha_k - 1) N_{k+1}(z) + z^{-1}\beta_k N_k(z) \right\} b(z)$$

or, taking into account the symmetry (C.2.68),

$$z^{-2n} \left\{ \bar{N}_n^* \left(1/z^*\right) - \sum_{k=-\infty}^{+\infty} z^{2(n-k-1)} \bar{G}_{n-k}(z) \left[(\alpha_k - 1) \bar{N}_{k+1}^*(1/z^*) + z\beta_k \bar{N}_k^*(1/z^*) \right] \right\} b(z)$$

$$= 1 - a(z) + \frac{z^2}{1-z^2} \sum_{k=-\infty}^{+\infty} \left(1 - z^{-2(n-k)}\right) \left\{ (\alpha_k - 1) M_{k+1}(z) + z^{-1}\beta_k M_k(z) \right\}.$$

From the summation equation (C.2.36b) it follows that the term in curly brackets in the left-hand side is equal to 1, so that

$$a(z) = 1 + \frac{z^2}{1-z^2} \sum_{k=-\infty}^{+\infty} \left\{ (\alpha_k - 1) + z^{-1}\beta_{k+1} \right\} M_{k+1}(z) \tag{C.2.70a}$$

$$b(z) = -\frac{z^2}{1-z^2} \sum_{k=-\infty}^{+\infty} z^{2k} \left\{ (\alpha_k - 1) + z\beta_{k+1} \right\} M_{k+1}(z). \tag{C.2.70b}$$

From (C.2.70a)–(C.2.70b), $\bar{a}$, $\bar{b}$ can be reconstructed using the symmetries (C.2.69).

Note that from (C.2.36a) and (C.2.70a)–(C.2.70b) it also follows that when β_n is identically zero (which is the case for the nonlinear ladder network) a and b are even functions of z,

$$a(-z) = a(z), \qquad b(-z) = b(z), \tag{C.2.71}$$

and the same result holds for $\bar{a}$ and $\bar{b}$.

Finally, we remark that, taking into account (C.2.38), the summation representation (C.2.70a) yields

$$a(z) = 1 - \sum_{k=-\infty}^{+\infty} (\alpha_k - 1) \prod_{\ell=-\infty}^{k} \alpha_\ell^{-1} + O(z^{-2}), \tag{C.2.72}$$

and, taking into account that, since $\alpha_n \to 1$ for $n \to \pm\infty$,

$$\sum_{k=-\infty}^{+\infty} (\alpha_k - 1) \prod_{\ell=-\infty}^{k} \alpha_\ell^{-1} = 0,$$

we get

$$a(z) = 1 + O(z^{-2}). \tag{C.2.73}$$

The representation (C.2.70a) proves that $a(z)$ has the same analytic properties as $M_n(z)$, that is, it is analytic for $|z| > 1$ and continuous for $|z| = 1$, and $\bar{a}(z)$ inherits the analytic properties from $M_n(z^{-1})$, that is, it is analytic for $|z| < 1$ and continuous for $|z| = 1$. Note that $b(z)$ and $\bar{b}(z)$ cannot in general be prosecuted off the unit circle.

C.2.4 Eigenvalues and norming constants

We define a proper eigenvalue for the scattering problem (C.2.25) to be a value of z for which the scattering problem admits bounded solutions going to zero as $n \to \pm\infty$. From equations (C.2.61)–(C.2.62), it follows that, if z_j, with $|z_j| > 1$, is a zero of $a(z)$, then $W(\phi_n(z_j), \psi_n(z_j)) = 0$, that is, $\phi_n(z_j)$ and $\psi_n(z_j)$ are linearly independent, that is,

$$\phi_n(z_j) = b_j \psi_n(z_j) \tag{C.2.74}$$

for some complex constant b_j. Then, from the asymptotics (C.2.26a)–(C.2.26b) it follows that

$$\phi_n(z_j) \sim (z_j)^n \qquad n \to -\infty$$

$$\phi_n(z_j) = b_j \psi_n(z_j) \sim b_j (z_j)^{-n} \qquad n \to +\infty,$$

and since $|z_j| > 1$, this means that z_j is a proper eigenvalue. Vice versa, proper eigenvalues outside the unit circle need to be zeros of $a(z)$. Analogously, one can show that $\bar{z}_\ell$ with $|\bar{z}_\ell| < 1$ is a proper eigenvalue if, and only if, $\bar{a}(\bar{z}_\ell) = 0$ and

$$\bar{\phi}_n(\bar{z}_\ell) = \bar{b}_\ell \bar{\psi}_n(\bar{z}_\ell). \tag{C.2.75}$$

We will assume that neither $a(z)$ nor $\bar{a}(z)$ have zeros for $|z| = 1$.

From the symmetries (C.2.66a) and (C.2.69) it also follows that, for real potentials,

$$a(z) = \bar{a}(1/z) = a^*(z^*) \tag{C.2.76a}$$

$$\bar{a}(z) = a(1/z) = \bar{a}^*(z^*), \tag{C.2.76b}$$

which means that if z_j is an eigenvalue, so is z_j^*, that is, the eigenvalues are real or come into complex conjugate pairs. Moreover, from (C.2.66a) it follows that z_j is an eigenvalue with $|z_j| > 1$ if, and only if, $\bar{z}_j = \frac{1}{z_j}$ is an eigenvalue with $|\bar{z}_j| < 1$. Finally, note that when $\beta_n = 0$ for all n, the symmetry (C.2.71) yields that the eigenvalues also come in pairs $\pm z_j$ ($\pm \bar{z}_j$). In conclusion, if β_n is identically zero and α_n is real, the scattering problem (C.1.11) has the sets of eigenvalues $\left\{ \pm z_j, \pm z_j^*, \pm \frac{1}{z_j}, \pm \frac{1}{z_j^*} \right\}_{j=1}^{J}$.

Let us write the scattering problem (C.1.12) in the form

$$w_{n+1} + w_{n-1} + \beta_n p_n w_n = -h^2 k^2\, p_n w_n,$$

where

$$p_n = \gamma_n^{-1} = \left(1 + h^2 |q_n|^2\right)^{-1} \tag{C.2.77}$$

$$\tilde{\lambda} = \left(z + z^{-1}\right) = -h^2 k^2. \tag{C.2.78}$$

Then for real potentials we have the following relations:

$$w_n^* \left(w_{n+1} + w_{n-1}\right) + \beta_n p_n |w_n|^2 = -h^2 k^2 p_n |w_n|^2$$
$$w_n \left(w_{n+1}^* + w_{n-1}^*\right) + \beta_n p_n |w_n|^2 = -h^2 k^{*2} p_n |w_n|^2$$

or, subtracting one from another,

$$\left(w_n^* w_{n+1} - w_{n-1}^* w_n\right) + \left(w_n^* w_{n-1} - w_{n+1}^* w_n\right) = -h^2 \left(k^2 - k^{*2}\right) p_n |w_n|^2.$$

Summing over all n at an eigenvalue k_j yields

$$\left(k_j^2 - k_j^{*2}\right) \sum_{n=-\infty}^{+\infty} p_n \left|w_n(k_j)\right|^2 = 0,$$

and, since $0 < p_n < 1$ for all $n \in \mathbb{Z}$, if $w_n\left(z_j\right) \in \ell^2$,

$$k_j^2 = k_j^{*2} \tag{C.2.79}$$

that is, either k_j is real or it is purely imaginary. This implies that $\tilde{\lambda}_j = \left(z_j + z_j^{-1}\right)$ is real and then either z_j is real or it is such that $|z_j| = 1$. Since we assumed the potential to be such that $a(z) \neq 0$ for $|z| = 1$, we conclude that in general the proper eigenvalues come in pairs $\left\{z_j, z_j^{-1} : z_j \in \mathbb{R}, |z_j| > 1\right\}_{j=1}^{J}$. In the special case $\beta_n = 0$, there are quartets of eigenvalues $\left\{\pm z_j, \pm z_j^{-1} : z_j \in (1, +\infty)\right\}_{j=1}^{J}$.

C.3 Inverse scattering problem

C.3.1 Recovery of the Jost functions

Suppose that $a(z)$ has J simple zeros $\left\{z_j : |z_j| > 1\right\}_{j=1}^{J}$. Then the function $\mu_n(z)$ given by (C.2.59) is meromorphic outside the unit circle with a finite number of simple poles at the points $z_1, \ldots, z_J$. Taking into account the symmetry (C.2.65), we can write equation (C.2.58a) as

$$\mu_n(z) = \bar{N}_n(z) + z^{-2n}\rho(z)\bar{N}_n\left(z^{-1}\right), \tag{C.3.80}$$

which yields

$$\text{Res}\left(\mu_n; z_j\right) = \lim_{z \to z_j}\left(z - z_j\right)\mu_n(z) = z_j^{-2n}\bar{N}_n\left(z_j^{-1}\right)C_j, \tag{C.3.81}$$

where

$$C_j = \lim_{z \to z_j}\left(z - z_j\right)\rho(z) = \frac{b(z_j)}{a'(z_j)} \tag{C.3.82}$$

are referred to as norming constants. We showed that when β_n is identically zero, the eigenvalues indeed come in pairs $\pm z_j$. As far as the corresponding norming constants are concerned, we have the relation

$$C_{-j} = \frac{b(-z_j)}{a'(-z_j)} = -C_j, \tag{C.3.83}$$

where we have taken into account that in this case both $a(z)$ and $b(z)$ are even functions of the spectral parameter z (cf. (C.2.71)). Note that from (C.2.38) and (C.2.73) it follows that at large z

$$\mu_n(z) = \Delta_n + O(z^{-1}), \tag{C.3.84}$$

where

$$\Delta_n = \prod_{j=-\infty}^{n-1} \alpha_j^{-1}. \tag{C.3.85}$$

Equation (C.3.80) defines a Riemann–Hilbert boundary problem on the unit circle. However, from (C.3.84) it follows that the boundary conditions on μ_n depend on the potential, which is unknown in the inverse problem. Therefore, we introduce the following modified functions:

$$\mu'_n(z) = \frac{\mu_n(z)}{\Delta_n} \tag{C.3.86}$$

$$\bar{N}'_n(z) = \frac{\bar{N}_n(z)}{\Delta_n} \tag{C.3.87}$$

such that

$$\mu'_n(z) = 1 + O(z^{-1}) \qquad |z| \to \infty \tag{C.3.88}$$
$$\bar{N}'_n(z) = \Delta_n^{-1} + O(z) \qquad z \to 0. \tag{C.3.89}$$

Note that μ'_n and $\bar{N}'_n$ are properly defined, according to the condition we assumed, $\alpha_n \neq 0$ for all $n \in \mathbb{Z}$.

Equation (C.3.80) then becomes

$$\mu'_n(z) = \bar{N}'_n(z) + z^{-2n}\rho(z)\bar{N}'_n\left(z^{-1}\right), \tag{C.3.90}$$

and consequently

$$\text{Res}\left(\mu'_n; z_j\right) = z_j^{-2n} \bar{N}'_n\left(z_j^{-1}\right) C_j. \tag{C.3.91}$$

Let us recall that for the IDNLS (Chapter 3) we introduced the following projection operator:

$$\bar{P}(f)(z) = \lim_{\substack{\zeta \to z \\ |\zeta|<1}} \frac{1}{2\pi i} \oint_{|w|=1} \frac{f(w)}{w-\zeta} dw, \tag{C.3.92a}$$

which is defined for $|z| \leq 1$ for any function $f(z)$ continuous on $|z| = 1$ (we will refer to $\bar{P}$ as the inside projector), and

$$P(f)(z) = \lim_{\substack{\zeta \to z \\ |\zeta|>1}} \frac{1}{2\pi i} \oint_{|w|=1} \frac{f(w)}{w-\zeta} dw \tag{C.3.92b}$$

defined for $|z| \geq 1$ for any function $f(z)$ continuous on $|z| = 1$ (outside projector).

Applying $\bar{P}$ to both sides of equation (C.3.90) and taking into account (C.3.88), (C.3.91) and the analytic properties of μ'_n, $\bar{N}'_n$ yields

$$\bar{N}'_n(z) = 1 + \sum_{j=1}^{J} \frac{z_j^{-2n}}{z - z_j} \bar{N}'_n\left(z_j^{-1}\right) C_j - \frac{1}{2\pi i} \lim_{\substack{\zeta \to z \\ |\zeta|<1}} \oint_{|w|=1} \frac{w^{-2n} \rho(w) \bar{N}'_n\left(w^{-1}\right)}{w-\zeta} dw$$

or, equivalently, using (C.2.66b),

$$\bar{N}'_n(z) = 1 + \sum_{j=1}^{J} \frac{z_j^{-2n}}{z - z_j} \bar{N}'_n\left(z_j^{-1}\right) C_j - \frac{1}{2\pi i} \lim_{\substack{\zeta \to z \\ |\zeta|<1}} \oint_{|w|=1} \frac{w^{2n-1} \bar{\rho}(w) \bar{N}'_n(w)}{1-\zeta w} dw. \tag{C.3.93a}$$

The sum in the right-hand side is performed over all the discrete eigenvalues z_j outside the unit circle, that is, with $\left|z_j\right| > 1$, and gives the contribution of the discrete spectrum, while the integral term corresponds to the contribution of the continuous spectrum. Note that, as we proved earlier, the eigenvalues are real for real potentials, and in the following we will restrict ourselves to this case. Equation (C.3.93a) depends on the additional constants $\bar{N}'_n\left(z_j^{-1}\right)$; therefore, in order to close the system, we evaluate (C.3.93a) at z_ℓ^{-1} for all $\ell = 1, \ldots, J$,

$$\bar{N}'_n(z_\ell^{-1}) = 1 + \sum_{j=1}^{J} \frac{z_j^{-2n}}{z_\ell^{-1} - z_j} \bar{N}'_n\left(z_j^{-1}\right) C_j - \frac{1}{2\pi i} \oint_{|w|=1} \frac{w^{2n-1} \bar{\rho}(w) \bar{N}'_n(w)}{1-z_\ell^{-1} w} dw, \tag{C.3.93b}$$

so that (C.3.93a)–(C.3.93b) constitute a linear algebraic–integral system on $|z| = 1$ that allows us, in principle, given the scattering data $\{z_j, C_j\}_{j=1}^{J} \cup \{\rho(w) : |w| = 1\}$ (i.e., eigenvalues and associated norming constants and reflection coefficient) to recover the modified Jost functions $\bar{N}'_n(z)$. Then, taking into account the asymptotics (C.3.89), from (C.3.93a) it follows that

$$\Delta_n^{-1} = 1 - \sum_{j=1}^{J} z_j^{-2n-1} \bar{N}'_n\left(z_j^{-1}\right) C_j - \frac{1}{2\pi i} \oint_{|w|=1} w^{2n-1} \bar{\rho}(w) \bar{N}'_n(w) dw, \tag{C.3.94}$$

which allows us to obtain $\bar{N}_n(z)$ from $\bar{N}'_n(z)$ (cf. (C.3.87)).

Note that for reflectionless potentials, that is, when $\rho(w) = 0$ for all w such that $|w| = 1$, the system (C.3.93a)–(C.3.93b) reduces to a linear–algebraic system.

Equations (C.3.93a)–(C.3.93b) can be written so that one takes explicitly into account the additional symmetry that corresponds to requiring $\beta_n \equiv 0$. Indeed, we showed that in this case the eigenvalues come in quartets $\left\{ \pm z_j, \ \pm z_j^{-1} \right\}_{j=1}^{J}$, and for the corresponding norming constants the relation (C.3.82) holds. Then the system (C.3.93a)–(C.3.93b) with the symmetries explicitly taken into account becomes

$$\bar{N}'_n(z) = 1 + \sum_{j=1}^{J} z_j^{-2n} \left[\frac{1}{z - z_j} \bar{N}'_n\left(z_j^{-1}\right) - \frac{1}{z + z_j} \bar{N}'_n\left(-z_j^{-1}\right) \right] C_j$$
$$- \frac{1}{2\pi i} \lim_{\substack{\zeta \to z \\ |\zeta| < 1}} \oint_{|w|=1} \frac{w^{2n-1} \bar{\rho}(w) \bar{N}'_n(w)}{1 - \zeta w} dw$$

$$\bar{N}'_n\left(z_\ell^{-1}\right) = 1 + \sum_{j=1}^{J} z_j^{-2n} \left[\frac{1}{z_\ell^{-1} - z_j} \bar{N}'_n\left(z_j^{-1}\right) - \frac{1}{z_\ell^{-1} + z_j} \bar{N}'_n\left(-z_j^{-1}\right) \right] C_j$$
$$- \frac{1}{2\pi i} \oint_{|w|=1} \frac{w^{2n-1} \bar{\rho}(w) \bar{N}'_n(w)}{1 - z_\ell^{-1} w} dw$$

$$\bar{N}'_n\left(-z_\ell^{-1}\right) = 1 + \sum_{j=1}^{J} z_j^{-2n} \left[\frac{1}{-z_\ell^{-1} - z_j} \bar{N}'_n\left(z_j^{-1}\right) - \frac{1}{-z_\ell^{-1} + z_j} \bar{N}'_n\left(-z_j^{-1}\right) \right] C_j$$
$$- \frac{1}{2\pi i} \oint_{|w|=1} \frac{w^{2n-1} \bar{\rho}(w) \bar{N}'_n(w)}{1 + z_\ell^{-1} w} dw,$$

which is a linear integro-algebraic system of $2J + 1$ equations in the $2J + 1$ unknowns ($\bar{N}'_n\left(\pm z_\ell^{-1}\right)$ for $\ell = 1, \ldots, J$ and $\bar{N}'_n(z)$), where J is the number of eigenvalues z_ℓ in the interval $(1, +\infty)$. We exploit the symmetry

$$\bar{N}'_n(z) = \bar{N}'_n(-z),$$

in order to reduce the above system to

$$\bar{N}'_n(z) = 1 + 2 \sum_{j=1}^{J} \frac{z_j^{-2n+1}}{z^2 - z_j^2} \bar{N}'_n\left(z_j^{-1}\right) C_j - \frac{1}{2\pi i} \lim_{\substack{\zeta \to z \\ |\zeta| < 1}} \oint_{|w|=1} \frac{w^{2n-1} \bar{\rho}(w) \bar{N}'_n(w)}{1 - \zeta w} dw$$

(C.3.95a)

$$\bar{N}'_n\left(z_\ell^{-1}\right) = 1 + 2 \sum_{j=1}^{J} \frac{z_j^{-2n+1}}{z_\ell^{-2} - z_j^2} \bar{N}'_n\left(z_j^{-1}\right) C_j - \frac{1}{2\pi i} \oint_{|w|=1} \frac{w^{2n-1} \bar{\rho}(w) \bar{N}'_n(w)}{1 - z_\ell^{-1} w} dw.$$

(C.3.95b)

Moreover, equation (C.3.94) becomes

$$\Delta_n^{-1} = 1 - 2 \sum_{j=1}^{J} z_j^{-2n-1} \bar{N}'_n\left(z_j^{-1}\right) C_j - \frac{1}{2\pi i} \oint_{|w|=1} w^{2n-1} \bar{\rho}(w) \bar{N}'_n(w) \, dw. \quad \text{(C.3.95c)}$$

C.3.2 Recovery of the potential

Since the eigenfunction $\bar{N}_n(z)$ is analytic inside the unit circle, it admits a convergent Taylor series about $z = 0$. Hence, let us write

$$\bar{N}_n(z) = 1 + z\bar{N}_n^{(1)} + z^2 \bar{N}_n^{(2)} + \cdots \quad \text{(C.3.96)}$$

and substitute this expansion into the scattering equation (C.2.28). Equating the powers of z, we find

$$O(1): \qquad \bar{N}^{(1)}_{n-1} + \beta_n = \bar{N}^{(1)}_n$$

$$O(z): \qquad \alpha_n + \bar{N}^{(2)}_{n-1} + \beta_n \bar{N}^{(1)}_n = 1 + \bar{N}^{(2)}_n,$$

which allows us to reconstruct the potentials in terms of the coefficients of the Taylor series expansion of the eigenfunction $\bar{N}_n(z)$, namely,

$$\beta_n = \bar{N}^{(1)}_n - \bar{N}^{(1)}_{n-1} \tag{C.3.97a}$$

$$\alpha_n = 1 + \bar{N}^{(2)}_n - \bar{N}^{(2)}_{n-1} - \beta_n \bar{N}^{(1)}_n. \tag{C.3.97b}$$

Reflectionless potentials

In the reflectionless case, equations (C.3.93a)–(C.3.94) reduce to

$$\bar{N}'_n(z) = 1 + \sum_{j=1}^{J} \frac{z_j^{-2n}}{z - z_j} \bar{N}'_n\left(z_j^{-1}\right) C_j \tag{C.3.98a}$$

$$\bar{N}'_n(z_\ell^{-1}) = 1 + \sum_{j=1}^{J} \frac{z_j^{-2n}}{z_\ell^{-1} - z_j} \bar{N}'_n\left(z_j^{-1}\right) C_j \tag{C.3.98b}$$

$$\Delta_n^{-1} = 1 - \sum_{j=1}^{J} z_j^{-2n-1} \bar{N}'_n\left(z_j^{-1}\right) C_j, \tag{C.3.98c}$$

and for $J = 1$ one gets

$$\bar{N}'_n(z_1^{-1}) = \left(1 + C_1(z_1 - z_1^{-1})^{-1} z_1^{-2(n+1)}\right)^{-1} \tag{C.3.99a}$$

$$\bar{N}'_n(z) = 1 + \frac{C_1 z_1^{2n}}{z - z_1}\left(1 + C_1(z_1 - z_1^{-1})^{-1} z_1^{-2(n+1)}\right)^{-1} \tag{C.3.99b}$$

$$\Delta_n^{-1} = \frac{1 + C_1(z_1 - z_1^{-1})^{-1} z_1^{-2(n+1)}}{1 + C_1(z_1 - z_1^{-1})^{-1} z_1^{-2n}}. \tag{C.3.99c}$$

Taking into account (C.3.87), (C.3.99a)–(C.3.99c) provide for the coefficients of the series expansion of $\bar{N}_n(z)$ about $z = 0$,

$$\bar{N}^{(1)}_n = -\frac{C_1 z_1^{-2(n+1)}}{1 + C_1(z_1 - z_1^{-1})^{-1} z_1^{-2(n+1)}} \tag{C.3.100}$$

$$\bar{N}^{(2)}_n = -\frac{C_1 z_1^{-2(n+1)-1}}{1 + C_1(z_1 - z_1^{-1})^{-1} z_1^{-2(n+1)}}, \tag{C.3.101}$$

and substituting into (C.3.97a)–(C.3.97b) yields

$$\beta_n = \frac{C_1(z_1 - z_1^{-1})z_1^{-2n-1}}{\left[1 + C_1(z_1 - z_1^{-1})^{-1}z_1^{-2(n+1)}\right]\left[1 + C_1(z_1 - z_1^{-1})^{-1}z_1^{-2n}\right]} \tag{C.3.102a}$$

$$\alpha_n = 1 + \frac{C_1(z_1 - z_1^{-1})z_1^{-2(n+1)}}{\left[1 + C_1(z_1 - z_1^{-1})^{-1}z_1^{-2(n+1)}\right]^2}. \tag{C.3.102b}$$

Note that α_n can be written as

$$\alpha_n = 1 + \frac{z_1^2 + z_1^{-2} - 2}{\left[Az_1^{n+1} + (Az_1^{n+1})^{-1}\right]^2}, \tag{C.3.103}$$

where

$$A = C_1^{-1/2}\left|z_1 - z_1^{-1}\right|^{1/2}. \tag{C.3.104}$$

When β_n is identically zero (e.g., the case of the nonlinear ladder network), the inverse problem is given by (C.3.95a)–(C.3.95c), which, in the reflectionless case, reduce to

$$\bar{N}_n'(z) = 1 + 2\sum_{j=1}^J \frac{z_j^{-2n+1}}{z^2 - z_j^2}\bar{N}_n'\left(z_j^{-1}\right)C_j \qquad j = 1, \ldots, J \tag{C.3.105a}$$

$$\bar{N}_n'\left(z_\ell^{-1}\right) = 1 + 2\sum_{j=1}^J \frac{z_j^{-2n+1}}{z_\ell^{-2} - z_j^2}\bar{N}_n'\left(z_j^{-1}\right)C_j \tag{C.3.105b}$$

$$\Delta_n^{-1} = 1 - 2\sum_{j=1}^J z_j^{-2n-1}\bar{N}_n'\left(z_j^{-1}\right)C_j. \tag{C.3.105c}$$

Moreover, (C.3.87) gives

$$\bar{N}_n(z) \sim 1 + \Delta_n \bar{N}_n'^{(2)}z^2 + \cdots,$$

which implies that

$$\bar{N}_n^{(2)} = -2\sum_{j=1}^J z_j^{-2n-3}\bar{N}_n\left(z_j^{-1}\right)C_j. \tag{C.3.106}$$

Therefore, taking into account (C.3.97b),

$$\alpha_n = 1 + 2\sum_{j=1}^J z_j^{-2n-1}\left[\bar{N}_{n-1}\left(z_j^{-1}\right) - z_j^{-2}\bar{N}_n\left(z_j^{-1}\right)\right]. \tag{C.3.107}$$

For $J = 1$, equations (C.3.105a)–(C.3.105b) give

$$\bar{N}_n'\left(z_1^{-1}\right) = \frac{1}{1 + 2\left(z_1^2 - z_1^{-2}\right)^{-1}C_1 z_1^{-2n+1}}$$

$$\Delta_n^{-1} = \frac{1 + 2C_1\left(z_1^2 - z_1^{-2}\right)^{-1}z_1^{-2n-3}}{1 + 2C_1\left(z_1^2 - z_1^{-2}\right)^{-1}z_1^{-2n+1}};$$

therefore, from (C.3.87) we get

$$\bar{N}_n \left(z_1^{-1}\right) = \frac{1}{1 + 2C_1 \left(z_1^2 - z_1^{-2}\right)^{-1} z_1^{-2n-3}},$$

and then from (C.3.107) it follows that

$$\alpha_n = 1 + \frac{2C_1 \left(1 - z_1^{-2}\right) z_1^{-n-1}}{\left[1 + 2C_1 \left(z_1^2 - z_1^{-2}\right)^{-1} z_1^{-2n-2}\right]\left[1 + 2C_1 \left(z_1^2 - z_1^{-2}\right)^{-1} z_1^{-2n-3}\right]}. \qquad \text{(C.3.108)}$$

In order to get a real and nonsingular potential, the norming constant C_1 has to be real and of the same sign as z_1, that is, $C_1 > 0$ if $z_1 \in (1, +\infty)$. If we write z_1 in the form $z_1 = e^\omega$ for some $\omega > 0$, then the potential becomes

$$\alpha_n = 1 + \frac{2C_1 \left(e^{-(2n+1)\omega} - e^{-(2n+3)\omega}\right)}{\left[1 + \frac{C_1}{\sinh 2\omega} e^{-(2n+1)\omega}\right]\left[1 + \frac{C_1}{\sinh 2\omega} e^{-(2n+3)\omega}\right]}$$

or

$$\alpha_n = 1 + \frac{\sinh \omega \sinh 2\omega}{\cosh\left[\left(n + \frac{1}{2}\right)\omega + x_0\right]\cosh\left[\left(n + \frac{3}{2}\right)\omega + x_0\right]}, \qquad \text{(C.3.109)}$$

where

$$e^{x_0} = \left(\frac{\sinh 2\omega}{C_1}\right)^{1/2}. \qquad \text{(C.3.110)}$$

C.3.3 Gel'fand–Levitan–Marchenko equations

We can also provide a reconstruction for the potentials by means of Gel'fand–Levitan–Marchenko integral equations. Indeed, let us represent the eigenfunctions ψ_n and $\bar{\psi}_n$ in terms of triangular kernels,

$$\psi_n(z) = \sum_{j=n}^{+\infty} z^{-j} K(n, j) \qquad |z| > 1 \qquad \text{(C.3.111a)}$$

$$\bar{\psi}_n(z) = \sum_{j=n}^{+\infty} z^j \bar{K}(n, j) \qquad |z| < 1. \qquad \text{(C.3.111b)}$$

Note that, due to the symmetry (C.2.64), $\bar{K}(n, j) = K(n, j)$. Moreover, taking into account (C.2.27b) and the asymptotics (C.2.40), we get $\bar{K}(n, n) = K(n, n) = 1$. We write equation (C.2.56a) in the form

$$\phi_n(z)a^{-1}(z) - \bar{\psi}_n(z) = \psi_n(z)\rho(z), \qquad \text{(C.3.112)}$$

with ρ given by (C.2.60), and apply the operator $\frac{1}{2\pi i}\oint_{|z|=1} dz \, z^{-m-1}$ for $m > n$ to equation (C.3.112). Taking into account the asymptotics (C.2.26a), (C.2.26b) and the analytic properties of the eigenfunctions and of the transmission coefficient $a^{-1}(z)$, as well as the triangular representations (C.3.111a)–(C.3.111b), we obtain the discrete integral equation (Gel'fand–Levitan–Marchenko equation)

$$K(n, m) + F(n + m) + \sum_{j=n+1}^{+\infty} K(n, j)F(m + j) = 0 \qquad m > n, \qquad \text{(C.3.113)}$$

where

$$F(n) = \sum_{j=1}^{J} z_j^{-n-1} C_j + \frac{1}{2\pi i} \oint_{|z|=1} z^{-n-1} \rho(z) dz. \qquad (C.3.114)$$

Comparing the representation (C.3.111a) for the eigenfunctions with (C.3.96) and recalling (C.3.97a) and (C.3.97b), we obtain the reconstruction of the potentials in terms of the kernels of GLM equations, that is,

$$\beta_n = K(n, n+1) - K(n-1, n) \qquad (C.3.115a)$$
$$\alpha_n = 1 + K(n, n+2) - K(n-1, n+1) - \beta_n K(n, n+1). \qquad (C.3.115b)$$

C.4 Time evolution and solitons for the Toda lattice and nonlinear ladder network

C.4.1 Toda lattice

Let us consider the time differential-difference equation relative to the Toda lattice (C.1.13), namely, equation (C.1.18),

$$\frac{d}{dt} v_n = A_n v_{n+1} + B_n v_n \qquad (C.4.116)$$

with A_n, B_n given by (C.1.21a)–(C.1.21b) and $A_\infty^{(0)} = 1$, $A_\infty^{(1)} = B_\infty^{(0)} = B_\infty^{(1)} = 0$, that is,

$$A_n = a_n \qquad B_n = \sum_{k=-\infty}^{n} \partial_t \log a_{k-1}. \qquad (C.4.117)$$

We rewrite (C.4.116) in terms of the eigenfunction φ_n of the scattering problem (C.1.11), that is, by means of the transformations (C.1.14)–(C.1.16), thus obtaining

$$\frac{d}{dt} \varphi_n = \alpha_n \varphi_{n+1}, \qquad (C.4.118)$$

where we took into account that, due to (C.1.16), $B_n = -\partial_t \log S_n$ and $\alpha_n = a_n^2$. Equation (C.4.118) determines the evolution of the Jost functions. From this we deduce the time-dependence of the scattering data.

First we observe that, since $\alpha_n \to 1$ as $|n| \to \infty$, the eigenfunctions fixed by the boundary conditions (C.2.26a)–(C.2.26b) are not compatible with the time evolution (C.4.118). Therefore we define the time-dependent functions

$$\Phi_n(z, t) = e^{zt} \phi_n(z, t), \qquad \Psi_n(z, t) = e^{z^{-1}t} \psi_n(z, t) \qquad (C.4.119)$$

to be solutions of the time-dependence equation (C.4.118). Such solutions are compatible with the boundary conditions (C.2.26a)–(C.2.26b) and by virtue of (C.2.56a), (C.2.64) satisfy the relation

$$\Phi_n(z, t) = b(z, t) e^{(z-z^{-1})t} \Psi_n(z, t) + a(z, t) \Psi_n(z^{-1}, t). \qquad (C.4.120)$$

To find the expression for the evolution of the scattering data, we first differentiate this equation with respect to t to obtain

$$\frac{d}{dt}\Phi_n(z,t) = e^{(z-z^{-1})t}\left\{\left[b_t(z,t) + \left(z - z^{-1}\right)b(z,t)\right]\Psi_n(z,t) + b(z,t)\frac{d}{dt}\Psi_n(z,t)\right\}$$

$$+ a_t(z,t)\Psi_n(z^{-1},t) + a(z,t)\frac{d}{dt}\Psi_n(z^{-1},t).$$

Taking into account that both Φ_n and Ψ_n solve (C.4.118), we get

$$b_t(z,t) = -\left(z - z^{-1}\right)b(z,t) \qquad a_t(z,t) = 0,$$

that is,

$$b(z,t) = e^{-\left(z-z^{-1}\right)t}b(z,0) \qquad a(z,t) = a(z,0), \tag{C.4.121}$$

and, consequently, for the reflection coefficient defined in (C.2.60)

$$\rho(z,t) = e^{-\left(z-z^{-1}\right)t}\rho(z,0). \tag{C.4.122}$$

Analogously, one obtains the evolution of the norming constants (C.3.82),

$$C_j(t) = e^{-\left(z_j - z_j^{-1}\right)t}C_j(0). \tag{C.4.123}$$

Finally, in the case of a purely discrete spectrum, the soliton solutions are computable in closed form. A one-soliton solution corresponding to a single eigenvalue z_1 is obtained by introducing the explicit time-dependence into (C.3.102b). Indeed, recalling (C.1.16) and (C.1.22c), we get from (C.3.102b)

$$e^{-(Q_n - Q_{n-1})} = 1 + \frac{z_1^2 + z_1^{-2} - 2}{A(t)z_1^{n+1} + \left(A(t)z_1^{n+1}\right)^{-1}}, \tag{C.4.124}$$

where, due to (C.4.123), $A(t) = C_1^{-1/2}(0)\exp\left(\left(z_1 - z_1^{-1}\right)t/2\right)|z_1 - z_1^{-1}|^{1/2}$. Setting $z_1 = \sigma e^\omega$, $\sigma = \pm 1$, reduces this to

$$e^{-(Q_n - Q_{n-1})} = 1 + \sinh^2\omega \operatorname{sech}^2\left(\omega(n - n_0) + \sigma\sinh\omega t\right), \tag{C.4.125}$$

where n_0 is a constant depending only on C_1, z_1. We note that, since $\sigma = \pm 1$, this soliton solution may travel in either the positive or the negative n direction.

C.4.2 Nonlinear ladder network

The time differential-difference equation relative to the nonlinear ladder network (C.1.24) is obtained from (C.1.18) with A_n, B_n given by (C.1.21a)–(C.1.21b) and $A_\infty^{(1)} = 1$, $B_\infty^{(0)} = -4$, $A_\infty^{(0)} = B_\infty^{(1)} = 0$ and $b_n = 0$, that is,

$$A_n = a_n\tilde{\lambda} \qquad B_n = -a_n^2 + \sum_{k=-\infty}^{n}\partial_t\log a_{k-1}. \tag{C.4.126}$$

We rewrite (C.1.18) in terms of the eigenfunction φ_n of the scattering problem (C.1.11), that is, by means of the transformations (C.1.14)–(C.1.16), thus obtaining

$$\frac{d}{dt}\varphi_n = \tilde{\lambda}\alpha_n\varphi_{n+1} - \alpha_n\varphi_n. \tag{C.4.127}$$

As before, we define the time-dependent functions

$$\Phi_n(z, t) = A(t)\phi_n(z), \qquad \Psi_n(z, t) = B(t)\psi_n(z) \qquad \text{(C.4.128)}$$

and determine $A(t)$, $B(t)$ such that Φ_n and Ψ_n solve the time-dependence equation (C.4.127) and are compatible, respectively, with the boundary conditions (C.2.26a)–(C.2.26b). Since $\alpha_n \to 1$ as $|n| \to \infty$, we get

$$\Phi_n(z, t) = e^{z^2 t}\phi_n(z), \qquad \Psi_n(z, t) = e^{z^{-2}t}\psi_n(z). \qquad \text{(C.4.129)}$$

Moreover, from (C.2.56a) and (C.2.64) it follows that Φ_n and Ψ_n satisfy the relation

$$\Phi_n(z, t) = b(z, t)e^{(z^2 - z^{-2})t}\,\Psi_n(z, t) + a(z, t)\Psi_n(1/z, t). \qquad \text{(C.4.130)}$$

To find the expression for the evolution of the scattering data, we first differentiate this equation with respect to t to obtain

$$\frac{d}{dt}\Phi_n(z, t) = e^{(z^2 - z^{-2})t}\left\{\left[b_t(z, t) + 2\left(z^2 - z^{-2}\right)b(z, t)\right]\,\Psi_n(z, t) + b(z, t)\frac{d}{dt}\Psi_n(z, t)\right\}$$

$$+ a_t(z, t)\Psi_n(1/z, t) + a(z, t)\frac{d}{dt}\Psi_n(1/z, t).$$

Taking into account that both Φ_n and Ψ_n solve (C.4.127), we get

$$b_t(z, t) = -\left(z^2 - z^{-2}\right)b(z, t) \qquad a_t(z, t) = 0, \qquad \text{(C.4.131)}$$

that is,

$$b(z, t) = e^{-(z^2 - z^{-2})t}b(z, 0) \qquad a(z, t) = a(z, 0) \qquad \text{(C.4.132a)}$$

and, consequently, for the reflection coefficient defined in (C.2.60)

$$\rho(z, t) = e^{-(z^2 - z^{-2})t}\rho(z, 0). \qquad \text{(C.4.132b)}$$

Analogously, one obtains the evolution of the norming constants (C.3.82),

$$C_j(t) = e^{-\left(z_j^2 - z_j^{-2}\right)t}C_j(0). \qquad \text{(C.4.132c)}$$

Finally, introducing the explicit time-dependence of the norming constant into (C.3.109), we obtain the expression for the pure one-soliton solution of the ladder network. Indeed, recalling (C.1.16) and (C.1.23), we get from (C.3.109)

$$e^{-u_n} = 1 + \frac{\sinh \omega \sinh 2\omega}{\cosh\left[\left(n + \frac{1}{2}\right)\omega + x_0 - 2(\sinh \omega)t\right]\cosh\left[\left(n + \frac{3}{2}\right)\omega + x_0 - 2(\sinh \omega)t\right]}.$$

$$\text{(C.4.133)}$$

Appendix D

Nonlinear Schrödinger systems with a potential term

D.1 Continuous NLS systems with a potential term

The modified NLS equation with a generic potential term

$$iq_t = q_{xx} + 2\left[|q|^2 + V(x, t)\right]q \tag{D.1.1}$$

was studied by Chen and Liu [52] and Balakrishnan [26]. It describes, for instance, the propagation of envelope solitons in inhomogeneous media – an example being that of electromagnetic waves in an inhomogeneous plasma. $V(x, t) = 0$ corresponds to a homogeneous plasma. This case is the standard NLS equation, which is known (see Chapter 2) to support a soliton traveling with constant velocity. The presence of $V(x, t)$ has the effect of introducing a potential barrier in the path of the soliton. The equation is also relevant in the context of Davydov's alpha-helix solitons [61], which are responsible for energy transport along molecular chains. $V(x, t)$ would then represent inhomogeneities in the arrangement of molecules along the chain.

Chen and Liu showed that, if $V(x, t)$ is a linear function of x, then the IST is applicable and soliton solutions exist. The presence of the potential forces the spectral parameter k occurring in the corresponding scattering problem to develop a dependence on both x and t. $k(x, t)$ itself satisfies a certain nonlinear evolution equation involving $V(x, t)$.

We note that if $q(x, t)$ satisfies the modified NLS equation (D.1.1) with a real potential term of the form considered below, namely,

$$V(x, t) = \mathcal{E}(t)x + \mathcal{F}(t),$$

the transformation of variables

$$u(y, t) = e^{-i\theta(x,t)}q(x, t) \qquad y = x - v(t), \tag{D.1.2}$$

with $\theta(x, t)$ a real function linearly depending on the space variable x and v a function of t, yields for u the standard NLS

$$iu_t = u_{yy} + 2|u|^2 u$$

provided

$$v' = -2\theta_x, \qquad \theta_t = \theta_x^2 - 2(\mathcal{E}x + \mathcal{F}), \tag{D.1.3}$$

where $'$ denotes derivative with respect to t. Therefore, writing

$$\theta(x, t) = -\tfrac{1}{2}v'x + \delta(t),$$

229

one has to determine v and δ such that

$$v'' = 4\mathcal{E} \qquad \delta' = \tfrac{1}{4}\left(v'\right)^2 - 2\mathcal{F},$$

that is,

$$v(t) = 4 \int_0^t d\tau \int_0^\tau d\tau' \mathcal{E}(\tau') \tag{D.1.4}$$

$$\delta(t) = 2 \int_0^t \left[2\left(\int_0^\tau \mathcal{E}(\tau')d\tau' \right)^2 - \mathcal{F}(\tau) \right] d\tau. \tag{D.1.5}$$

Let us consider the system

$$iq_t = q_{xx} - 2\left[qr - V(x,t)\right]q \tag{D.1.6}$$

$$-ir_t = r_{xx} - 2\left[qr - V(x,t)\right]r, \tag{D.1.7}$$

which reduces to the single PDE (D.1.1) under the reduction $r = -q^*$, as the compatibility condition of the following linear equations:

$$v_x = \begin{pmatrix} -ik & q \\ r & ik \end{pmatrix} v \tag{D.1.8}$$

and

$$v_t = \begin{pmatrix} A & B \\ C & -A \end{pmatrix} v. \tag{D.1.9}$$

The conventional formalism (cf. [21]) assumes k is a constant. Allowing for the possibility $k = k(x,t)$ leads to the constraints

$$-ik_t = A_x - qC + rB \tag{D.1.10a}$$

$$q_t = B_x + 2ikB + 2qA \tag{D.1.10b}$$

$$r_t = C_x - 2ikC - 2rA. \tag{D.1.10c}$$

The comparison with the Lax pair for the standard NLS system (2.2.3)–(2.2.4) suggests introducing functions W, Y, Z such that

$$A = iqr + 2ik^2 - iV(x,t) + W \tag{D.1.11a}$$

$$B = -iq_x - 2kq + Y \tag{D.1.11b}$$

$$C = ir_x - 2kr + Z, \tag{D.1.11c}$$

where $W = W(x,t,k)$ and likewise for Y and Z. Inserting (D.1.11a)–(D.1.11c) into (D.1.10a)–(D.1.10c), we find

$$W_x - qZ + rY = 0 \tag{D.1.12a}$$

$$Y_x + 2ikY + 2qW + 2k_xq = 0 \tag{D.1.12b}$$

$$Z_x - 2ikZ - 2rW - 2k_xr = 0 \tag{D.1.12c}$$

and also a nonlinear evolution equation for the spectral parameter

$$k_t = -2\left(k^2\right)_x + V_x. \tag{D.1.12d}$$

We see that $k_x = 0$ implies $V_x = 0$ or $V_x =$const, that is, $V(x,t)$ is a linear function of x. These cases can be handled directly without modifying the conventional IST formalism, and therefore we will restrict ourselves to them. If we only allow the spectral parameter to be time-dependent, that is, $k_x = 0$, equations (D.1.12a)–(D.1.12c) are satisfied by simply taking $W = Y = Z = 0$. Therefore, for potentials of the form

$$V(x,t) = \mathcal{E}(t)x + \mathcal{F}(t), \tag{D.1.13}$$

the auxiliary spectral problem (D.1.9) for the system of equations (D.1.6)–(D.1.7) can be chosen as

$$v_t = \begin{pmatrix} 2ik^2 + iqr - iV(x,t) & -2kq - iq_x \\ -2kr + ir_x & -2ik^2 - iqr + iV(x,t) \end{pmatrix} v \tag{D.1.14}$$

and (D.1.12d) fixes the time-dependence of the spectral parameter as

$$k(t) = k_0 + \int_0^t \mathcal{E}(t')dt'. \tag{D.1.15}$$

Since $k = k(t)$, the direct and inverse problems for the system (D.1.6)–(D.1.7) are therefore the ones described in Chapter 2 for the standard NLS. For potentials q, r decaying as $|x| \to \infty$, the time-dependence asymptotically satisfies

$$\partial_t v = \begin{pmatrix} 2ik^2 - iV & 0 \\ 0 & -2ik^2 + iV \end{pmatrix} v \qquad \text{as } x \to \pm\infty. \tag{D.1.16}$$

However, the fixed (in time) boundary conditions of the Jost functions (2.2.5)–(2.2.6) are not compatible with the time-dependence (D.1.16). Therefore, we introduce the time-dependent functions

$$\Phi(x,t) = e^{ig(t)}\phi(x,t), \qquad \bar{\Phi}(x,t) = e^{-ig(t)}\bar{\phi}(x,t) \tag{D.1.17a}$$

$$\Psi(x,t) = e^{-ig(t)}\psi(x,t), \qquad \bar{\Psi}(x,t) = e^{ig(t)}\bar{\psi}(x,t), \tag{D.1.17b}$$

where the function $g(t)$ has to be determined such that (D.1.17a)–(D.1.17b) solve asymptotically the time-differential equation (D.1.16). Making use of the boundary conditions (2.2.5)–(2.2.6) for the asymptotic form of v in (D.1.16), one obtains

$$g_t = 2k^2 - V + k_t x,$$

which, taking into account (D.1.15) and (D.1.13), yields

$$g(t) = \int_0^t (2k^2(t') - \mathcal{F}(t'))dt'. \tag{D.1.18}$$

From (2.2.24a)–(2.2.24b) one gets

$$\Phi(x,t) = \bar{\Psi}(x,t)a(t) + e^{2ig(t)}\Psi(x,t)b(t) \tag{D.1.19a}$$

$$\bar{\Phi}(x,t) = \Psi(x,t)\bar{a}(t) + e^{-2ig(t)}\bar{\Psi}(x,t)\bar{b}(t), \tag{D.1.19b}$$

and differentiating with respect to t yields

$$\frac{\partial\Phi}{\partial t} = \frac{\partial\bar{\Psi}}{\partial t}a + \bar{\Psi}\frac{\partial a}{\partial t} + 2ig_t e^{2ig(t)}\Psi b + e^{2ig(t)}\frac{\partial\Psi}{\partial t}b + e^{2ig(t)}\Psi\frac{\partial b}{\partial t}$$

$$\frac{\partial\bar{\Phi}}{\partial t} = \frac{\partial\Psi}{\partial t}\bar{a} + \Psi\frac{\partial\bar{a}}{\partial t} - 2ig_t e^{-2ig(t)}\bar{\Psi}\bar{b} + e^{-2ig(t)}\frac{\partial\bar{\Psi}}{\partial t}\bar{b} + e^{-2ig(t)}\bar{\Psi}\frac{\partial\bar{b}}{\partial t}.$$

Hence, taking into account (D.1.16),

$$\partial_t a(k) = 0, \qquad \partial_t \bar{a}(k) = 0$$

$$\partial_t b(k) = \left(-4ik^2 + 2i\mathcal{F}\right)b(k), \qquad \partial_t \bar{b}(k) = \left(4ik^2 - 2i\mathcal{F}\right)\bar{b}(k)$$

or, explicitly,

$$a(k, t) = a(k(0), 0), \qquad \bar{a}(k, t) = \bar{a}(k(0), 0) \qquad (D.1.20a)$$

$$b(k, t) = e^{-2i \int_0^t (2k^2 - \mathcal{F}) d\tau} b(k(0), 0), \qquad \bar{b}(k, t) = e^{2i \int_0^t (2k^2 - \mathcal{F}) d\tau} \bar{b}(k(0), 0). \qquad (D.1.20b)$$

The evolution of the reflection coefficients is given by

$$\rho(k, t) = \rho(k(0), 0) e^{-2i \int_0^t (2k^2 - \mathcal{F}) d\tau}, \qquad \bar{\rho}(k, t) = \bar{\rho}(k(0), 0) e^{2i \int_0^t (2k^2 - \mathcal{F}) d\tau}, \qquad (D.1.20c)$$

and this also gives the evolution of the norming constants,

$$C_j(t) = C_j(0) e^{-2i \int_0^t (2k_j^2 - \mathcal{F}) d\tau}, \qquad \bar{C}_j(t) = \bar{C}_j(0) e^{2i \int_0^t (2k_j^2 - \mathcal{F}) d\tau}. \qquad (D.1.20d)$$

Note that the Gel'fand–Levitan–Marchenko integral equations are the same as in (2.2.62) and (2.2.64), but the time-dependence of the scattering data is now given by (D.1.20c)–(D.1.20d). For instance, equations (2.2.60a)–(2.2.60b), which reconstruct the potential, read

$$r(x, t) = \frac{1}{\pi} \int_{\mathcal{C}_t} \rho(k(0), 0) e^{2ik(t)x + 2ig(k(t), t)} N^{(2)}(x, k(t)) dk \qquad (D.1.21a)$$

$$q(x, t) = \frac{1}{\pi} \int_{\bar{\mathcal{C}}_t} \bar{\rho}(k(0), 0) e^{-(2ik(t)x + 2ig(k(t), t))} \bar{N}^{(1)}(x, k(t)) dk, \qquad (D.1.21b)$$

where $\mathcal{C}_t$ (resp. $\bar{\mathcal{C}}_t$) is a contour from $-\infty$ to $+\infty$ that passes above all the zeros of $a(k(t), t)$ (resp. below all the zeros of $\bar{a}(k(t), t)$) at time t. Note that the latter equations depend on both $\mathcal{E}(t)$ and $\mathcal{F}(t)$ (via $k(t)$ and $g(k(t), t)$).

In order to obtain the one-soliton solution for the equation (D.1.1), we first note that a discrete eigenvalue evolves in time according to (D.1.15). Assuming the potential term $V(x, t)$ is real, that is, both $\mathcal{E}(t)$ and $\mathcal{F}(t)$ in (D.1.13) are real, $k_1(t) \equiv \xi(t) + i\eta(t) = \xi(t) + i\eta_0$, where $k_1(0) = \xi_0 + i\eta_0$ and

$$\xi(t) = \xi_0 + \int_0^t \mathcal{E}(\tau) d\tau.$$

Taking into account the time-dependence of C_1 as given by (D.1.20d), we obtain from (2.3.86)

$$q(x, t) = 2\eta_0 e^{-2i\xi(t)x + 4if(t) - i\psi_0} \operatorname{sech} \left[2\eta_0 x - 8\eta_0 \left(\xi_0 t + \int_0^t d\tau \int_0^\tau \mathcal{E}(\tau') d\tau' \right) - 2\delta_0 \right] \qquad (D.1.22)$$

with

$$f(t) = \int_0^t \left[\xi(\tau)^2 - \eta_0^2 - \frac{1}{2} \mathcal{F}(\tau) \right] d\tau$$

and ψ_0 and δ_0 given by (2.3.89).

Note that this is related to the one-soliton solution of the NLS by means of the variable transformation (D.1.2).

The previous results are generalized straightforwardly to the multicomponent case, that is, for a modified matrix nonlinear Schrödinger system,

$$i\mathbf{Q}_t = \mathbf{Q}_{xx} - 2[\mathbf{QR} + V(x, t)\mathbf{I}_N]\mathbf{Q} \qquad (D.1.23a)$$

$$-i\mathbf{R}_t = \mathbf{R}_{xx} - 2[\mathbf{RQ} + V(x, t)\mathbf{I}_M]\mathbf{R}, \qquad (D.1.23b)$$

where $\mathbf{Q} = \mathbf{Q}(x, t)$ is an $N \times M$ matrix; $\mathbf{R} = \mathbf{R}(x, t)$ is an $M \times N$ matrix; and $\mathbf{I}_N$, $\mathbf{I}_M$ are $N \times N$ and $M \times M$ identity matrices, respectively.

The Lax pair for such a system is naturally obtained from the matrix generalization of equations (D.1.8) and (D.1.14), that is,

$$\mathbf{v}_x = \begin{pmatrix} -ik\mathbf{I}_N & \mathbf{Q} \\ \mathbf{R} & ik\mathbf{I}_M \end{pmatrix} \mathbf{v}, \qquad (D.1.24a)$$

and

$$\mathbf{v}_t = \begin{pmatrix} \left[2ik^2 - iV(x, t)\right]\mathbf{I}_N + i\mathbf{QR} & -2k\mathbf{Q} - i\mathbf{Q}_x \\ -2k\mathbf{R} + i\mathbf{R}_x & -\left[2ik^2 - iV(x, t)\right]\mathbf{I}_M - i\mathbf{RQ} \end{pmatrix} \mathbf{v} \quad (D.1.24b)$$

again with

$$V(x, t) = \mathcal{E}(t)x + \mathcal{F}(t)$$

and the time-dependence of the scattering data results to be the same as in (D.1.20a)–(D.1.20d).

The Gel'fand–Levitan–Marchenko integral equations are the same as in (4.2.61a) and (4.2.61c), but the time-dependence of the scattering data is now given by (D.1.20c)–(D.1.20d). For instance, equations (4.2.59a)–(4.2.59b), which reconstruct the potential, read

$$\mathbf{R}(x, t) = \frac{1}{\pi} \int_{\mathcal{C}_t} \rho(k(0), 0) e^{2ik(t)x + 2ig(k(t), t)} \mathbf{N}^{(dn)}(x, k(t)) dk \qquad (D.1.25a)$$

$$\mathbf{Q}(x, t) = \frac{1}{\pi} \int_{\bar{\mathcal{C}}_t} \bar{\rho}(k(0), 0) e^{-(2ik(t)x + 2ig(k(t), t))} \bar{\mathbf{N}}^{(up)}(x, k(t)) dk, \qquad (D.1.25b)$$

where $\mathcal{C}_t$ (resp. $\bar{\mathcal{C}}_t$) is a contour from $-\infty$ to $+\infty$ in the complex k-plane that passes above all the zeros of $\det \mathbf{a}(k(t), t)$ (resp. below all the zeros of $\det \bar{\mathbf{a}}(k(t), t)$) at time t. Note that the latter equations depend on both $\mathcal{E}(t)$ and $\mathcal{F}(t)$ (via $k(t)$ and $g(k(t), t)$). The vector–soliton solution for the modified matrix NLS equation is given by (4.3.84), once the explicit time-dependence is taken into account, that is,

$$\mathbf{q}(x, t) = 2\eta_0 e^{-2i\xi(t)x + 4if(t) - i\frac{\pi}{2}} \text{sech}\left[2\eta_0 x - 8\eta_0 \left(\xi_0 t + \int_0^t d\tau \int_0^\tau \mathcal{E}(\tau')d\tau'\right) - 2\delta_0\right]\mathbf{p}$$

$$(D.1.26)$$

with

$$\mathbf{p} = \frac{\mathbf{C}_1^H(0)}{\|\mathbf{C}_1\|} \qquad f(t) = \int_0^t \left[\xi(\tau)^2 - \eta_0^2 - \frac{1}{2}\mathcal{F}(\tau)\right]d\tau \qquad (D.1.27)$$

and δ_0 given by (4.3.85).

D.2 Discrete NLS systems with a potential term

In [46] and [110], the dynamics of a nonlinear Schrödinger chain in a time-varying spatially uniform electric field is shown to be integrable. In the limit of a static electric field, the system exhibits a periodic evolution that is a nonlinear counterpart of Bloch oscillations.

The governing differential-difference equation for the chain system is

$$i\frac{d}{d\tau}Q_n = Q_{n+1} + Q_{n-1} - 2Q_n + \mu\left(Q_{n+1} + Q_{n-1}\right)|Q_n|^2 - 2\nu Q_n |Q_n|^2 + V_n Q_n,$$
(D.2.28)

where $V_n(\tau)$ is a potential. The μ term can be viewed as the first-order correction to the intersite overlap integral taking into account the nonlinearity as the induced self-interaction.

In general, the system (D.2.28) is not integrable, and when the potential term is time-dependent, the system is no longer conservative. In [46], Cai et al. analyzed the system with a potential of the form $V_n = \mathcal{E}(\tau)n$, where $\mathcal{E}(\tau)$ is any function of time. This potential corresponds to a time-dependent, spatially uniform electric field along the chain direction. When $\nu = 0$ and the potential is of this form, it can be shown that equation (D.2.28) is exactly integrable. The general case with $\nu \neq 0$ can be treated perturbatively.

We consider the system (D.2.28) with $\nu = 0$, $\mu = 1$, and V_n of the form

$$V_n(\tau) = \mathcal{E}(\tau)n + \mathcal{F}(\tau),$$
(D.2.29)

which admits the Lax pair

$$v_{n+1} = \begin{pmatrix} z & Q_n \\ R_n & z^{-1} \end{pmatrix} v_n$$
(D.2.30)

$$\frac{d}{d\tau}v_n = \begin{pmatrix} i\left[Q_n R_{n-1} - \frac{1}{2}(z^2 + z^{-2}) + f_n\right] & -izQ_n + iz^{-1}Q_{n-1} \\ iz^{-1}R_n - izR_{n-1} & -i\left[R_n Q_{n-1} - \frac{1}{2}(z^2 + z^{-2}) + f_n\right] \end{pmatrix} v_n$$
(D.2.31)

with

$$f_n(\tau) = -\frac{1}{2}\mathcal{E}(\tau)n + \mathcal{G}(\tau), \qquad \mathcal{G}(\tau) = \frac{1}{4}\mathcal{E}(\tau) - \frac{1}{2}\mathcal{F}(\tau) + 1.$$
(D.2.32)

Indeed, as in Section D1, allowing the spectral parameter z to be time-dependent, the compatibility condition between (D.2.30) and (D.2.31) gives

$$i\frac{d}{d\tau}Q_n = Q_{n+1} + Q_{n-1} - R_n Q_n(Q_{n+1} + Q_{n-1}) - (f_{n+1} + f_n)Q_n,$$
(D.2.33a)

$$-i\frac{d}{d\tau}R_n = R_{n+1} + R_{n-1} - Q_n R_n(R_{n+1} + R_{n-1}) - (f_{n+1} + f_n)R_n,$$
(D.2.33b)

and

$$\frac{d}{d\tau}z = iz(f_{n+1} - f_n).$$
(D.2.34)

Therefore the spectral parameter in the above IST problem depends on time according to

$$z = z_0\Lambda(\tau), \qquad \Lambda(\tau) = e^{-\frac{i}{2}\int_0^\tau \mathcal{E}(t)dt},$$
(D.2.35)

and this requires some modifications to the usual isospectral theory of the IST. It can be solved by generalizing the approach formulated in [26] and [52] to the lattice.

The operator (D.2.31) determines the evolution of the eigenfunctions. From this we deduce the time evolution of the scattering data. Since we have assumed that

$Q_n, R_n \to 0$ as $n \to \pm\infty$, then the time-dependence is asymptotically of the form

$$\partial_\tau v_n = \begin{pmatrix} -\frac{i}{2}(z^2 + z^{-2} - 2f_n) & 0 \\ 0 & \frac{i}{2}(z^2 + z^{-2} - 2f_n) \end{pmatrix} v_n \qquad \text{as } n \to \pm\infty, \quad \text{(D.2.36)}$$

where both f_n and the spectral parameter z depend on time.

However, the fixed (in time) boundary conditions of the eigenfunctions (3.2.8a)–(3.2.8b) are not compatible with the time-dependence (D.2.36), and therefore we introduce the time-dependent functions

$$\Phi_n(z, \tau) = e^{-ig(\tau)}\phi_n(z, \tau), \qquad \bar{\Phi}_n(z, \tau) = e^{ig(\tau)}\bar{\phi}_n(z, \tau) \qquad \text{(D.2.37a)}$$
$$\Psi_n(z, \tau) = e^{ig(\tau)}\psi_n(z, \tau), \qquad \bar{\Psi}_n(z, \tau) = e^{-ig(\tau)}\bar{\psi}_n(z, \tau), \qquad \text{(D.2.37b)}$$

where $g(\tau)$ has to be determined in such a way that (D.2.37a)–(D.2.37b) are solutions of the time-dependence equation (D.2.36). Making use of the boundary conditions (3.2.8a)–(3.2.8b) for the asymptotic form of v in (D.2.36), one obtains

$$\partial_\tau g = -inz^{-1}\partial_\tau z + \frac{1}{2}(z^2 + z^{-2} - 2f_n),$$

which, taking into account (D.2.29), (D.2.32), and (D.2.34), yields

$$g(\tau) = \frac{1}{2}\int_0^\tau \left[z^2(\tau') + z^{-2}(\tau') - 2\mathcal{G}(\tau') \right] d\tau'. \qquad \text{(D.2.38)}$$

These τ-dependent functions satisfy the relations

$$\Phi_n(z, \tau) = e^{-2ig(\tau)}\Psi_n(z, \tau)b(z, \tau) + \bar{\Psi}_n(z, \tau)a(z, \tau) \qquad \text{(D.2.39a)}$$
$$\bar{\Phi}_n(z, \tau) = e^{2ig(\tau)}\bar{\Psi}_n(z, \tau)\bar{b}(z, \tau) + \Psi_n(z, \tau)\bar{a}(z, \tau), \qquad \text{(D.2.39b)}$$

which are obtained from equations (3.2.59a)–(3.2.59b). Differentiating (D.2.39a)–(D.2.39b) with respect to τ and using (D.2.36) yields for the time evolution of the scattering data

$$\frac{d}{d\tau}a(z(t), t) = 0, \qquad \frac{d}{d\tau}b(z(\tau), \tau) = 2ig_\tau b(z(\tau), \tau)$$
$$\frac{d}{d\tau}\bar{a}(z(t), t) = 0, \qquad \frac{d}{d\tau}\bar{b}(z(\tau), \tau) = -2ig_\tau \bar{b}(z(\tau), \tau)$$

or, explicitly,

$$a(z(\tau), \tau) = a(z(0), 0), \qquad b(z(\tau), \tau) = b(z(0), 0)\,\Omega(\tau) \qquad \text{(D.2.40a)}$$
$$\bar{a}(z(\tau), \tau) = \bar{a}(z(0), 0), \qquad \bar{b}(z(\tau), \tau) = \bar{b}(z(0), 0)\,\bar{\Omega}(\tau), \qquad \text{(D.2.40b)}$$

where

$$\Omega(\tau) = e^{i\int_0^\tau dt(z^2 + z^{-2} - 2\mathcal{G})}, \qquad \bar{\Omega}(\tau) = e^{-i\int_0^\tau dt(z^2 + z^{-2} - 2\mathcal{G})}. \qquad \text{(D.2.41)}$$

Consequently, the reflection coefficients and the norming constants evolve in time according to

$$\rho(z(\tau), \tau) = \rho(z(0), 0)\,\Omega(\tau), \qquad \bar{\rho}(z(\tau), \tau) = \bar{\rho}(z(0), 0)\,\bar{\Omega}(\tau) \qquad \text{(D.2.42a)}$$

$$C_j(\tau) = C_j(0)\Lambda(\tau)\Omega(z_j(\tau), \tau) \qquad \text{(D.2.42b)}$$

$$\bar{C}_\ell(\tau) = \bar{C}_\ell(0)\Lambda(\tau)\bar{\Omega}(\bar{z}_\ell(\tau), \tau), \qquad \text{(D.2.42c)}$$

where z_j is the j-th zero of $a(z, t)$ and $\bar{z}_\ell$ is the ℓ-th zero of $\bar{a}(z, t)$. When $R_n = -Q_n^*$, according to (3.2.86)–(3.2.87), $\bar{z}_j = \left(z_j^*\right)^{-1}$ and $\bar{C}_j = C_j^*/(z_j^*)^2$.

Note that the Gel'fand–Levitan–Marchenko equations are the same as in (3.2.120a)–(3.2.120b), but the time-dependence of the scattering data is now given by (D.2.42a)–(D.2.42b). For instance, equations (3.2.104a)–(3.2.104b), which reconstruct the potential, are given by

$$R_n(\tau) = 2 \sum_{j=1}^{J} C_j(\tau) z_j(\tau)^{-2(n+1)} N_n'^{(2)}(z_j(\tau))$$

$$+ \frac{1}{2\pi i} \oint_{|w|=1} w^{-2(n+1)} \rho(w) \Omega(\tau) N_n'^{(2)}(w) dw \qquad \text{(D.2.43a)}$$

$$Q_{n-1}(\tau) = -2 \sum_{j=1}^{J} \bar{C}_j(\tau) \bar{z}_j(\tau)^{2(n-1)} \bar{N}_n'^{(1)}(\bar{z}_j(\tau))$$

$$+ \frac{1}{2\pi i} \oint_{|w|=1} w^{2(n-1)} \bar{\rho}(w) \bar{\Omega}(\tau) \bar{N}_n'^{(1)}(w) dw. \qquad \text{(D.2.43b)}$$

As far as the one-soliton solution is concerned, from (D.2.35) and (D.2.42b),

$$z_1(\tau) = e^{\alpha_1(\tau) + i\beta_1(\tau)} = z_1(0) e^{-\frac{i}{2} \int_0^\tau \mathcal{E}(t) dt}, \qquad \text{(D.2.44a)}$$

$$\bar{z}_1(\tau) = e^{-\alpha_1(\tau) + i\beta_1(\tau)} = \bar{z}_1(0) e^{-\frac{i}{2} \int_0^\tau \mathcal{E}(t) dt}, \qquad \text{(D.2.44b)}$$

$$\alpha_1(\tau) = \alpha_1(0) \qquad \beta_1(\tau) = \beta_1(0) - \frac{1}{2} \int_0^\tau \mathcal{E}(t) dt, \qquad \text{(D.2.44c)}$$

and

$$C_1(\tau) = C_1(0) e^{2i \int_0^\tau \left[\cosh 2\alpha_1(0) \cos 2\beta_1(t) - 1 + i \sinh 2\alpha_1(0) \sin 2\beta_1(t) + \frac{1}{2}\mathcal{F}(t) - \frac{1}{2}\mathcal{E}(t) \right] dt}.$$

Hence, substituting this explicit time-dependence into (3.2.109a) with $D_1 = C_1^*$ yields

$$Q_n(\tau) = -\sinh(2\alpha_1(0)) e^{2i(n+1)\beta_1(\tau) - if(\tau)} \operatorname{sech}\left[2\alpha_1(0)(n+1) - 2v(\tau) - \delta(0)\right],$$
$$\text{(D.2.45)}$$

where

$$v(\tau) = -\sinh 2\alpha_1(0) \int_0^\tau \sin 2\beta_1(t) dt \qquad \text{(D.2.46)}$$

$$f(\tau) = \psi_0 + 2 \int_0^\tau \left[\cosh 2\alpha_1(0) \cos 2\beta_1(t) - 1 + \frac{1}{2}\mathcal{F}(t) - \frac{1}{2}\mathcal{E}(t) \right] dt. \quad \text{(D.2.47)}$$

Note that, for $\mathcal{F} \to 2h^2\mathcal{F}$, $\mathcal{E} \to 2h^3\mathcal{E}$, $Q_n = hq_n$, and $t = h^2\tau$, in the limit $h \to 0$, $nh \to x$, one obtains the one-soliton solution of the continuous problem (D.1.22).

Now let us consider the system of equations

$$i\frac{d}{d\tau}\mathbf{Q}_n = \mathbf{Q}_{n+1} - 2\mathbf{Q}_n + \mathbf{Q}_{n-1} - \mathbf{Q}_{n+1}\mathbf{R}_n\mathbf{Q}_n - \mathbf{Q}_n\mathbf{R}_n\mathbf{Q}_{n-1} + V_n\mathbf{Q}_n \quad \text{(D.2.48a)}$$

$$-i\frac{d}{d\tau}\mathbf{R}_n = \mathbf{R}_{n+1} - 2\mathbf{R}_n + \mathbf{R}_{n-1} - \mathbf{R}_{n+1}\mathbf{Q}_n\mathbf{R}_n - \mathbf{R}_n\mathbf{Q}_n\mathbf{R}_{n-1} + V_n\mathbf{R}_n, \quad \text{(D.2.48b)}$$

where $\mathbf{Q}_n$ and $\mathbf{R}_n$ are, respectively, $N \times M$ and $M \times N$ matrices and V_n is a real, time-varying scalar function. In the absence of the field V_n, the system corresponds to (5.1.1a)–(5.1.1b).

We consider potentials of the form (D.2.29). The system (D.2.48a)–(D.2.48b) with such V_n given by (D.2.29) results from the compatibility condition of the following linear equations:

$$\mathbf{v}_{n+1} = \begin{pmatrix} z\mathbf{I}_N & \mathbf{Q}_n \\ \mathbf{R}_n & z^{-1}\mathbf{I}_M \end{pmatrix} \mathbf{v}_n \tag{D.2.49}$$

$$\frac{d}{d\tau}\mathbf{v}_n =$$
$$\begin{pmatrix} i\left[\mathbf{Q}_n\mathbf{R}_{n-1} - \frac{1}{2}(z^2 + z^{-2})\mathbf{I}_N + f_n\mathbf{I}_N\right] & -iz\mathbf{Q}_n + iz^{-1}\mathbf{Q}_{n-1} \\ iz^{-1}\mathbf{R}_n - iz\mathbf{R}_{n-1} & -i\left[\mathbf{R}_n\mathbf{Q}_{n-1} - \frac{1}{2}(z^2 + z^{-2})\mathbf{I}_M + f_n\mathbf{I}_M\right] \end{pmatrix} \mathbf{v}_n, \tag{D.2.50}$$

where, as usual, $\mathbf{I}_N$ is the $N \times N$ identity matrix, $\mathbf{I}_M$ is the $M \times M$ identity matrix, and f_n is given by (D.2.32) provided that the spectral parameter z depends on time according to (D.2.35).

As was discussed in Chapter 5, the system (D.2.48a)–(D.2.48b) does not, in general, admit the reduction $\mathbf{R}_n = \mp\mathbf{Q}_n^H$. However, if one takes $M = N$ and restricts $\mathbf{R}_n$ and $\mathbf{Q}_n$ to be such that

$$\mathbf{R}_n\mathbf{Q}_n = \mathbf{Q}_n\mathbf{R}_n = \alpha_n\mathbf{I}_N \tag{D.2.51}$$

and α_n is real when $\mathbf{R}_n = \mp\mathbf{Q}_n^H$, then this symmetry is a consistent reduction of (D.2.48a)–(D.2.48b), which reduces to the single (matrix) equation

$$i\frac{d}{d\tau}\mathbf{Q}_n = \mathbf{Q}_{n+1} - 2\mathbf{Q}_n + \mathbf{Q}_{n-1} - \alpha_n\left(\mathbf{Q}_{n+1} + \mathbf{Q}_{n-1}\right) + V_n\mathbf{Q}_n. \tag{D.2.52}$$

For instance, in the case $N = 2$, the matrices

$$\mathbf{Q}_n = \begin{pmatrix} Q_n^{(1)} & Q_n^{(2)} \\ (-1)^n R_n^{(2)} & (-1)^{n+1} R_n^{(1)} \end{pmatrix}, \qquad \mathbf{R}_n = \begin{pmatrix} R_n^{(1)} & (-1)^n Q_n^{(2)} \\ R_n^{(2)} & (-1)^{n+1} Q_n^{(1)} \end{pmatrix} \tag{D.2.53}$$

satisfy the condition (D.2.53) with

$$\mathbf{R}_n\mathbf{Q}_n = \mathbf{Q}_n\mathbf{R}_n = \alpha_n\mathbf{I} = \left(R_n^{(1)} Q_n^{(1)} + R_n^{(2)} Q_n^{(2)}\right)\mathbf{I},$$

and equation (D.2.52) gives the vector generalization of (D.2.28) with $Q_n \to \left(Q_n^{(1)}, Q_n^{(2)}\right)$ and $|Q_n|^2 \to \|Q_n\|^2$.

The scattering theory for the system of equations (D.2.48a)–(D.2.48b) is therefore the same as in the absence of the external potential, and we only need to determine the time-dependence of the scattering data.

The operator (D.2.50) determines the evolution of the eigenfunctions. From this we deduce the time evolution of the scattering data. Since we have assumed that $\mathbf{Q}_n$, $\mathbf{R}_n \to 0$ as $n \to \pm\infty$, then the time-dependence is asymptotically of the form

$$\partial_\tau \mathbf{v}_n = \begin{pmatrix} -\frac{i}{2}(z^2 + z^{-2} - 2f_n)\mathbf{I}_N & 0 \\ 0 & \frac{i}{2}(z^2 + z^{-2} - 2f_n)\mathbf{I}_M \end{pmatrix} \mathbf{v}_n \qquad \text{as } n \to \pm\infty, \tag{D.2.54}$$

where both f_n and the spectral parameter z depend on time. Like in the scalar case, this yields for the time evolution of the scattering data,

$$\mathbf{b}(z, \tau) = \Omega(\tau)\mathbf{b}(z(0), 0) \qquad \mathbf{a}(z, \tau) = \mathbf{a}(z(0), 0) \tag{D.2.55}$$
$$\bar{\mathbf{a}}(z, \tau) = \bar{\mathbf{a}}(z, 0) \qquad \bar{\mathbf{b}}(z, \tau) = \bar{\Omega}(\tau)\bar{\mathbf{b}}(z(0), 0) \tag{D.2.56}$$

with Ω and $\bar{\Omega}$ given in (D.2.41).

The evolution of the norming constants follows from the definitions (5.2.75), (5.2.77) and the relations (D.2.55)–(D.2.56), that is,

$$\mathbf{C}_j(\tau) = \Lambda(\tau)\Omega(\tau)\mathbf{C}_j(0)|_{z=z_j}, \qquad \bar{\mathbf{C}}_j(\tau) = \Lambda(\tau)\bar{\Omega}(\tau)\bar{\mathbf{C}}_j(0)|_{z=\bar{z}_j}. \tag{D.2.57}$$

The discrete vector–soliton solution, for instance, in the case of fundamental solitons, is obtained from (5.2.139a), once the explicit time-dependence is taken into account, that is,

$$\begin{pmatrix} Q_n^{(1)}(\tau) \\ Q_n^{(2)}(\tau) \end{pmatrix} = -\frac{\gamma_1(0)}{||\gamma_1(0)||} \sinh(2\alpha_1(0))$$

$$\times\, e^{2i(n+1)\beta_1(\tau)-if(\tau)} \operatorname{sech}\left[2\alpha_1(0)\,(n+1) - 2v(\tau) - \delta(0)\right]$$

$$\tag{D.2.58}$$

with the same $v(\tau)$ and $f(\tau)$ as in (D.2.46)–(D.2.47).

Appendix E

NLS systems in the limit of large amplitudes

The limit of large amplitude (also called the strong coupling limit) of the system (4.1.1a)–(4.1.1b) is obtained by letting $\mathbf{q}$, $\mathbf{r}$ become arbitrary large. For instance, positing $\mathbf{q} = \tilde{\mathbf{q}}/\epsilon$, $\mathbf{r} = \tilde{\mathbf{r}}/\epsilon$, $t = \epsilon^2 \tilde{t}$, where ϵ is a small real parameter, and dropping the $\tilde{\ }$, one obtains the following equations:

$$i\mathbf{q}_t = -2\mathbf{qrq} \tag{E.1.1a}$$

$$-i\mathbf{r}_t = -2\mathbf{rqr}. \tag{E.1.1b}$$

The general solution is readily obtained since, multiplying the first equation from the right by $\mathbf{r}$, the second from the left by $\mathbf{q}$, or vice versa, and subtracting, yields

$$\frac{\partial}{\partial t}(\mathbf{qr}) = 0, \qquad \frac{\partial}{\partial t}(\mathbf{rq}) = 0,$$

and hence

$$(\mathbf{qr})(x, t) = (\mathbf{qr})(x, 0), \qquad (\mathbf{rq})(x, t) = (\mathbf{rq})(x, 0).$$

Following [20], the inverse scattering transform for the system (4.1.1a)–(4.1.1b) can be shown to reduce to the obtained solution via perturbation.

We consider now the IDNLS system (3.1.2a)–(3.1.2b) in the limit of large amplitude, that is,

$$i\frac{d}{d\tau}Q_n = -Q_n R_n (Q_{n+1} + Q_{n-1}) \tag{E.1.2a}$$

$$-i\frac{d}{d\tau}R_n = -Q_n R_n (R_{n+1} + R_{n-1}). \tag{E.1.2b}$$

In this case, it is convenient to consider the quantities $Q_n R_{n-1}$ and $Q_{n-1} R_n$, which, according to (E.1.2a)–(E.1.2b), satisfy the following evolution equations:

$$i\frac{d}{d\tau}(Q_n R_{n-1}) = -Q_n R_{n-1} (Q_{n+1} R_n - Q_{n-1} R_{n-2}) \tag{E.1.3a}$$

$$-i\frac{d}{d\tau}(Q_{n-1} R_n) = -Q_{n-1} R_n (Q_n R_{n+1} - Q_{n-2} R_{n-1}). \tag{E.1.3b}$$

239

In their turn, these equations can be reduced to

$$i\frac{d}{d\tau}s_n = e^{s_{n-1}} - e^{s_{n+1}} \tag{E.1.4a}$$

$$i\frac{d}{d\tau}\tilde{s}_n = -e^{\tilde{s}_{n-1}} + e^{\tilde{s}_{n+1}}, \tag{E.1.4b}$$

where

$$s_n = \log(Q_n R_{n-1}), \qquad \tilde{s}_n = \log(Q_{n-1} R_n), \tag{E.1.5}$$

that is, the nonlinear ladder network equation (C.1.24) up to the change $\tau \to i\tau$ and a trivial change of sign in the potentials. Then, from (E.1.2a)–(E.1.2b) it follows that

$$i\frac{d}{d\tau}Q_n = -Q_n(s_{n+1} + \tilde{s}_n) \tag{E.1.6a}$$

$$-i\frac{d}{d\tau}R_n = -R_n(\tilde{s}_{n+1} + s_n), \tag{E.1.6b}$$

which implies that the initial-value problem for the IDNLS system in the strong coupling limit can in principle be solved in terms of the solutions of two decoupled nonlinear ladder networks.

In Appendix C we gave a detailed description of the IST of the discrete linear Schrödinger equation (C.1.9), which is associated to the nonlinear ladder network. We notice that the potentials s_n and $\tilde{s}_n$ in (E.1.5) are not necessarily real, which has to be taken into account when applying the IST machinery. In particular, for certain initial data, the solution could blow up in finite time.

We can write explicitly the one-soliton solution for this system. According to (C.4.133), we have

$$s_n(\tau) = -1 - \frac{\sinh(\omega + i\theta)\sinh 2(\omega + i\theta)}{p_n(\tau)p_{n+1}(\tau)}, \tag{E.1.7}$$

where

$$\begin{aligned}
p_n(\tau) = {}& \cosh\left[\left(n + \tfrac{1}{2}\right)\omega + x_0 + 2(\cosh\omega)(\sin\theta)\tau\right] \\
& \times \cos\left[\left(n + \tfrac{1}{2}\right)\theta - 2(\sinh\omega)(\cos\theta)\tau\right] \\
& + i\sinh\left[\left(n + \tfrac{1}{2}\right)\omega + x_0 + 2(\cosh\omega)(\sin\theta)\tau\right] \\
& \times \sin\left[\left(n + \tfrac{1}{2}\right)\theta - 2(\sinh\omega)(\cos\theta)\tau\right]
\end{aligned}$$

and

$$\tilde{s}_n(\tau) = -s_n(-\tau),$$

where we have taken into account that, for complex potentials, the discrete eigenvalues for the discrete linear Schrödinger scattering problem (C.2.25) are not necessarily real, that is, $\omega \to \omega + i\theta$ (see Section C.1.3), and performed the substitution $t \to i\tau$ that is required to reconduct (E.1.4a)–(E.1.4b) to the nonlinear ladder network equation (C.1.24). From (E.1.6a), we finally have

$$Q_n(\tau) = Q_n(0)e^{i\sinh(\omega+i\theta)\sinh 2(\omega+i\theta)}$$

$$\times \exp\left\{i\int_0^\tau\left[\frac{1}{p_{n+2}(\tau')p_{n+1}(\tau')} - \frac{1}{p_n(-\tau')p_{n+1}(-\tau')}\right]d\tau'\right\}. \tag{E.1.8}$$

We now turn to the discrete matrix NLS system. In the matrix case, the corresponding equations (5.1.1a)–(5.1.2b) in the limit of large amplitude are reduced to

$$i\frac{d}{d\tau}\mathbf{Q}_n = -\mathbf{Q}_{n+1}\mathbf{R}_n\mathbf{Q}_n - \mathbf{Q}_n\mathbf{R}_n\mathbf{Q}_{n-1} \qquad (\text{E.1.9a})$$

$$-i\frac{d}{d\tau}\mathbf{R}_n = -\mathbf{R}_{n+1}\mathbf{Q}_n\mathbf{R}_n - \mathbf{R}_n\mathbf{Q}_n\mathbf{R}_{n-1}, \qquad (\text{E.1.9b})$$

which, in terms of the quantities

$$\mathbf{s}_n = \mathbf{Q}_n\mathbf{R}_{n-1}, \qquad \tilde{\mathbf{s}}_n = \mathbf{R}_n\mathbf{Q}_{n-1}, \qquad (\text{E.1.10})$$

become

$$i\frac{d}{d\tau}\mathbf{s}_n = \mathbf{s}_n\mathbf{s}_{n-1} - \mathbf{s}_{n+1}\mathbf{s}_n \qquad (\text{E.1.11a})$$

$$i\frac{d}{d\tau}\tilde{\mathbf{s}}_n = -\tilde{\mathbf{s}}_n\tilde{\mathbf{s}}_{n-1} + \tilde{\mathbf{s}}_{n+1}\tilde{\mathbf{s}}_n, \qquad (\text{E.1.11b})$$

that is, a matrix (nonabelian) generalization of the nonlinear ladder equation.

In analogy with the scalar case, one can relate this matrix nonlinear ladder to a discrete analog of the matrix Schrödinger spectral problem,

$$\mathbf{s}_n\mathbf{v}_{n+1} + \mathbf{v}_{n-1} = \lambda\mathbf{v}_n, \qquad (\text{E.1.12})$$

with an associated time evolution equation of the form

$$\frac{d}{dt}\mathbf{v}_n = \mathbf{A}_n\mathbf{v}_{n+1} + \mathbf{B}_n\mathbf{v}_n. \qquad (\text{E.1.13})$$

Taking the time derivative of (E.1.12) and using (E.1.12) itself to solve for $\mathbf{v}_{n+2}$ and $\mathbf{v}_{n-1}$, we find two equations by setting the coefficients of the terms $\mathbf{v}_n$ and $\mathbf{v}_{n+1}$, respectively, to zero (and assuming the spectral parameter λ to be time-independent):

$$\frac{d\mathbf{s}_n}{dt} + \lambda\mathbf{s}_n\mathbf{A}_{n+1}\mathbf{s}_{n+1}^{-1} + \mathbf{s}_n\mathbf{B}_{n+1} - \mathbf{B}_{n-1}\mathbf{s}_n = \lambda\mathbf{A}_n \qquad (\text{E.1.14a})$$

$$-\mathbf{s}_n\mathbf{A}_{n+1}\mathbf{A}_{n+1}^{-1} + \mathbf{A}_{n-1} + \lambda\mathbf{B}_{n-1} = \lambda\mathbf{B}_n. \qquad (\text{E.1.14b})$$

Expanding $\mathbf{A}_n$ and $\mathbf{B}_n$ in the form

$$\mathbf{A}_n = \mathbf{A}_n^{(1)}\lambda + \mathbf{A}_n^{(0)} \qquad \mathbf{B}_n = \mathbf{B}_n^{(1)}\lambda + \mathbf{B}_n^{(0)},$$

and requiring the coefficients of the different powers in λ to vanish, we obtain

$$\frac{d\mathbf{s}_n}{dt} = -\mathbf{s}_n\mathbf{B}_{n+1}^{(0)} + \mathbf{B}_{n-1}^{(0)}\mathbf{s}_n \qquad (\text{E.1.15a})$$

$$\mathbf{A}_n^{(0)} = \mathbf{s}_n\mathbf{A}_{n+1}^{(0)}\mathbf{s}_{n+1}^{-1} + \mathbf{s}_n\mathbf{B}_{n+1}^{(1)} - \mathbf{B}_{n-1}^{(1)}\mathbf{s}_n \qquad (\text{E.1.15b})$$

$$\mathbf{A}_n^{(1)} = \mathbf{s}_n\mathbf{A}_{n+1}^{(1)}\mathbf{s}_{n+1}^{-1} \qquad (\text{E.1.15c})$$

$$\mathbf{A}_{n-1}^{(0)} = \mathbf{s}_n\mathbf{A}_{n+1}^{(0)}\mathbf{s}_{n+1}^{-1} \qquad (\text{E.1.15d})$$

$$\mathbf{B}_n^{(0)} = -\mathbf{s}_n\mathbf{A}_{n+1}^{(1)}\mathbf{s}_{n+1}^{-1} + \mathbf{A}_{n-1}^{(1)} + \mathbf{B}_{n-1}^{(0)}, \qquad (\text{E.1.15e})$$

which are satisfied by

$$\mathbf{A}_n^{(1)} = -\mathbf{s}_n, \qquad \mathbf{A}_n^{(0)} = \mathbf{0} \qquad (\text{E.1.16})$$

$$\mathbf{B}_n^{(1)} = -\mathbf{0}, \qquad \mathbf{B}_n^{(0)} = \mathbf{s}_n, \qquad (\text{E.1.17})$$

that is, one can take as an associated time evolution

$$\frac{d}{dt}\mathbf{v}_n = -\lambda\mathbf{s}_n\mathbf{v}_{n+1} + \mathbf{s}_n\mathbf{v}_n. \qquad (E.1.18)$$

The IST for the discrete matrix Schrödinger spectral problem was given in [44]. In a subsequent paper, Bruschi et al. also found the hierarchy of evolution equations associated with the discrete matrix Schrödinger equation [45]. From (E.1.9a)–(E.1.9b) it then follows that

$$i\frac{d}{d\tau}\mathbf{Q}_n = -\mathbf{s}_{n+1}\mathbf{Q}_n - \mathbf{Q}_n\tilde{\mathbf{s}}_n \qquad (E.1.19a)$$

$$-i\frac{d}{d\tau}\mathbf{R}_n = -\tilde{\mathbf{s}}_{n+1}\mathbf{R}_n - \mathbf{R}_n\mathbf{s}_n, \qquad (E.1.19b)$$

where $\mathbf{s}_n$ and $\tilde{\mathbf{s}}_n$ are considered to be known in terms of their initial value once the IST for the two decoupled matrix nonlinear ladder network equations have been solved via the IST.

Bibliography

[1] Ablowitz M. J., *Nonlinear evolution equations – continuous and discrete*, SIAM Rev. **19** (1977) 663

[2] Ablowitz M. J., Biondini G., Chakravarty S., Jenkins R. B., and Sauer J. R., *Four-wave mixing in wavelength-division multiplexed soliton systems: Damping and amplification*, Opt. Lett. **21** (1996) 1646

[3] Ablowitz M. J., Biondini G., Chakravarty S., and Horne R., *On timing jitter in wavelength-division multiplexed soliton systems*, Opt. Commun. **150** (1998) 305–318

[4] Ablowitz M. J. and Biondini G., *Multiscale pulse dynamics in communication systems with strong dispersion management*, Opt. Lett. **23** (1998) 1668–1670

[5] Ablowitz M. J., Chakravarty S., and Halburd R., *The generalized Chazy equation from the self-duality equations*, Stud. Appl. Math. **103** (1999) 287–304

[6] Ablowitz M. J. and Clarkson P. A., *Solitons, Nonlinear Evolution Equations and Inverse Scattering*, London Mathematical Society Lecture Notes Series **149**, Cambridge University Press (1991)

[7] Ablowitz M. J. and Fokas A. S., *Complex Variables*, Cambridge Texts in Applied Mathematics, Cambridge University Press (1997)

[8] Ablowitz M. J. and Hirooka T., *Managing nonlinearity in strongly dispersion-managed optical pulse transmission*, JOSAB **19** (2002) 425–439

[9] Ablowitz M. J., Hirooka T., and Biondini G., *Quasi-linear optical pulses in strongly dispersion managed transmission systems*, Opt. Lett. **26** (2001) 459–461

[10] Ablowitz M. J., Kaup D. J., Newell A. C., and Segur H., *The inverse scattering transform – Fourier analysis for nonlinear problems*, Stud. Appl. Math. **53** (1974) 249–315

[11] Ablowitz M. J. and Ladik J. F., *Nonlinear differential-difference equations*, J. Math. Phys. **16** (1975) 598–603

[12] Ablowitz M. J. and Ladik J. F., *Nonlinear differential-difference equations and Fourier Analysis*, J. Math. Phys. **17** (1976) 1011–1018

[13] Ablowitz M. J. and Ladik J. F., *A nonlinear difference scheme and inverse scattering*, Stud. Appl. Math. **55** (1976) 213

[14] Ablowitz M. J. and Musslimani Z. H., *Discrete diffraction managed spatial solitons*, Phys. Rev. Lett. **87** (2001) 254102

[15] Ablowitz M. J. and Musslimani Z. H., *Discrete vector spatial solitons in a nonlinear waveguide array*, Phys. Rev. E **65** (2002) 056618

[16] Ablowitz M. J. and Musslimani Z. H., *Discrete spatial solitons in a diffraction managed nonlinear waveguide array: A unified approach*, To appear in *Physica D* (2003)

[17] Ablowitz M. J., Z. H. Musslimani Z. H., and Biondini G., *Methods for discrete solitons in nonlinear lattices*, Phys. Rev. E **65**, (2002) 026602

[18] Ablowitz M. J., Ohta Y., Trubatch A. D., *On discretizations of the vector nonlinear Schrödinger equation*, Phys. Lett. A **253** (1999) 253–287

[19] Ablowitz M. J., Ohta Y., and Trubatch A. D., *On integrability and chaos in discrete systems*, Chaos, Solitons and Fractals **11** (2000) 159–169

[20] Ablowitz M. J. and Schultz L., *Strong coupling of certain multidimensional nonlinear wave equations*, Stud. Appl. Math. **80** (1989) 229–238

[21] Ablowitz M. J. and Segur H., *Solitons and the inverse scattering transform*, SIAM **4** (1981)

[22] Aceves A. B., DeAngelis C., Trillo S., and Wabnitz S., *Storage and steering of self-trapped discrete solitons in nonlinear waveguide arrays*, Opt. Lett. **19** (1994) 332

[23] Aceves A. B., DeAngelis C., Peschel T., Muschall R. Lederer F., Trillo S., and Wabnitz S., *Discrete self-trapping, soliton interactions and beam steering in nonlinear waveguide arrays*, Phys. Rev. E **53** (1996) 1172

[24] Agrawal G. P., *Nonlinear Fiber Optics*, Academic Press (1995)

[25] Atiyah M. F. and Ward R. S., *Instantons and geometry*, Commun. Math. Phys. **55** (1977) 117

[26] Balakrishnan R., *Soliton propagation in nonuniform media*, Phys. Rev. A **32** (1985) 1145–1149

[27] Beals R. and Coifman R. R., *Scattering and inverse scattering for first order systems*, Commun. Pure Appl. Math. **37** (1984) 39

[28] Beals R. and Coifman R. R., *Inverse scattering and evolution equations*, Commun. Pure Appl. Math. **38** (1985) 29

[29] Belavin A. A. and Zakharov V. E., *Yang-Mills equations as an inverse scattering problem*, Phys. Lett. B, **73** (1978) 53–57

[30] Benney D. J. and Newell A. C., *The propagation of nonlinear wave envelopes*, J. Math. Phys. **46** (1967) 133–139

[31] Benney D. J. and Roskes G. J., *Wave instabilities*, Stud. Appl. Math. **48** (1969) 377–385

[32] Black W., Weideman J. A. C., and Herbst B. M., *A note on an integrable discretization of the nonlinear Schrödinger equation*, Inv. Probl. **15** (1999) 807–810

[33] Bobenko A., Pinkall U., *Discrete surfaces with constant negative Gaussian curvature and the Hirota equation*, J. Diff. Geom. **43** (1996) 527–611

[34] Bobenko A. I., Hoffmann T., and Suris Y. B., *Hexagonal circle patterns and integrable systems: patterns with the multi-ratio property and Lax equations on the regular triangular lattice*, Intern. Math. Research Notices **3** (2002) 111–164

[35] Boiti M., M. Bruschi, Pempinelli F., and Prinari B., *A discrete Schrödinger spectral problem and associated evolution equations*, J. Phys. A: Math. Gen. **36** (2003) 139

[36] Boiti M., Leon J., and Pempinelli F., *Nonlinear discrete systems with nonanalytic dispersion relations*, J. Math. Phys. **37** (1996) 2824

[37] Boiti M., Pempinelli F., Prinari B., and Spire A., *An integrable discretization of KdV at large times*, Inv. Probl. **17** (2001) 515

[38] Boussinesq J., *Theorie de l'intumescence liquid appelée onde solitaire ou de translation, se propagente dans un canal rectangulaire*, Compte Rendus Acad. Sci. Paris, **72** (1871) 755–759

[39] Boussinesq J., *Theorie des ondes et de remous qui se propagent le long d'un canal rectangulaire horizontal, en communiquant au liquide contenu dans ce canal des vitesses sensiblemant parielles de la surface au fond*, J. Math. Pure Appl. Ser. 2 **17** (1872) 55–108

[40] Braun O. M., Kivshar Y. S., *Nonlinear dynamics of the Frenkel-Kontorova model*, Phys. Rep. **306** (1998) 1

[41] Briggs C. S., *Topics in Nonlinear Mathematics*, PhD Thesis, Clarkson College of Technology (1983)

[42] Bruschi M., *On a new integrable hamiltonian system with nearest-neighbor interaction*, Inv. Probl. **5** (1989) 983–998

[43] Bruschi M., Levi D., and Ragnisco O., *Discrete version of the Nonlinear Schrödinger with linearly x-dependent coefficients*, Nuovo Cimento **53A** (1979) 21

[44] Bruschi M., Manakov S. V., Ragnisco O., and Levi D., *The nonabelian Toda lattice – Discrete analog of the matrix Schrödinger spectral problem*, J. Math. Phys. **21** (1980) 2749–2753

[45] Bruschi M., Manakov S. V., Ragnisco O., and Levi D., *Evolution equations associated with the discrete analog of the matrix Schrödinger spectral problem solvable by IST*, J. Math. Phys. **22** (1981) 2463–2471

[46] Cai D., Bishop A. R., Grønbech-Jensen N., and Salerno M., *Electric field-induced nonlinear Bloch oscillations and dynamical localizations*, Phys. Rev. Lett. **74** (1995) 1186

[47] Calogero F., Degasperis A., *Spectral Transform and Solitons I* North Holland (1982)

[48] Case K. M., *On discrete inverse scattering problems II*, J. Math. Phys. **14** (1973) 916

[49] Case K. M. and Kac M., *A discrete version of the inverse scattering problem*, J. Math. Phys. **14** (1973) 594

[50] Chakravarty S., Ablowitz M. J., and Clarkson P. A., *Reductions of self-dual Yang-Mills fields and classical equations*, Phys. Rev. Lett. **65** (1990) 1085–1087

[51] Chakravarty S. and Ablowitz M. J., *On reductions of self-dual Yang Mills equations*, in: *Painlevé Transcendents*, Ed. D. Levi and P. Winternitz, Plenum (1992)

[52] Chen H. H., and Liu C. S., *Solitons in nonuniform media*, Phys. Rev. Lett. **37** (1976) 693–697

[53] Chiu S. C. and Ladik J. F., *Generating exactly soluble nonlinear discrete evolution equations by a generalized Wronskian technique*, J. Math. Phys. **18** (1977) 690

[54] Christodoulides D. N. and Joseph R. J., *Discrete self-focusing in nonlinear arrays of coupled waveguides*, Opt. Lett. **13** (1988) 794

[55] Claude Ch., Kishvar Y. S., Kluth O., and Spatscheck K. H., *Moving modes in localized nonlinear lattices*, Phys. Rev. **47** (1993) 14228

[56] Common A. K., *A solution of the initial value problem for half infinite integrable lattice systems*, Inverse Problems **8** (1992) 393

[57] Darmanyan S., Kobyakov A., and Lederer F., *Strongly localized modes in discrete systems with quadratic nonlinearity*, Phys. Rev. E **57** (1998) 2344–2349

[58] Darmanyan S., Kobyakov A., Schmidt E., and Lederer F., *Strongly localized vectorial modes in nonlinear waveguide arrays*, Phys. Rev. E **57** (1998) 3520–3530

[59] Davey A. and Stewartson K., *On three-dimensional packets of surface waves*, Proc. Roy. Soc. London Ser. A **338** (1974) 101–110

[60] Davydov A. S., *Theory of contraction of proteins under their excitation*, J. Theor. Biol. **38** (1973) 559–569

[61] Davydov A. S., *The role of solitons in the energy and electron transfer in one-dimensional molecular systems*, Physica 3 D (1981) 1

[62] Davydov A. S., *Solitons in Molecular Systems*, Kluwer Academic (1991)

[63] Deift P. and Trubowitz E., *Inverse scattering on the line*, Commun. Pure Appl. Math. **32** (1979) 121

[64] Desurvivre E. J., Simpson J. R., and Becker P. C., *High-gain Erbium-doped traveling-wave fiber amplifier*, Opt. Lett. **12** (1987) 888

[65] Dodd R. K., Eilbeck J. C., Gibbon J. D., and Morris H. C., *Solitons and Nonlinear Waves*, Academic (1982)

[66] Doliwa A. and Santini P. M., *An elementary geometric characterization of the integrable motions of a curve*, Phys. Lett. A **185** (1994) 373

[67] Doliwa A. and Santini P. M., *Integrable dynamics of a discrete curve and the Ablowitz-Ladik hierarchy*, J. Math. Phys. **36** (1995) 1259

[68] Duncan D. B., Eilbeck J. C., Feddersen H., and Wattis J. A. D., *Solitons on lattices*, Physica D **68** (1993) 1–11

[69] Eilbeck J. C., Lombdahl P. S., and Scott A. C., *The discrete self-trapping equation*, Physica **16** (1985) 318–338

[70] Eisenberg H., Silberberg Y., Morandotti R., Boyd A., and Aitchison J., *Discrete spatial optical solitons in waveguide arrays*, Phys. Rev. Lett. **81** (1998) 3383

[71] Eisenberg H., Silberberg Y., Morandotti R., and Aitchison J., *Diffraction management*, Phys. Rev. Lett. **85** (2000) 1863

[72] Evangelidis S. G., Mollenauer L. F., Gordon J. P., and Bergano N. S., *Polarization multiplexing with solitons*, J. Lightwave Tech. **10** (1992) 28–35

[73] Faddeev L. D., *The inverse problem in the quantum theory of scattering*, J. Math. Phys. **4** (1963) 72

[74] Faddeev L. D. and Takhtajan L. A., *Hamiltonian Methods in the Theory of Solitons*, Springer-Verlag (1980)

[75] Fermi E., Pasta J., and Ulam S., *Studies of nonlinear problems*, Los Alamos Rep. LA1940 (1955)

[76] Flach S., Willis C. R., *Discrete breathers*, Phys. Rep. **295** (1998) 181

[77] S. Flach S., Zolotaryuk, K., and Kladko K., *Moving lattice kinks and pulses: An inverse method*, Phys. Rev. E **59** (1999) 6105–6115

[78] Flaschka H., *The Toda lattice. I. Existence of integrals*, Phys. Rev. **B9** (1974) 24

[79] Flaschka H., *On the Toda lattice II. Inverse scattering solutions*, Progr. Theo. Phys. **51** (1974) 703–706

[80] Gabitov I. and Turitsyn S., *Averaged pulse dynamics in a cascaded transmission system with passive dispersion compensation*, Opt. Lett. **21** (1996) 327

[81] Gantmacher F. R., *Theory of Matrices*, Vol. 1, Chelsea (1959)

[82] Gardner C. S., Greene J. M., Kruskal M. D., and Miura R. M., *Method for solving the Korteweg-de Vries equation*, Phys. Rev. Lett. **19** (1967) 1095–1097; *The Kortewey-de Vries equation and generalizations. VI. Methods for exact solution*, Comm. Pure Appl. Math. **21** (1974) 97–133

[83] Gerdjikov V. S. and Ivanov M. I., *Block discrete Zakharov-Shabat system. I Generalized Fourier expansions*, Commun. JINR E2-81-811 (1981)

[84] Gerdjikov V. S. and Ivanov M. I., *Block discrete Zakharov-Shabat system. II. Hamiltonian structures*, Commun. JINR E2-81-812 (1981)

[85] Gerdjikov V. S. and Ivanov M. I., *The Hamiltonian structure of multicomponent nonlinear Schrödinger difference equations*, Theor. Math. Phys. **52** (1982) 89

[86] Gerdjikov V. S., Ivanov M. I., and Kulish P. P., *Expansions of the "squared" solutions and difference evolution equations* J. Math. Phys. **25** (1984) 25

[87] Gerdjikov V. S., Ivanov M. I., and Vaklev Y. S., *Gauge transformations and generating operators for the discrete Zakharov-Shabat system*, Inv. Probl. **2** (1986) 413–432

[88] Gordon J. P. and Haus H. A., *Random walk of coherently amplified solitons in optical fiber transmission*, Opt. Lett. **11** (1986) 665

[89] Hasegawa A. Ed., *Massive WDM and TDM Soliton Transmission Systems*, Kluwer (2000)

[90] Hasegawa A. and Kodama Y., *Solitons in Optical Communications*, Oxford University Press (1995)

[91] Hasegawa A. and Tappert F., *Transmission of stationary nonlinear optical pulses in dispersive dielectric fibers I. Anomalous dispersion*, Appl. Phy. Lett. **23** (1973) 142

[92] Hasegawa A. and Tappert F., *Transmission of stationary nonlinear optical pulses in dispersive dielectric fibers II. Normal dispersion*, Appl. Phy. Lett. **23** (1973) 171

[93] Herbst B. M. and Weideman J. A. C., *Finite difference methods for an AKNS eigenproblem*, Math. Comp. Sim. **43** (1997) 77–88

[94] Hirota R., *Nonlinear partial difference equations I. A difference analog of the Korteweg-de Vries equation*, J. Phys. Soc. Japan **43** (1977) 1429

[95] Hirota R., *Nonlinear partial difference equations II. Discrete time Toda equation*, J. Phys. Soc. Japan **43** (1977) 2074

[96] Hirota R., *Nonlinear partial difference equations III. Discrete sine-Gordon equations*, J. Phys. Soc. Japan **43** (1977) 2079

[97] Its A. R., Izergin A. G., Korepin V. E., and Slavnov N. A., *Temperature correlations of quantum spins*, Phys. Rev. Lett. **70** (11) (1993) 1704–1706

[98] Kac M. and van Moerbeke P., *On an explicit soluble system of nonlinear differential equations related to certain Toda lattices*, Advances Math. **16** (1975) 160

[99] Kac M. and van Moerbeke P., *A complete solution of the periodic Toda problem*, Proc. Nat. Acad. Sci. **72** (1975) 2879

[100] Kadomtsev B. B. and Petviashvili V. I., *On the stability of solitary waves in weakly dispersive media*, Sov. Phys. Dokl. **15** (1970) 539–541

[101] Kajinaga Y., Tsuchida T., and Wadati M., *Coupled nonlinear Schrödinger equations for two-component wave systems*, J. Phys. Soc. Japan **67** (1998) 1565

[102] Kenkre V. M. and Campbell D. K., *Self-trapping on a dimer: Time-dependent solutions of a discrete nonlinear Schrödinger equation*, Phys. Rev. B **35** (1987) 1473–1484

[103] Kenkre V. M. and Tsironis D. K., *Nonlinear effects in quasi-linear neutron scattering: Exact line calculation for a dimer*, Phys. Rev. B **34** (1986) 4959–4961

[104] Kevrekidis P. G., Kevrekidis I. G., and Malomed B. G., *Stability of solitary waves in finite Ablowitz-Ladik lattices*, J. Phys. A **35** (2002) 267–283

[105] Kevrekidis P. G., Rasmussen K. Ø., and Bishop A. R., *The discrete nonlinear Schrödinger equation: A survey of recent results*, Int. J. Mod. Phys. B **15** (2001) 2833–2900

[106] Kivshar Y. S., *Self-localization in arrays of defocusing waveguides*, Opt. Lett. **18** (1993) 1147–1149

[107] Kivshar Y. S. and Luther-Davies B., *Dark optical solitons: physics and applications*, Phys. Rep. **298** (1998) 81

[108] Kodama Y. and Hasegawa A., *Generation of asymptotically stable optical solitons and suppression of the Gordon-Haus effect*, Opt. Lett. **17** (1992) 31

[109] Konotop V. V., *Lattice dark solitons in the linear potential*, Theor. Math. Phys. **99** (1994) 687

[110] Konotop V. V., Chubykalo O. A., and Vasquez L., *Dynamics and interaction of solitons on an integrable inhomogeneous lattice*, Phys. Rev. E **48** (1993) 563

[111] Konotop V. V. and Vekslerchik V. E., *On soliton creation in the nonlinear Schrödinger models: Discrete and continuous versions*, J. Phys. A: Math. Gen. **25** (1992) 4037

[112] Korteweg D. J. and de Vries G., *On the change of form of long waves advancing in a rectangular canal, and on a new type of long stationary waves*, Philos. Mag. Ser. 5 **39** (1895) 422–443

[113] Krolikowski W. and Kivshar Y. S., *Soliton-based optical switching in waveguide arrays*, J. Opt. Soc. Am. B **13** (1996) 876–887

[114] Kulish P. P., *Quantum difference nonlinear Schrödinger equation* Lett. Math. Phys. **5** (1981) 191–197

[115] Lax P. D., *Integrals of nonlinear equations of evolution and solitary waves*, Commun. Pure Appl. Math. **21** (1968) 467–490

[116] Lederer F. and Aitchison J. S., *Discrete solitons in nonlinear waveguide arays*, in: *Les Houches Workshop on Optical Solitons*, Ed. Zakharov V. E. and Wabnitz S., Springer-Verlag (1999)

[117] Lederer F., Darmanyan S., and Kobyakov A., *Discrete solitons*, in: *Spatial Solitons*, Ed. S. Trillo and W. Torruellas, Springer-Verlag (2001) 267–290

[118] Malomed B. and Weinstein M. I., *Soliton dynamics in the discrete nonlinear Schrödinger equations*, Phys. Lett. A **220** (1996) 91–96

[119] Mamyshev P. V. and Mollenauer L. F., *Pseudo-phase-matched four-wave mixing in soliton wavelength-division multiplexed transmission*, Opt. Lett. **21** (1996) 396

[120] Manakov S. V., *On the theory of two-dimensional stationary self-focusing of electromagnetic waves*, Sov. Phys. JETP **38** (1974) 248–253

[121] Manakov S. V., *Complete integrability and stochastization of discrete dynamical systems*, Sov. Phys. JETP **40** (1975) 269–274

[122] Marcuse D., Menyuk C. R., and Wai P. K. A., *Applications of the Manakov-pmd equations to studies of signal propagation in fibers with randomly-varying birefringence*, J. Lightwave Tech. **15** (1997) 1735–1745

[123] Marquii P., Bilbaut J. M., and Remoissenet M., *Observation of nonlinear localized modes in an electrical lattice*, Phys. Rev. E **51** (1995) 6127–6133

[124] Mason L. J. and Sparling G. A. J., *Nonlinear Schrödinger and Korteweg-de Vries are reductions of self-dual Yang Mills*, Phys. Lett. A **137** (1989) 29–33

[125] Mason L. J. and Woodhouse N. M. J., *Integrability, Self-Duality and Twistor Theory*, LMS Monograph, New Series 15, Oxford University Press (1996)

[126] Maszczyk R., Mason L. J., and Woodhouse N. M. J., *Self-dual Bianchi metrics and the Painlevé transcendents*, Classical Quant. Gravity **11** (1994) 65–71

[127] Mears R. J., Reekie L., Jauncey I. M., and Payne D. N., *Low-noise Erbium-doped fiber amplifier operating at 1.54 μm*, Electron. Lett. **23** (1987) 1026

[128] Mecozzi A. and Haus H. A., *Effect of filters on soliton interactions in wavelength-division multiplexing systems*, Opt. Lett. **17** (1992) 988

[129] Mecozzi A., Moores J. D., Haus H. A., and Lai Y., *Soliton transmission control*, Opt. Lett. **16** (1991) 1841

[130] Mecozzi A., *Timing jitter in wavelength-division-multiplexed filtered soliton transmission*, J. Opt. Soc. Am. B **15** (1998) 152

[131] Menyuk C. R., *Nonlinear pulse propagation in birefringent optical fibers*, IEEE J. Quant. Electron. **23** (1987) 174–176

[132] Menyuk C. R., *Pulse propagation in an elliptically birefringent Kerr Medium*, IEEE J. Quant. Electron. **25** (1989) 2674–2682

[133] Mollenauer L. F., Evangelides S. G., and Gordon J. P., *Wavelength division multiplexing with solitons in ultra-long distance transmission using lumped amplifiers*, J. Lightwave Tech. **9** (1991) 362

[134] Mollenauer L. F., Gordon J. P., and Evangelides S. G., *The sliding-frequency guiding filter: An improved form of soliton jitter control*, Opt. Lett. **17** (1992) 1575

[135] Mollenauer L. F., Stolen L. F., and Gordon J. P., *Experimental observation of picoseconds pulse narrowing and solitons in optical fibers*, Phys. Rev. Lett. **45** (1980) 1095

[136] Morandotti R., Peschel U., Aitchison J., Eisenberg H., and Silberberg Y., *Dynamics of discrete solitons in optical waveguide arrays*, Phys. Rev. Lett. **83** (1999) 2726

[137] Morandotti R., Peschel U., Aitchison J., Eisenberg H., and Silberberg Y., *Experimental observation of linear and nonlinear optical Bloch oscillations*, Phys. Rev. Lett. **83** (1999) 4756

[138] Morandotti R., Eisenberg H., Silberberg Y., Sorel M., and Aitchison J., *Self-focusing and defocusing in waveguide arrays*, Phys. Rev. Lett. **86** (2001) 3296

[139] Nihoff F. and Capel H., *The discrete Korteweg-de Vries equation*, Acta Appl. Math. **39** (1995) 133

[140] Novikov S. P., Manakov S. V., Pitaevskii L. P., and Zakharov V. E., *Theory of Solitons. The Inverse Scattering Method*, Plenum (1984)

[141] Orfanidis S. J., *Sine-Gordon equations and nonlinear σ model on a lattice*, Phys. Rev. D **18** (1978) 3828

[142] Radakrishnan R., Lakshmanan M., and Hietarinta J., *Inelastic collision and switching of coupled bright solitons in optical fibers*, Phys. Rev. E **56** (1997) 2213

[143] Ramani A., Grammaticos B., and Tanizhmani K. M., *An integrability test for differential-difference systems*, J. Phys. A – Math. Gen. **25** (1992) 14

[144] Roskes G. J., *Some nonlinear multiphase interactions*, Stud. Appl. Math. **55** (1976) 231

[145] Ruijsenaars S. N. M., *Relativistic Toda systems*, Comm. Math. Phys. **133** (1990) 217–247

[146] Russell J. S., *Report of the committee on waves*, Report of the 7th Meeting of British Association for the Advancement of Science, Liverpool (1838) 417–469

[147] Russell J. S., *Report on waves*, Report of the 14th Meeting of British Association for the Advancement of Science, John Murray, London (1844) 311–390

[148] Scott A. C. and Macneil L., *Binding-energy versus nonlinearity for a small stationary soliton*, Phys. Lett. A **98** (1983) 87–88

[149] Sievers A. J. and Takeno S., *Intrinsic localized modes in anharmonic crystals*, Phys. Rev. Lett. **61** (1988) 970–973

[150] Sophocleus C. and Parker D. F., *Pulse collisions and polarization conversion for optical fibers*, Opt. Commun. **112** (1994) 214

[151] Su W. P., Schieffer J. R., and Heeger A. J., *Solitons in polyacetylene*, Phys. Rev. Lett. **42** (1979) 698

[152] Suris Y. B., *On the bi-Hamiltonian structure of Toda and relativistic Toda lattices*, Phys. Lett. A **188** (1994) 419–429

[153] Suris Y. B., *A discrete-time relativistic Toda lattice*, J. Phys. A – Math. Gen. **29** (1996) 451–465

[154] Suris Y. B., *New integrable systems related to the relativistic Toda lattice*, J. Phys. A – Math. Gen. **30** (1997) 1745–1761

[155] Suris Y. B., *On some integrable systems related to the Toda lattice*, J. Phys. A – Math. Gen. **30** (1997) 2235–2249

[156] Suris Y. B., *A note on an integrable discretization of the nonlinear Schrödinger equation*, Inv. Probl. **13** (1997) 1121

[157] Suris Y. B., *On an integrable discretization of the modified Korteweg-de Vries equation*, Phys. Lett. A **234** (1997) 91–102

[158] Suris Y. B., *Integrable discretizations for lattice systems: Local equations of motion and their Hamiltonian properties*, Rev. Math. Phys. **11** (1999) 727

[159] Suris Y. B., *A reply to a comment: A note on an integrable discretization of the nonlinear Schrödinger equation*, Inv. Probl. **16** (2000) 1071–1077

[160] Suris Y. B., *The Problem of Integrable Discretization: Hamiltonian Approach*, Birkhäuser Verlag, Cambridge, MA (2003)

[161] Taha T. R. and Ablowitz M. J., *Analytical and numerical aspects of certain nonlinear equations. I. Analytical*, J. Comp. Phys. **55** (1984) 192

[162] Taha T. R. and Ablowitz M. J., *Analytical and numerical aspects of certain nonlinear equations. II. Numerical, nonlinear Schrödinger equation*, J. Comp. Phys. **55** (1984) 203

[163] Taha T. R. and Ablowitz M. J., *Analytical and numerical aspects of certain nonlinear equations. II. Numerical, Korteweg-de Vries equation*, J. Comp. Phys. **55** (1984) 231

[164] Taha T. R. and Ablowitz M. J., *Analytical and numerical aspects of certain nonlinear equations. II. Numerical, modified Korteweg-de Vries equation*, J. Comp. Phys. **77** (1984) 540

[165] Toda M., *Waves in nonlinear lattices*, Prog. Theor. Phys. Suppl. **45** (1970) 174–200

[166] Trombettoni A. and Smerzi A., *Discrete solitons and breathers with dilute Bose-Einstein condensates*, Phys. Rev. Lett. **86** (2001) 2353–2356

[167] Trubatch A. D., PhD Thesis, University of Colorado at Boulder (2000)

[168] Tsuchida T., *Integrable discretizations of coupled nonlinear Schrödinger equations*, Rep. Math. Phys. **46** (2000) 269

[169] Tsuchida T., Ujino H., and Wadati M., *Integrable semi-discretization of the coupled modified KdV equations*, J. Math. Phys. **39** (1998) 4785

[170] Tsuchida T., Ujino H., and Wadati M., *Integrable semi-discretization of the coupled nonlinear Schrödinger equations*, J. Phys. A – Math. Gen. **32** (1999) 2239–2262

[171] Tsuchida T. and Wadati M., *New integrable systems of derivative nonlinear Schrödinger equation with multiple components*, Phys. Lett. A **257** (1999) 53

[172] Tsuchida T. and Wadati M., *Complete integrability of derivative nonlinear Schrödinger-type equations*, Inv. Probl. **15** (1999) 1363

[173] Vakhnenko A. A. and Gadidei Y. B., *The motion of solitons in discrete moleular chains*, Theor. Math. Phys. **68** (1987) 873

[174] Vakhnenko O. O. and Velgakis M. J., *Multimode soliton dynamics in perturbed ladder lattices*, Phys. Rev. E **63** (2000) 1

[175] Vekslerchik V. E., *Finite nonlinear Schrödinger chain*, Phys. Lett. A **174** (1993) 285

[176] Vekslerchik V. E. and Konotop V. V., *Discrete nonlinear Schrödinger equation under nonvanishing boundary conditions*, Inv. Probl. **8** (1992) 889

[177] Veselov A. P. and Shabat A. B., *Dressing chains and the spectral theory of the Schrödinger operator*, Func. Anal. Appl. **27** (1993) 81

[178] Wadati M., Iizuka T., and Hisekado M., *A coupled nonlinear Schrödinger equation and optical solitons*, J. Phys. Soc. Japan. **61** (1992) 2241

[179] Wai P. K. A. and Menyuk C. R., *Polarization mode dispersion, decorrelation and diffusion in optical fibers with randomly-varying birefringence*, J. Lightwave Tech. **14** (1996) 148

[180] Wei P. K. A., Menyuk C. R., and Chen H. H., *Stability of solitons in randomly varying birefrengent fibers* Opt. Lett. **16** (1991) 1231–1233

[181] Ward R. S., *On self-dual gauge fields*, Phys. Lett. A **61** (1977) 81

[182] Ward R. S., *Ansatze for self-dual Yang-Mills fields*, Commun. Math. Phys. **80** (1981) 563

[183] Ward R. S., *Integrable and solvable systems, and relations among them*, Proc. Roy. Soc. Lond. A **315** (1985) 451

[184] Yang J., *Classification of the solitary waves in coupled nonlinear Schrödinger equations*, Physica D **108** (1997) 92–112

[185] Yang J., *Multiple permanent-wave trains in nonlinear systems*, Stud. Appl. Math. **100** (1998) 127

[186] Yang J., *Multisoliton perturbation theory for the Manakov equations and its applications to nonlinear optics*, Phys. Rev. E **59** (1999) 240

[187] Yang J., *Interactions of vector solitons*, Phys. Rev. E **64** (2001) 026607

[188] Yariv A., *Optical Electronics in Modern Communications*, 5th Ed., Oxford University Press (1997)

[189] Zabusky N. J. and Kruskal M. D., *Interaction of solitons in a collisionless plasma and the recurrence of initial states*, Phys. Rev. Lett., **15** (1965) 240–243

[190] Zakharov V. E., *Stability of periodic waves of finite amplitude on the surface of a deep fluid*, Sov. Phys. J. Appl. Mech. Tech. Phys. **4** (1968) 190–194

[191] Zakharov V. E., *Collapse of Langmuir waves*, Sov. Phys. JETP **35** (1972) 908–914

[192] Zakharov V. E. and Shabat A. B., *Exact theory of two-dimensional self-focusing and one-dimensional self-modulation of waves in nonlinear media*, Sov. Phys. JETP **34** (1972) 62–69

[193] Zakharov V. E. and Shabat A. B., *Interaction between solitons in a stable medium*, Sov. Phys. JETP **37** (1973) 823–828

[194] Zvezdin A. K. and Popkov A. F., *Contribution to the nonlinear theory of magnetostatic spin waves*, Sov. Phys. JETP **2** (1983) 350

References for further reading

[1] Ablowitz M.J., Halburd R., and Herbst B., *On the extension of the Painlevé property to difference equations*, Nonlinearity **13** (2000) 889–905

[2] Ablowitz M.J., Ramani A., and Segur H., *A connection between nonlinear evolution equations and ordinary differential equations of P-type. I, II*, J. Math. Phys. **21** (1980) 715–721, 1006–1015

[3] Akhmediev N.N., Krolikowski W., and Snyder A.W., *Partially coherent solitons of variable shape*, Phys. Rev. Lett. **81** (1998) 4632

[4] Albanese C. and Frölich J., *Periodic solutions of some infinite-dimensional Hamiltonian systems associated with nonlinear partial difference equations. I, II*,Commun. Math. Phys. **116, 119** (1988) 475–502, 677–699

[5] Anastassiou C., Segev M., Steiglitz K., Giordmaine J.A., Mitchell M., Shih M., Lan S., and Martin S., *Energy-exchange interactions between colliding vector solitons*, Phys. Rev. Lett. **83** (1999) 2332

[6] Anastassiou C., Fleischer J.W., Carmon T., Segev M., and Steiglitz K., *Information transfer via cascaded collisions of vector solitons*, Opt. Lett. **26** (2001) 1498

[7] Ankiewicz K., W. Krolikowski, and Akhmediev N.N., *Partially coherent solitons of variable shape in a slow Kerr-like medium: Exact solutions*, Phys. Rev. E. **59** (1999) 6079

[8] Belokolos E.D., Bobenko A.I., Enol'skii V.Z., Its A.R., and Matveev V.B., *Algebro-geometric approach to nonlinear integrable equations*, Springer Series in Nonlinear Dynamics (1994)

[9] Boiti M., Leon J.J-P., Martina L., and Pempinelli F., *Scattering of localized solitons in the plane*, Phys. Lett. **132A** (1988) 432–439

[10] Boiti M., Leon J.J-P., and Pempinelli F., *Multidimensional solitons and their spectral transform*, J. Math. Phys. **31** (1990) 2612–2618

[11] Bourgain J., *Fourier restriction phenomena for certain lattice subsets and applications to nonlinear evolution equations*, Geom. Funct. Anal. **3** (1993) 107–156, 209–262

[12] Bourgain J., *Periodic nonlinear Schrödinger equation and invariant measures*, Commun. Math. Phys. **166** (1994) 1–26

[13] Bronski J., *Semiclassical eigenvalue distribution of the Zakharov-Shabat eigenvalue problem*, Physica D **97** (1996) 376–397

[14] Bruschi M., Levi D., and Ragnisco O., *Discrete version of the modified Korteweg-de Vries equation with x-dependent coefficients*, Nuovo Cim. **48A** (1978) 213

[15] Burzlaff J., *The soliton number of optical soliton bound states for two special families of input pulses*, J. Phys. A: Math. Gen. **21** (1988) 561–566

[16] Cai D., Bishop A.R., Grønbech-Jensen N., *Spatially localized, temporally quasiperiodic, discrete nonlinear excitations*, Phys. Rev. E **52** (1995) R5784–R5787

[17] Cascaval R.C., Gesztesy F., Holden H., and Latushkin Y. *Spectral analysis of Darboux transformations for the focusing NLS hierarchy*, to appear in J. Analyse Math. (2003)

[18] Christiansen P.L., Eilbeck J.C., Enolskii V.Z., and Kostov N.A., *Quasi-periodic solutions of coupled nonlinear Schrödinger equations of Manakov type*, Proc. R. Soc. A **451** (1995) 685–700

[19] Christiansen P.L., Eilbeck J.C., Enolskii V.Z., and Kostov N.A., *Quasi-periodic and periodic solutions for coupled nonlinear Schrödinger equations of Manakov type*, Proc. R. Soc. A **456** (2000) 2263–2281

[20] Conte R. and Musette M., *A new method to test discrete Painlevé equations*, Phys. Lett. A **223** (1996) 439–448

[21] Deift P. and Zhou X., *Long-time asymptotics for solutions of the NLS equation with initial data in a weighted Sobolev space*, Comm. Pure Appl. Math. **56** (2003) 1029–1077

[22] Dubrovin B.A. and Novikov S.P., *Periodic and conditionally periodic analogues of the many-soliton solutions of the Korteweg-de Vries equation*, Sov. Phys. JETP **40** (1974) 1058–1063

[23] Eilbeck J.C., *The discrete nonlinear Schrödinger equation – 20 Years on*, Proceedings of the "Third Conference Localization and Energy Transfer in Nonlinear Systems," June 17–21, San Lorenzo de El Escorial, Madrid, eds Vázquez L., MacKay R.S., and Zorzano M.P., World Scientific, Singapore (2003) 44–67

[24] Eilbeck J.C., Enolskii V.Z., and Kostov N.A., *Quasi-periodic and periodic solutions for vector nonlinear Schrödinger equations*, J. Maths. Phys **41** (2000) 8236–8248

[25] Fokas A. S. and Zakharov V.E., eds., *Important developments in soliton theory*, Springer-Verlag (1993)

[26] Fokas A.S., *Integrable nonlinear evolution equations on the half-line*, Commun. Math. Phys. **230** (2002) 1–39

[27] Fokas A.S., *A new transformation method for evolution PDE's*, IMA J. Appl. Math. **67** (2002) 559–590

[28] Fokas A.S., Grammaticos B., and Ramani A., *From continuous to discrete Painlevé equations*, J. Math. Anal. Appl. **180** (1993) 342–360

[29] Goncharenko V.M. and Veselov A.P., *Yang-Baxter maps and matrix solitons*, arXiv:math-ph/0303032 (2003)

[30] Hirota R, *Discrete analogue of a generalized Toda equation*, J. Phys. Soc. Jap. **50** (1981) 3785–3791

[31] Its A.R., *Asymptotics of nonlinear Schrödinger equation solutions and isomonodromic deformations of the linear-differential equation systems*, Dokl. Akad. Nauk. SSSR **261** (1981) 14–18

[32] Its A.R. and Kotlyarov V.P., *Explicit formulas for solutions of nonlinear Schrödinger equation*, Dokl. Akad. Nauk. SSSR **11** (1976) 965–968

[33] Its A.R. and Matveev V.B., *The periodic Korteweg-de Vries equation*, Func. Anal. Appl. **9** (1975) 67

[34] Izergin A.G. and Korepin V.E., *A lattice model related to the nonlinear Schrödinger equation*, Sov. Phys. Dokl. **26** (1981) 653–654

[35] Jakubowsky M.H., Steiglitz K., and Squier R., *State transformations of colliding optical solitons and possible applications to computation in bulk media*, Phys. Rev. E **58** (1998) 6752

[36] Kang J.U., Stegeman G.I., Aitchison J.S., and Akhmediev N., *Observation of Manakov soliton in AlGaAs planar waveguides*, Phys. Rev. Lett **76** (1996) 3699

[37] Kang J.U., Stegeman G.I., Aitchison J.S., and Ahkmediev N.N., *Spatial solitons in AlGaAs waveguides*, Opt. Quant. Elect. **30** (1998) 649–671

[38] Kanna T. and Lakshmanan M., *Exact soliton solutions, shape changing collisions, and partially coherent solitons in coupled nonlinear Schrödinger equations*, Phys. Rev. Lett. **86** (2001) 5043

[39] Kanna T. and Lakshmanan M., *Exact soliton solutions of coupled nonlinear Schrödinger equations: Shape changing collisions, logic gates and partially coherent solitons*, arXiv:nlin.SI/0303025 (2003)

[40] Klaus M. and Shaw K., *Purely imaginary eigenvalues of Zakharov-Shabat systems*, Phys. Rev. **65** (2002) 1–5

[41] Klaus M. and Shaw K., *On the eigenvalues of Zakharov-Shabat systems*, SIAM J. Math. Anal. **34** (2003) 750–773

[42] Kotlyarov V.P., *Asymptotic analysis of the Marchenko integral equation and soliton asymptotics of a solution of the nonlinear Schrödinger equation*, Physica D **87** (1995) 176–185

[43] Krichever I.M., *Algebraic curves and commuting matrix differential operators*, Func. Anal. Appl. **10** (1976) 144–146

[44] Krolikowski W., Akhmediev N.N., and Luther-Davies B., *Collision-induced shape transformations of partially coherent solitons*, Phys. Rev. E. **59** (1999) 4654

[45] Kivshar Y.S., *On the soliton generation in optical fibers*, J. Phys. A: Math. Gen. **22** (1989) 337–340

[46] Kulish P.P., *Factorization of the classical and quantum S-matrix and conservation laws*, Theor. Math. Phys. **26** (1976) 132–137

[47] Levi D. and Ragnisco O., *Extension of the spectral-transform method for solving nonlinear differential difference equations*, Lett. Nuovo Cim. **22** (1978) 691

[48] Levi D., Pilloni L., and Santini P.M., *Integrable three-dimensional lattices*, J. Phys. A **14** (1981) 1567

[49] Ma Y.C. and Ablowitz M.J., *The periodic cubic Schrödinger equation*, Stud. Appl. Math. **65** (1981) 113–158

[50] Maimistov A.I. and Sklyarov Y.M., *Influence of regular phase modulation on formation of optical solitons*, Sov. J. Quant. Electron. **17** (1987) 500–504

[51] Manakov S.V., *Nonlinear Fraunhofer diffraction*, Sov. Phys. JETP **38** (1974) 693–696

[52] McKean H.P. and Trubowitz E., *Hill's operator and hyper-elliptic function theory in the presence of infinitely many branch points*, Commun. Pure Appl. Math. **29** (1976) 143–226

[53] McKean H.P. and van Moerbeke P., *The spectrum of Hill's equation*, Invent. Math. **30** (1975) 217

[54] Mikhailov A.V., *Integrability of a two-dimensional generalization of the Toda chain*, JETP Lett. **30** (1979) 414

[55] Nijhoff F.W., Satsuma J., Kajiwara K., Grammaticos B., and Ramani A., *A study of the alternative discrete Painlevé-II equation* Inv. Probl. **12** (1996) 697–716

[56] Novikov S.P., *The periodic problem for the Korteweg-de Vries equation*, Func. Anal. Appl. **8** (1974) 236–246

[57] Novokshenov V.Y., *Asymptotic behavior as* $t \to \infty$ *of the solution of the Cauchy problem for a nonlinear Schrödinger equation*, Sov. Math. Dokl. **212** (1980) 529–533

[58] Porubov A.V. and Parker D.F., *Some general periodic solutions to coupled nonlinear Schrödinger equations*, Wave Motion **29** (1999) 97–109

[59] Ramani A., Grammaticos B., and Hietarinta J., *Discrete versions of the Painlevé equations*, Phys. Rev. Lett. **67** (1991) 1829–1832

[60] Salerno M., *Quantum deformations of the discrete nonlinear Schrödinger equation*, Phys. Rev. A **46** (1992) 6856–6859

[61] Satsuma J. and Yajima N., *Initial value problems of the one-dimensional self-modulation of nonlinear waves in dispersive media*, Progr. Theor. Phys. Suppl. **55** (1974) 284–306

[62] Solijačić M., Steiglitz K., Sears, S.M., Segev M., Jakubowski M.H., and Squier R., *Collisions of two solitons in an arbitrary number of coupled nonlinear Schrödinger equations*, Phys. Rev. Lett. **90** (2003) 254102

[63] Steiglitz K., *Time-gated Manakov solitons are computationally universal*, Phys. Rev. E **63** (2000) 016608

[64] Sukhorukov A.A. and Akhmediev N.N., *Coherent and incoherent contributions to multisoliton complexes*, Phys. Rev. Lett. **83** (1999) 4736

[65] Suris Y. and Veselov A.P., *Lax matrices for Yang-Baxter maps*, arXiv:math.QA/0304122 (2003)

[66] Tovbis A. and Venakides S., *The eigenvalue problem for the focusing nonlinear Schrödinger equation: New solvable cases*, Physica D **146** (2000) 150–164

[67] Wadati M., *Transformation theories for nonlinear discrete systems*, Progr. Theor. Phys. Suppl. **59** (1976) 33–63

[68] Wreszinski W.F., *Inverse scattering theory for systems of the Zakharov-Shabat type*, J. Math. Phys. **24** (1983) 1502–1508

[69] Zakharov V.E. and Manakov S.V., *On the complete integrability of the nonlinear Schrödinger equation*, Theor. Math. Phys. **19** (1974) 551–559

Index